PROBABILISTIC DESIGN TOOLS FOR VERTICAL BREAKWATERS

Probabilistic Design Tools for Vertical Breakwaters

HOCINE OUMERACI, ANDREAS KORTENHAUS
Technical University of Braunschweig, Leichtweiß-Institut für Wasserbau, Braunschweig, Germany

WILLIAM ALLSOP
HR Wallingford, Wallingford, U.K.

MAARTEN DE GROOT
Geodelft, Delft, The Netherlands

ROGER CROUCH
University of Sheffield, Department of Civil and Structural Engineering, Sheffield, U.K.

HAN VRIJLING, HESSEL VOORTMAN
Delft University of Technology, Hydraulic and Offshore Engineering Section, Delft, The Netherlands

Edited by
Andreas Kortenhaus and Hessel Voortman

CRC Press
Taylor & Francis Group
Boca Raton London New York

CRC Press is an imprint of the
Taylor & Francis Group, an **informa** business

Library of Congress Cataloging-in-Publication Data

Applied for

CRC Press
Taylor & Francis Group
6000 Broken Sound Parkway NW, Suite 300
Boca Raton, FL 33487-2742

ISBN-13: 978-9-058-09248-9 (hbk)
ISBN-13: 978-9-058-09249-6 (pbk)

Publisher's Note

The publisher has gone to great lengths to ensure the quality of this reprint but points out that some imperfections in the original copies may be apparent.

Cover design: Studio Jan de Boer, Amsterdam, The Netherlands.

Visit the Taylor & Francis Web site at
http://www.taylorandfrancis.com

and the CRC Press Web site at
http://www.crcpress.com

TABLE OF CONTENTS

PREFACE

The PROVERBS project (Probabilistic Design Tools for Vertical Breakwaters) is a large multi-disciplinary project comprising 23 institutes from 8 European countries and funded by the Marine Science and Technology (MAST) Research Programme of the European Union (EU) under its Fourth Framework Programme (1994-1998).

It has broken new frontiers by developing probability-based analysis and design methods for a wide range of harbour and coastal monolithic structures subject to wave attack.

It demonstrates how the "European added value" provided by RTD projects (Research and Development) funded by the European Union can best be used to successfully address highly complex problems (such as those related to breakwater stability and design), where a common platform for integration and combination of expertise, efforts and resources on a European-wide scale is needed.

The results from PROVERBS, numerous and impressive as they are, are expected to make a sound contribution towards making European marine construction industry more competitive world-wide. The potential end-users of the results are many - port authorities, coastal/harbour engineers, governments authorities, just to mention a few - and the benefits for the entire society considerable.

Coastal protection against flooding, erosion and pollution is a legitimate concern of the European citizen. As such, it has been given high priority in the newly launched Fifth Framework Programme of the EU under the *Sustainable Marine Ecosystem* Key Action, the latter being the successor activity of MAST.

Research in this context will be targeted towards making coastal structures *safer, more economic and environmentally more compatible.*

In the years to come, the validity of the methods and tools developed within PROVERBS will have to be tested by the international community (researchers, practitioners, public authorities). The contribution of the latter to improve them and enhance their applicability worldwide will undoubtedly be most invaluable and welcome.

In concluding, I would like to take this opportunity to thank all PROVERBS partners for their outstanding research efforts and their exemplary commitment and collaboration over the past six years.

Christos Fragakis
European Commission
Directorate General for Research

A GUIDE TO THIS BOOK

This book represents the final report of the MAST III PROVERBS project (Probabilistic Design Tools for Vertical Breakwaters) of the European Union (1996 to 1999). It covers the design of vertical breakwaters including the various aspects of hydraulic loading and foundation and structural strength. The book deals with the modelling of physical processes relevant for this type of structures as well as an overall probabilistic design concept in which the physics are integrated.

The structure of this book is briefly sketched in the figure overleaf and can be summarised as follows:

Chapter 1 first deals with the motivation for the project and describes the structure of the research project itself. It then provides an extensive overview of the key results that were achieved. The first chapter is intended for readers who want to quickly familiarise themselves with the contents of the project and the subjects covered. It may also serve as a starting point for more detailed reading.

Chapter 2 describes the design wave conditions and then goes further into wave load modelling. Quasi-static loads and impact loads, both landward and seaward are introduced and models for their description are given.

Chapter 3 deals with foundation modelling of vertical breakwaters. Several types of foundation failure are described on different levels of sophistication. The sets of models are explicitly aimed at different stages in the design process. Limit state equations and uncertainties of parameters and models for the foundation are given.

Chapter 4 gives an overview of structural failures of concrete caissons. Like in chapter 3, simple design formulae for preliminary design are given next to an overview of a more sophisticated approach, including finite element modelling.

Chapter 5 describes the overall concept of probabilistic design. After a general introduction of this concept the application to vertical breakwaters is shown. Two alternative methods for including probabilistic analysis into a design process are given, the first of which is an optimisation framework to find the optimal safety level and the second is a partial safety factor approach.

The book closes with some concluding remarks in *chapter 6*. Future research tasks of each of the aspects addressed are briefly outlined and some final conclusion for the work within the project are sketched.

Each chapter of the book contains references to either peer reviewed journals and conference proceedings or to additional papers and reports which were produced throughout the course of the project and which are referred to as Volume II (whereas this book is referred to as Volume I). A complete list of these documents is given at the end of each chapter and in Annex 3, respectively.

A brief overview of the key results of the project incl. possible end users of the results, main benefits of the project is given in Annex 4 of this book.

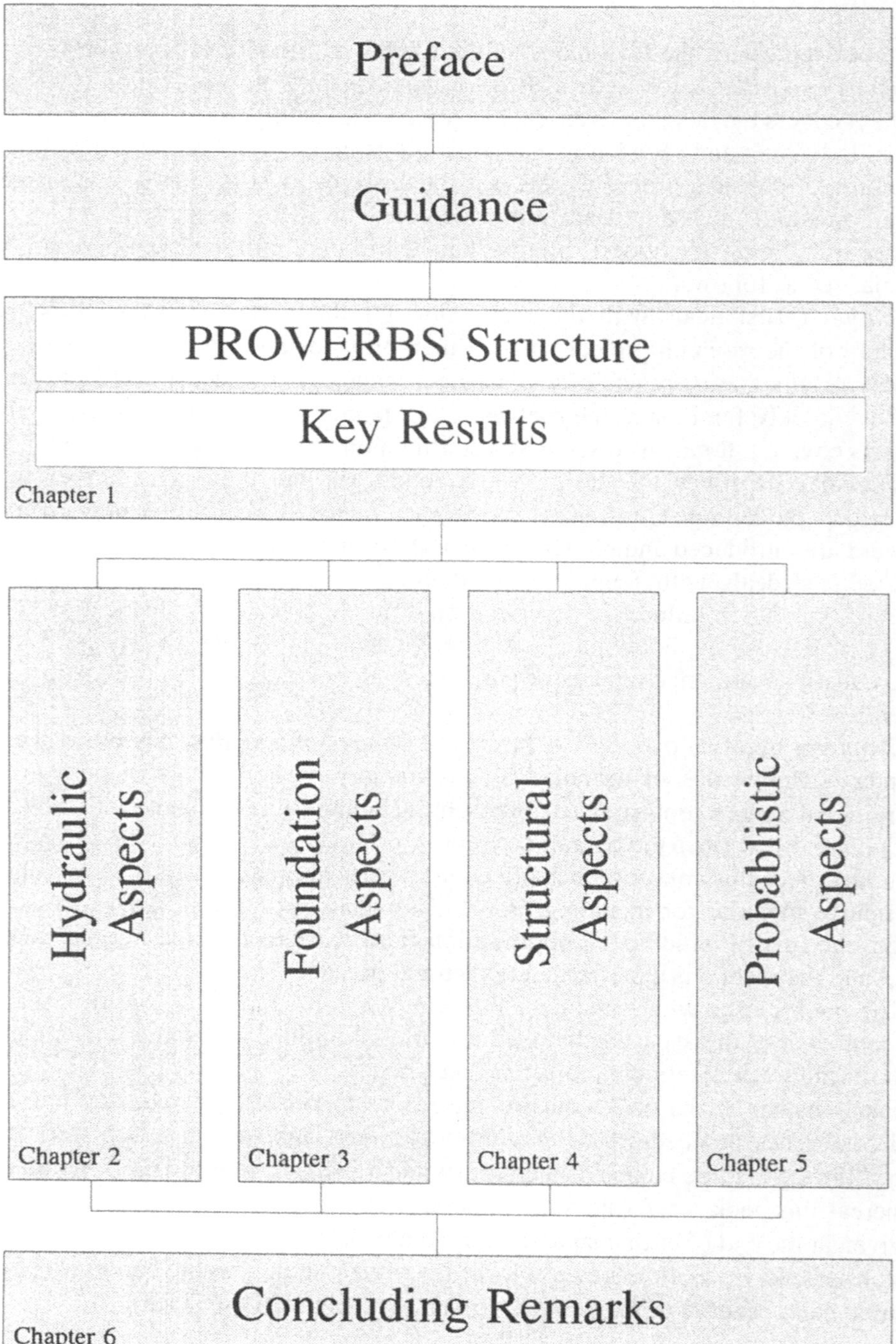
Preface
Guidance
PROVERBS Structure
Key Results
Chapter 1
Hydraulic Aspects
Chapter 2
Foundation Aspects
Chapter 3
Structural Aspects
Chapter 4
Probablistic Aspects
Chapter 5
Concluding Remarks
Chapter 6

CHAPTER 1

General introduction and selected key results of the overall project

Hocine Oumeraci

1.1 GENERAL BACKGROUND, OPPORTUNITY AND MOTIVATIONS

1.1.1 *General background and opportunity*

Due to the growing pressure exerted by human activities and the growing needs in the industrial and amenity sector within the coastal zone, interest in protective structures against the hydrodynamic actions of the sea is also expected to increase. A further important reason supporting this expectation is the increase of the magnitude, duration and frequency of storm surges observed in the last decades.

On the other hand, the construction of coastal structures is still essentially based on empirical design methods, as well as on trial and error approaches, thus making any optimisation almost impossible. The latter statement is particularly supported by the fact that most of the catastrophic failures experienced by coastal structures could not be predicted at the design stage and cannot yet be satisfactorily explained by present design methods and analyses. In fact, most of the failure modes which have been identified to date are associated with the dynamic nature of the wave loads and the highly transient phenomena involved in the structure-foundation interactions. With this background an urgent need arises for the development of integrated rational design approaches which are based on an increased understanding of the hydrodynamic, geotechnical and structural processes involved in the wave-structure-foundation interaction and which account for the uncertainties involved in the models used as well as in the input data. An opportunity to start with a research project based on this general background was given within MAST (Marine Science and Technology Research Programme of the European Union). In this programme two main groups of research topics directly related to Coastal Engineering have been addressed: Coastal Morphodynamics and Coastal Structures (Fig. 1-1). The MAST III project PROVERBS which is described here belongs to the second group which is more related to applied research but does not exclude substantial contributions to basic research.

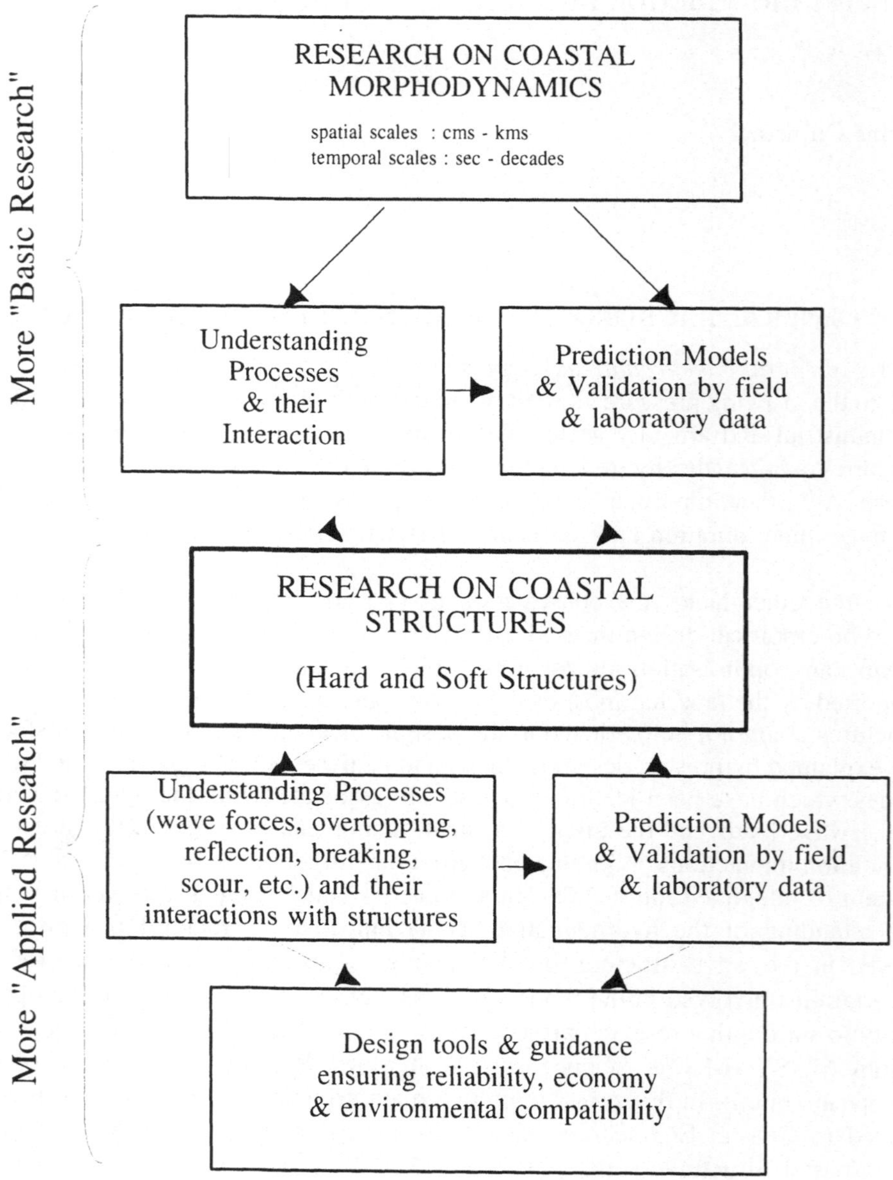

Figure 1-1. Research in coastal engineering within the EU / MAST-Programme.

PROVERBS started at 1st February 1996 and ended at 31st January 1999. In this introductory chapter, a brief presentation of PROVERBS will be given, including motivations, objectives, organisation structure, research strategy and methodology as well as a brief outline of the practical importance of the key scientific and engineering findings which are described within the next four chapters.

1.1.2 *Motivations and Position of the Design Problem*

The major reasons why monolithic structures and why probabilistic design methods have been selected as candidate research topics for a large European project are first outlined before the design problem is addressed in more details to better illustrate the necessity of an integrated probabilistic design approach.

1.1.2.1 *Motivations for Monolithic Coastal Structures / Breakwaters*

- *Catastrophic failures*: Numerous severe and catastrophic failures were experienced by vertical breakwaters in the 1930s. It should be stressed that major failures may cost 2-3 times more to re-build than the original construction costs. As a consequence, the vertical breakwater type was almost abandoned - except in some few countries - in favour of the rubble mound breakwater type. In Japan for instance, about 7 failures per year have been experienced by vertical breakwaters (Oumeraci, 1994). After a series of catastrophic failures experienced by large rubble mound breakwaters at the end of the 1970s and the beginning of the 1980s, a number of actions were started to promote the revival of vertical breakwaters and the development of new breakwater concepts (Oumeraci et al., 1991). In this respect, extensive research efforts at interdisciplinary and multinational level were urgently required.
- *Need for breakwaters at greater depth*: To suit the increasing draught of large vessels, breakwaters should be founded in increasingly deeper water, thus making the cost of such structures more prohibitive. Construction costs of 100 to 350 million €/km breakwater are not seldom. In this respect, a type of structure is needed which represents a better alternative not only in terms of technical performance and total costs, but also in terms of standardisation, quality control, environmental aspects, construction time and maintenance. Moreover, more focus should be put on the optimisation of the design.
- *Need for environmentally friendly structures*: Monolithic caisson type structures can easily be given any shape, perforations and any further constructional features to reduce the impact on the environment. Moreover, less material and less energy for material transportation than for further traditional breakwater types is required for construction. In fact, most of the material involved in a caisson breakwater is sand - dredged from deeper sea, thus minimising the energy required for transportation and maximising the conservation of scarce construction material.

- *Need for multi-purpose structures*: Due to the inherent considerable public investments, it is expected that the objective of coastal structures will not solely be limited to the damping of wave action. Taking the opportunity of such expensive structures, facilities for amenity and wave power extraction might also be integrated in the structure (cost sharing, acceptance by society etc.). In this respect, caisson structures are suited for this purpose, because of their flexibility to adapt to any requirements related to their shape, size and multi-purpose use. Although not yet based on rational design methods, the caisson type structures have already demonstrated their capability because they can easily be adapted to meet also:
  - further purposes like amenity, wave power extraction and further industrial needs;
  - environmental requirements by providing a suitable shape and further interesting features to reduce wave reflection, to increase the water exchange between open sea and protected area, to minimise the disturbance of the sea ground and to fit into the maritime scenery.

Examples from Japan (Tanimoto & Takahashi, 1994), Monaco (Bouchet et al., 1994) and Korea (Lee & Hong, 1994) have already shown that the potential of adapting caissons type structures to meet any requirement of technical, social and ecological nature is higher than for any other traditional type of structure. This however, requires a high level of knowledge and technology. The results of PROVERBS are expected to contribute to bringing the European maritime construction industry in a world leading position in this field by increasing the knowledge and skill levels as well as the competitiveness of European coastal / harbour engineers working world-wide.

- *Potential large-scale application for sea-walls*: Because of the competitiveness of caisson structures in terms of technical performance, total costs, environment, quality control, construction time and standardisation, it is believed that there is also a large potential for their use as sea walls to respond to the potential increase of storminess and sea level rise. This will help to react more rapidly and better protect the coastal zones of high economic, social and environmental values.

1.1.2.2 *Motivations for Probabilistic Design Methods*

- *Need for more and better optimisation*: Breakwaters and coastal structures represent considerable public investments (in the range of 1 billion € for a 5 km long breakwater in deep water) which encounter less and less acceptance by government and local authorities, due to the decreasing availability of funding for this kind of large projects and the increasing awareness of environmental impacts by governments and society. Beside the need to diversify the use of such structures (see multi-purpose use as described above), there is an urgent need to use more rational methods for design and more sophisticated tools for the optimisation of such structures. It is obvious that such methods can only be developed in an integrated manner, requiring a multi-national framework and a multi-disciplinary research strategy.

- *Complexity of physical problems involved*: The results of previous MAST I and MAST II-projects have highlighted the complexity and the integrated nature of the problems related to breakwater stability and design, including the considerable importance of the three-dimensional and stochastic nature of the processes involved in the wave-structure-foundation interactions (wave loading and dynamic response), as well as the large number of possible failure mechanisms and their complex interaction. This necessarily prescribes the use of probability-based analysis methods as the sole alternative for the design. In fact, this is the only alternative which may provide a systematic and comprehensive framework not only for optimisation procedures, but also for the application of engineering judgement.
- *Stimulating aspects*: Since the prospective probabilistic design methods will essentially be based on the feedback from prototype experience, the results are expected to stimulate a rapid and continuous feedback between researchers and practitioners, thus enhancing the world-wide applicability of these tools. This will also actively stimulate the collection of more and better data, because it is the essence of probability-based tools to use more and better information.
- *Sustainable development in coastal zones*: It is expected that within the next 10 - 15 years reliability tools will represent the standard methods in the design of engineering works as well as in the management of coastal zones. In fact, without a risk management framework based on reliability methods a sustainable protection of coastal zones will hardly be achieved (Fig. 1-2).

1.1.2.3 *Position of the Design Problem*

Given a sea wave climate the main function of a breakwater is to provide sufficient protection of the area behind it; i.e. the wave transmission around, through and over the breakwater should be reduced to some acceptable level which strongly depends on the purpose of the protected area (birthed or manoeuvring ships, beach protection etc.). Any situation where this hydraulic function fails to be fulfilled must be considered as a failure. Besides wave transmission, hydraulic performance aspects like wave reflection (important for ship manoeuvres and scour potential assessment) and wave overtopping (important for installations/operations on and behind the breakwater) might also be of importance.

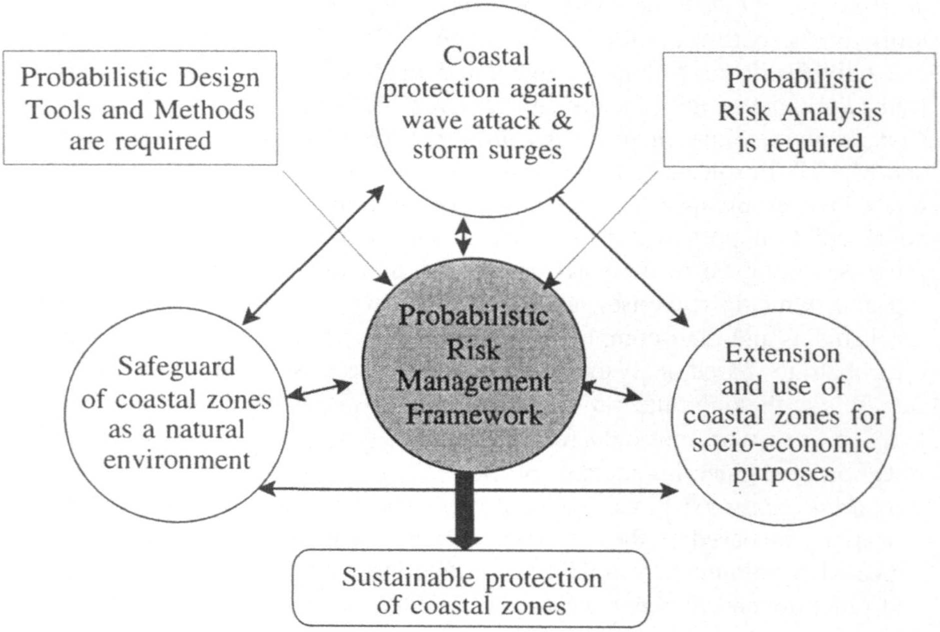

Figure 1-2. Probabilistic risk management framework: a prerequisite for a sustainable coastal protection.

However, before the breakwater will come to fail fulfilling its main function there are a number of previous failures associated with the loss of the stability of the structure components (structural failure modes) and that of the foundation (geotechnical failure modes). It is one of the main and most difficult task of the design process to properly describe and predict these failure modes (Fig. 1-3).

For this purpose the waves at the structure, the subsequent wave loading as well as the associated structural and foundation response should be determined. The stochastic nature of the waves and the wave loading, as well as the uncertainties associated with the prediction models and the input parameters together with the complex interaction of the failure mechanisms of hydraulic, structural and geotechnical nature require a probabilistic design approach which can address all relevant hydraulic, structural and geotechnical aspects of the design in an integrated manner.

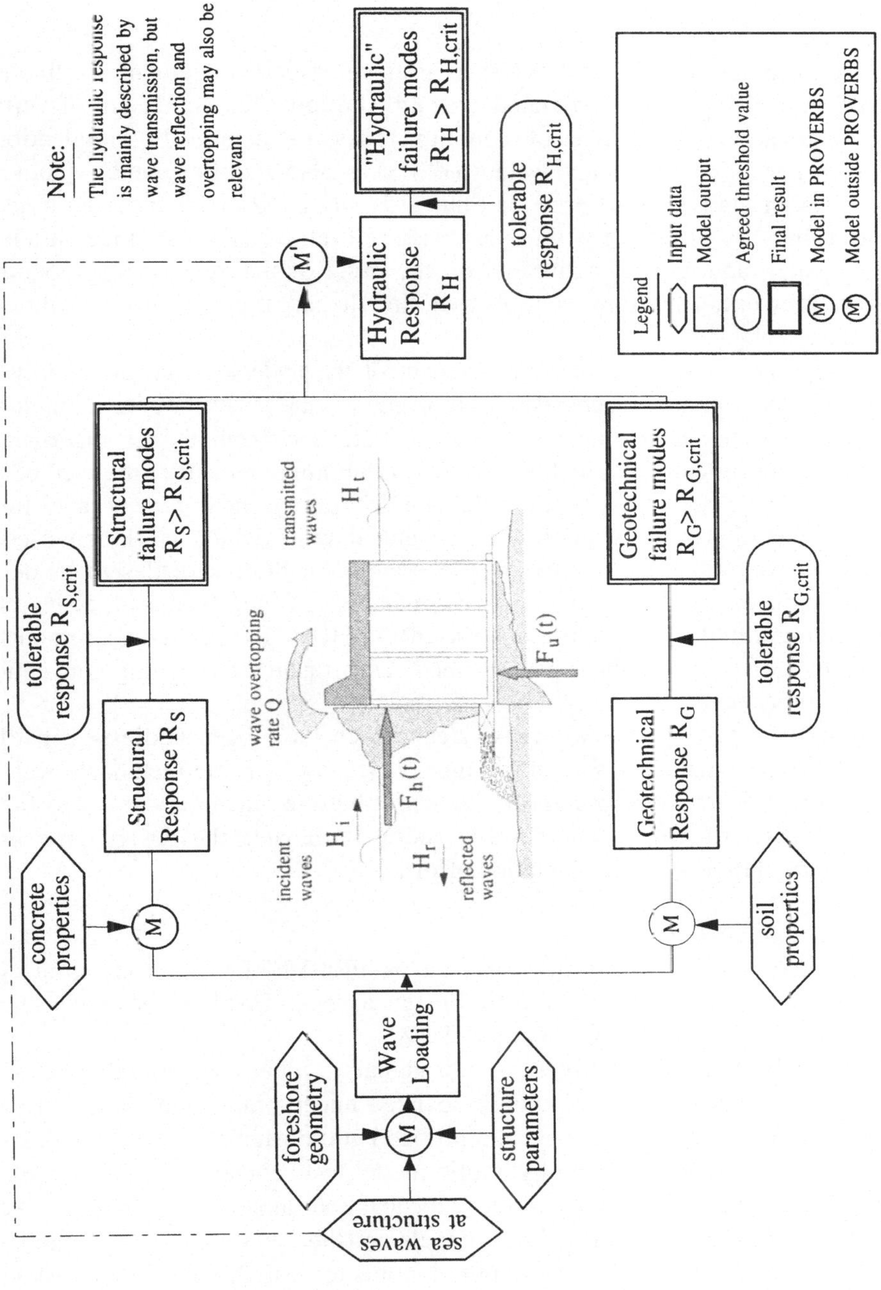

Figure 1-3. Problem definition in the design of vertical breakwaters.

1.2 BRIEF PRESENTATION OF PROVERBS

1.2.1 *Objectives*

The overall objective of PROVERBS is to develop and implement a reliability-based framework and the associated probabilistic tools for the design of vertical breakwaters and further classes of monolithic coastal structures. For application in design, a partial safety factor system (PSFS) is also required which should be coherent with the relevant Eurocode Standards, but is more advanced as it uses a target probability of failure within the structure lifetime as an entrance parameter to the system and as it is calibrated on the basis of the developed probability-based framework. Therefore the tools to define the target probability of failure are also addressed.

Further objectives are (i) the improvement of the understanding of the physical processes and failure mechanisms involved in the wave-structure-foundation interaction, (ii) the development of reliable data and methods for the dynamic analysis of the structure and its foundation, (iii) the improvement of the methods for the performance and analysis of field and laboratory measurement and finally (iv) the generation of a scientific and technical basis which might be used by authorised institutions and committees to establish authoritative design guidelines and by engineers to develop innovative structures fulfilling technical, economical, ecological and multi-purpose use criteria. PROVERBS also intends to demonstrate the advantages of probability based methods as compared to existing conventional design approaches.

The main types of vertical breakwaters which have been addressed in PROVERBS are schematically shown in Figure 1-4. However, the methods developed in PROVERBS are more generic in the sense that they also may apply to further similar coastal structures where wave action dominates the design process by means of a relatively small adaptation effort.

1.2.2 *Research Issues*

Four main issues have been addressed by PROVERBS: hydrodynamic, geotechnical, structural and probabilistic design aspects. Each of these aspects has been dealt with in a sub-project called "Task".

- *Task 1: Hydrodynamic Aspects*: Hydrodynamic inputs are provided by focusing on wave loadings, including associated uncertainties and statistical distributions required to implement the probabilistic design tools in Task 4. Prototype measurements and 3D-hydraulic model tests constitute the major part of this work, supplemented by further physical and numerical modelling. The results of these measurements and models are then analysed in detail and integrated together to provide the required inputs to Task 2, Task 3 and Task 4.

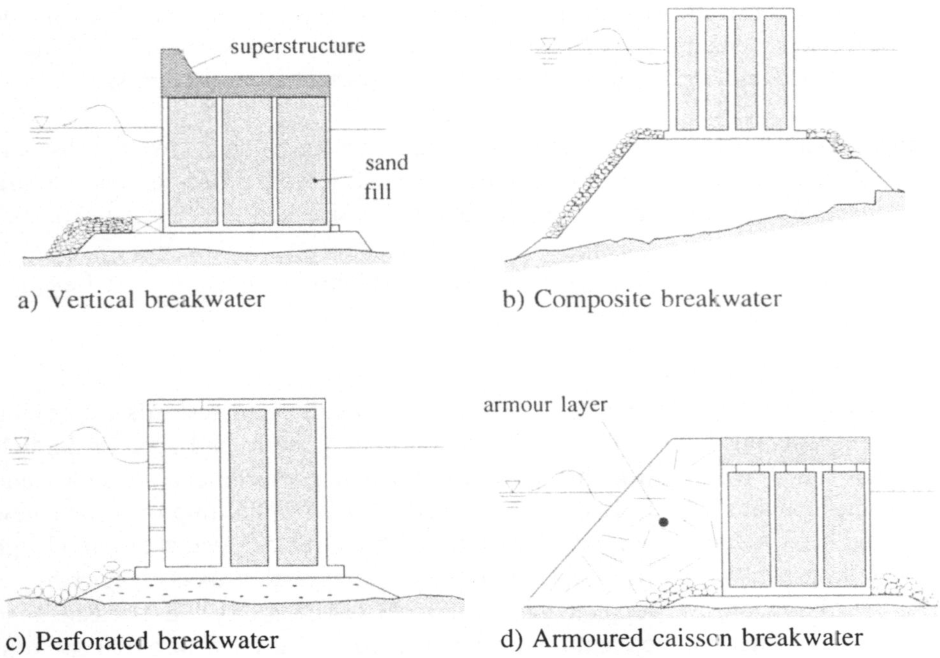

Figure 1-4. Main types of vertical breakwaters addressed in PROVERBS.

- *Task 2: Foundation Aspects*: Beside the development of new knowledge related to failure mechanisms and dynamic soil properties, information is generated on uncertainties for soil parameters and models, making them readily applicable for the implementation of probabilistic design tools in Task 4. Analysis, numerical modelling and laboratory tests build the major means for the investigations, supplemented by centrifuge modelling and prototype measurements for validation purposes.
- *Task 3: Structural Aspects*: This task provides the methods to assess the structural strength of breakwater walls during placing of caissons, under pulsating and extreme impact loading, as well as their durability. Uncertainties in the loading and resistance parameters are accounted for as needed for the implementation of the probabilistic design tools. Finite element analyses are principally used for the investigations.
- *Task 4: Probabilistic Design Tools*: In a first preparatory phase a probabilistic framework is developed by linking together all above aspects by describing the failure modes, by establishing the fault trees, by evaluating the associated uncertainties and by developing a reliability design philosophy. In a second phase, probabilistic design tools are developed. Based on reliability calcula-

tions at Level II and III, on the analysis of case studies and involvement of practitioners, a method to assess the overall reliability of vertical breakwaters is developed and applied to a set of selected representative structures.

Furthermore, Tasks 1-3 were also intended to enhance the knowledge associated with the hydrodynamic, geotechnical and structural processes and failure mechanisms involved in wave-structure-foundation-interaction.

1.2.3 *Research Strategy and Development Procedure for Probabilistic Design Tools*

1.2.3.1 *Overall Strategy*

The research strategy pursued in PROVERBS was directed towards improving the related existing knowledge and producing new generic and specific knowledge associated with hydrodynamic, structural and geotechnical processes and failure mechanisms within Tasks 1 - 3 of PROVERBS. This improved and new generated knowledge is then linked together with the related available knowledge from previous MAST-Projects and further national research projects (Fig. 1-5).

All this knowledge is supplemented within PROVERBS by the assessment of the associated uncertainties and then implemented in Task 4 to provide the reliability tools required for design. The overall strategy in the development of these tools may be summarised as follows:

- Although only vertical breakwaters are explicitly addressed in PROVERBS, the developed methods are generic, i.e. only moderate adaptation effort is required for application to further classes of coastal structures.
- Development of a probabilistic framework based on level II/III reliability analyses which is primarily used for the reassessment of existing structures and the calibration of Partial safety factors systems (PSFS), but can also be used for the design of new structures. This also includes the development of methods to define the economically optimal probability of failure (target reliability).
- Development of a Partial Safety Factor System (PSFS) (Level I) calibrated on the basis of the aforementioned probabilistic framework and primarily used for the design of new structure.

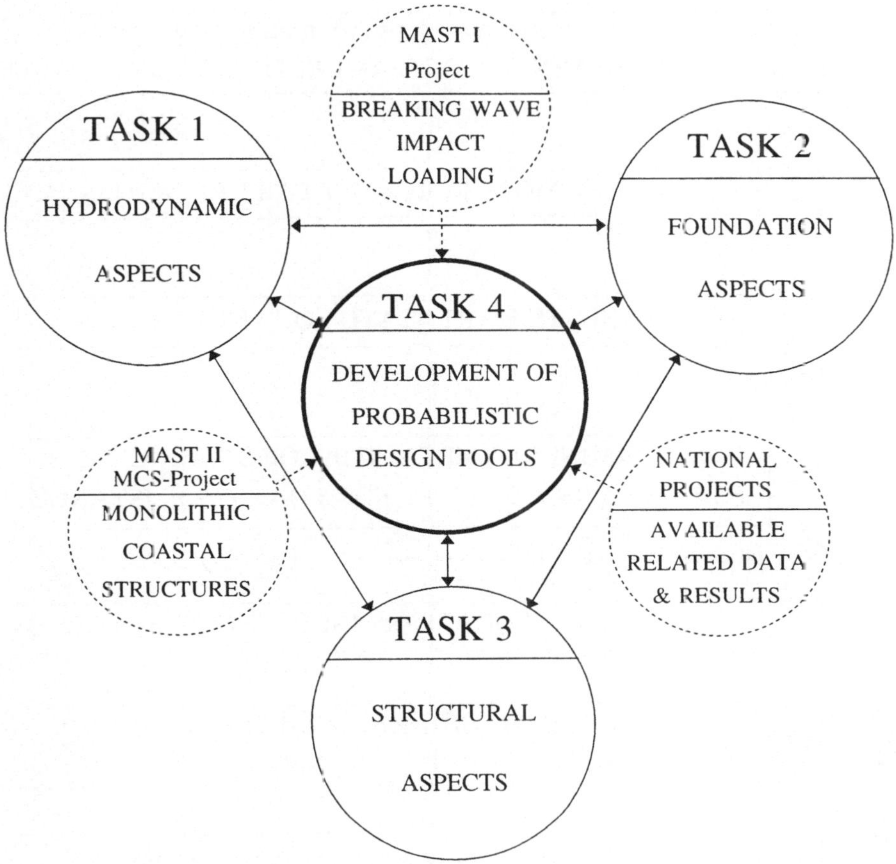

Figure 1-5. PROVERBS' research strategy: implementation of existing and new knowledge in probabilistic design framework.

1.2.3.2 *Development Procedure for Probabilistic Tools*

The developed probabilistic tools are based on level II/III reliability analyses using Bayesian updating, and include methods to account for maintenance as well as time dependent reliability methods.

The basic steps of the probabilistic procedure are illustrated by the flow chart in Figure 1-6. First, the dimensions and properties of the structure are determined on the basis of a common deterministic design procedure. Then, the most important failure modes are identified and the associated limit state equations are formulated without any consideration of the uncertainties involved (see Fig. 1-7).

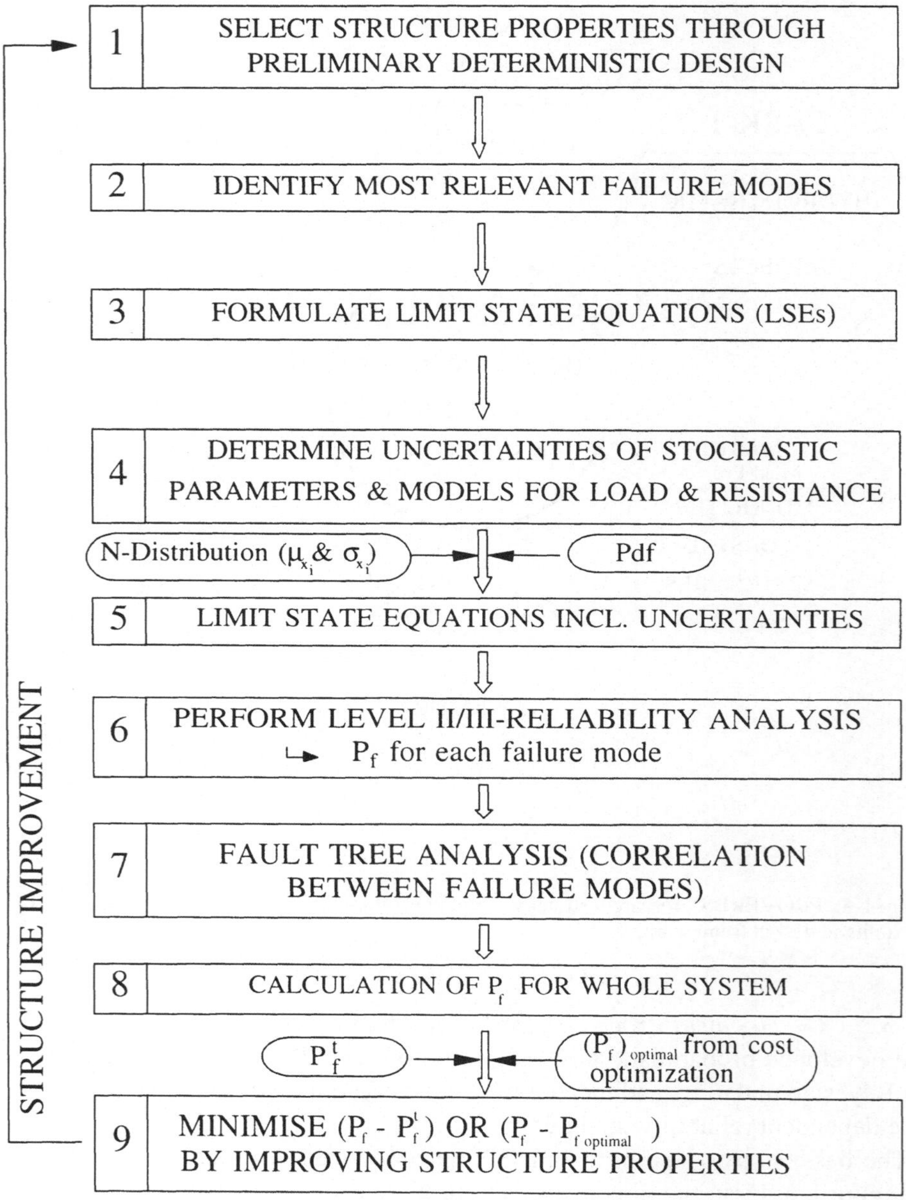

Figure 1-6. Development procedure for probabilistic design in PROVERBS (simplified principle flow chart).

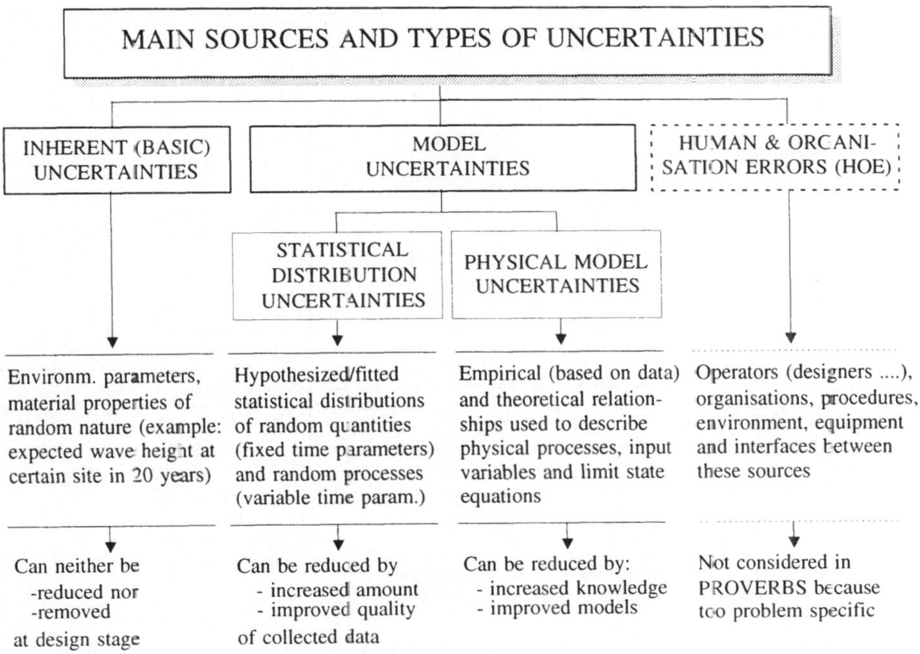

Figure 1-7. Sources and types of uncertainties.

Two types of limit states are generally considered in PROVERBS: Ultimate Limit State (ULS) and Serviceability Limit State (SLS). The uncertainties in the models and the associated input variables are determined within Step 4. The sources and types of uncertainties considered in PROVERBS are summarised in Figure 1-7. Further details are given in chapter 5.

However, Human and Organisation Errors (HOE) have been omitted in PROVERBS although they generally represent a considerable part of all uncertainties involved in the design process. There are two reasons for this omission. First, HOE are too problem specific to be treated in general terms within PROVERBS. Second, there are techniques available to account additionally for HOE, once the probability of failure has been determined according to the PROVERBS tools. One of these techniques which is used for offshore structures is shown exemplarily in Figure 1-8.

At Step 5 the related uncertainties are included into the limit state equations for each of the hydraulic, geotechnical and structural failure modes (Fig. 1-9). A level II/III reliability analysis is then performed within Step 6 (see also chapter 5). As a result the probability of failure P_f is obtained for each failure mode.

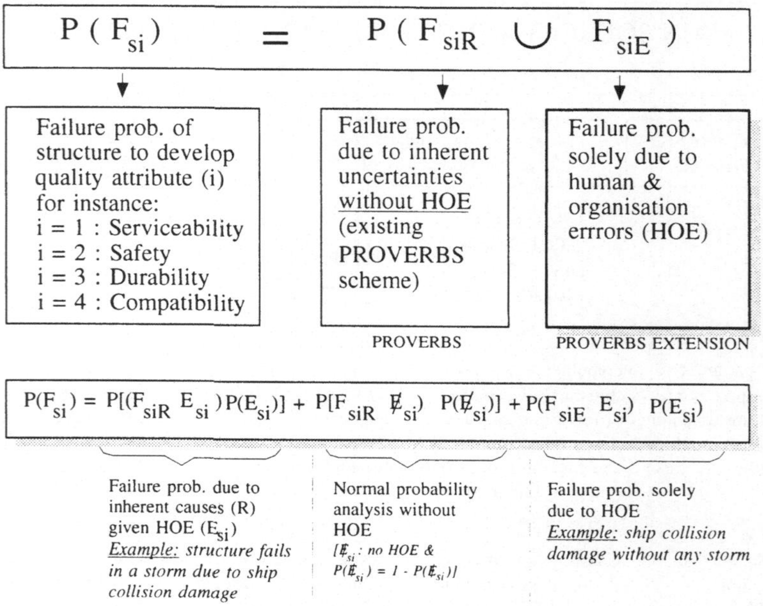

Figure 1-8. Method to account for human and organisation errors (HOE) as a possible extension of PROVERBS.

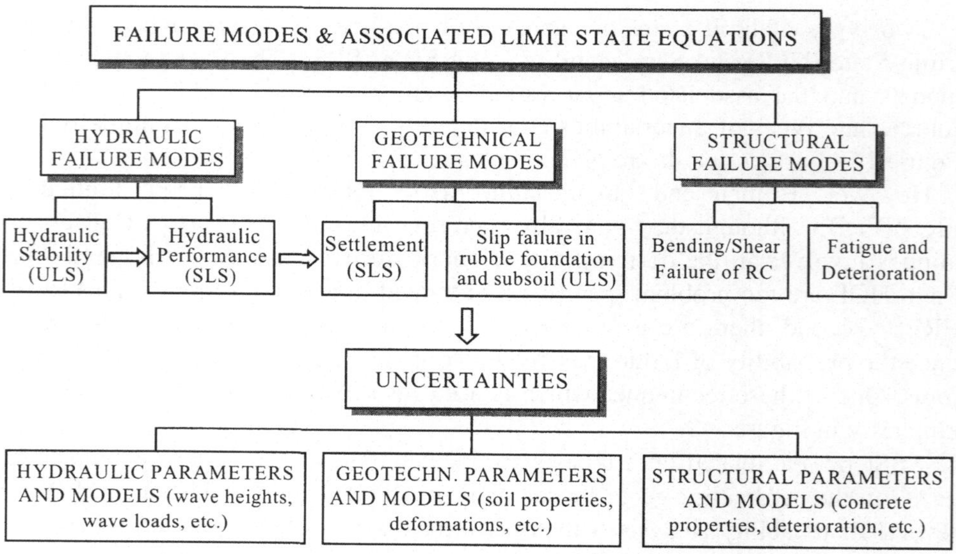

Figure 1-9. Failure modes, limit state equation and associated uncertainties.

Generally, P_f does not represent the actual probability, but only a nominal value. In Step 7 a fault tree analysis is performed in order to identify the interactions and the possible critical paths of the failure modes (see chapter 5). As a result, a probability of failure P_f for the whole system is obtained in Step 8. At this step, one of the most difficult and important tasks has to be solved: the selection of a target probability of failure P_f^t. Generally, P_f^t strongly depends on the expected consequences of failure (risk to human injury and life, cost of damage) as well as on the efforts and costs to achieve the required safety level. The criteria generally adopted in existing codes and regulations to select P_f^t are summarised in Figure 1-10 which well illustrates the degree of complexity of the problem (Example figures for P_f^t see chapter 5 and section 4.2 of Volume IId).

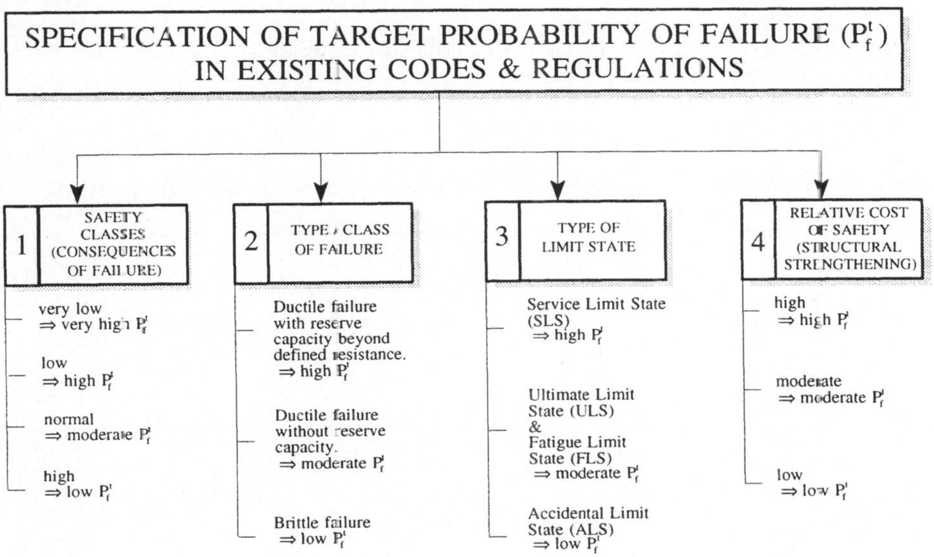

Figure 1-10. Criteria for adopting target probabilities of failure in existing codes and regulations.

The three methods which are generally used to select P_f^t are summarised in Figure 1-11. Since breakwater failures generally imply no risk to human injury and life, cost optimisation over the life time of the structure is considered as an important means to select P_f^t. The optimisation procedure is schematically illustrated in Figure 1-12 by considering the ultimate limit state related to sliding stability and the required caisson width B_c.

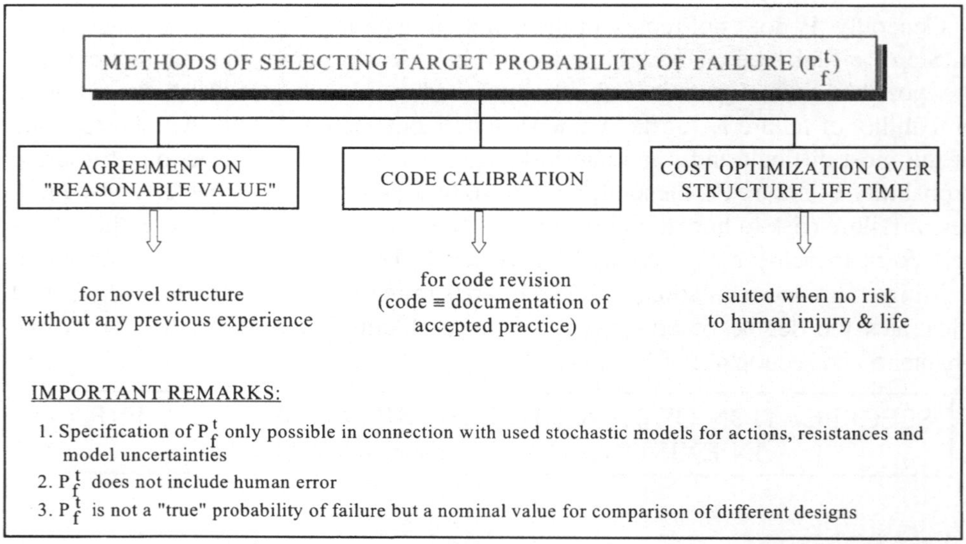

Figure 1-11. Selection criteria for target probability of failure in existing codes and regulations.

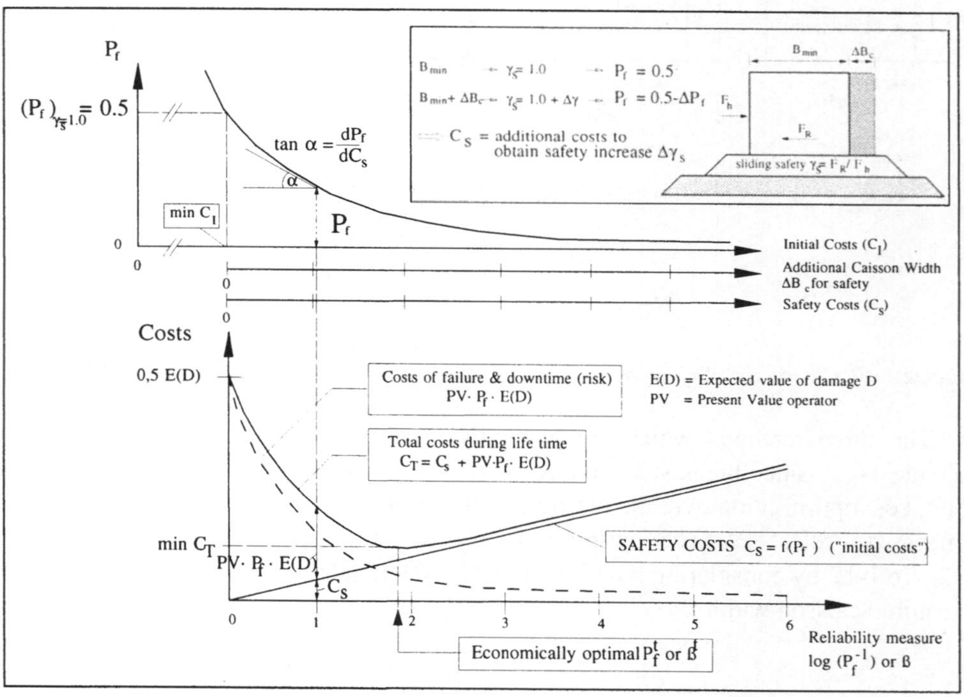

Figure 1-12. Selection of target probability of failure P_f^t based on coast optimisation (principle sketch).

Minimum initial (investment) costs (min C_I) are achieved with a minimum width B_{min} which corresponds to a safety against sliding $\gamma_s = 1.0$ associated with a probability of failure $P_f = 0.5$. Every additional investment costs (ΔC_I) by increasing the caisson width (ΔB_c) will solely serve to increase sliding safety ($\Delta\gamma$), thus decreasing the probability of failure (ΔP_f). Therefore, the additional initial costs (ΔC_I) resulting from this strengthening measure (ΔB_c) represent safety costs C_s. Increasing the safety costs will decrease the probability of failure P_f (or increase the reliability index β), but will decrease the cost of failure and down time resulting from the expected damages D. The target probability of failure P_f^t or the target reliability index β^t is then obtained at the point of minimum total costs min C_T. A method for economic optimisation has been developed to select the target probability of failure, which uses three design variables (see chapter 5 and section 4.1 of Volume IId).

1.2.3.3 *Development Procedure for Partial Safety Factor System (Level I)*

Since a level II/III reliability analysis is more difficult to handle in the design practice and since most design codes and regulations are based on safety factors systems, it is necessary to develop a Partial Safety Factor System (PSFS). In order to enhance its acceptance and application by design engineers, the following requirements were specified: (i) coherence with the Eurocode framework, (ii) use of a target probability of failure P_f^t as an entrance parameter. The general procedure for the development of PSFS in PROVERBS is schematically shown in Figure 1-13. Further details on the procedure are given in chapter 5 as well as in section 4.2 of Volume IId and Ditlevsen & Madsen (1996).

1.2.3.4 *Representative Example Structures for Application*

The reliability design tools developed in PROVERBS are applicable to a variety of vertical breakwaters and similar classes of structures subject to wave action. Nevertheless, 5 representative example structures have been selected from a set of more than 20 structures in and outside Europe to illustrate the application of the developed reliability design tools and to identify at an early stage of the project potential simplifications of the approaches applied and further unexpected problems. The choice of these 5 structures has been based on the collection and analysis of data related to hydraulic, geotechnical, structural design conditions and further information on experienced failures/successes as well as on a set of selection criteria developed in the project (Kortenhaus & Oumeraci, 1997b). The five selected structures are shown in Figure 1-14.

1	DEFINITION OF CLASSES OF STRUCTURES FOR WHICH PSFS IS AIMED
⇩	
2	DEFINITION OF SCOPE AND OBJECTIVE OF PSFS
⇩	
3	SELECTION OF TYPICAL EXAMPLE STRUCTURES WITH TYPICAL RANGES OF PARAMETER VARIATION (importance weighting)
⇩	
4	SELECTION OF TARGET RELIABILITY INDEX LEVELS
⇩	
5	CALCULATION OF CALIBRATED PSF BASED ON PERFORMANCE OF LEVEL II/III RELIABILITY LEVELS FOR WHOLE RANGES OF PARAMETER VARIATIONS IN STEP 3
⇩	
6	VERIFICATION OF PSFS (comparison with existing structures, variability of P_f over selected parameter space)

Figure 1-13. Development procedure for Partial Safety Factor System (PSFS) (principle flow chart).

The Easchel breakwater is a fictitious combination of the existing Gela breakwater/Italy and the existing sandy subsoil in Eastern Scheldt/Netherlands. For the three Japanese example structures two extreme alternatives are considered for the subsoil in the calculations: a strong and a weak subsoil, because no precise soil data could be made available by Japanese authorities.

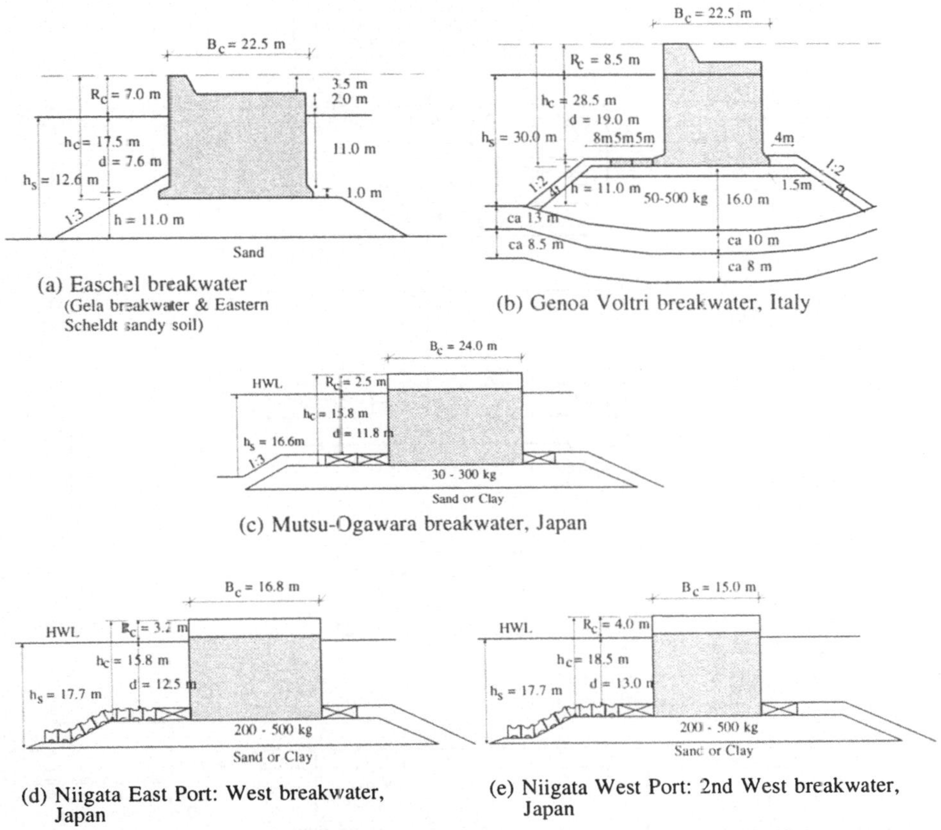

(*) since no detailed data available on subsoil two extreme cases are examined: (a) drained (sand) and (b) undrained (clay) subsoil

Figure 1-14. Representative example structures for the application of probabilistic design tools developed in PROVERBS.

1.3 KEY RESULTS AND THEIR PRACTICAL IMPORTANCE

1.3.1 *Hydrodynamic Aspects (Task 1)*

Task 1 provides for the other Tasks the hydraulic input parameters including uncertainties, as well as the models and the associated uncertainties to predict the hydraulic performance and loads needed to perform the foundation (Task 2), structural (Task 3), and reliability (Task 4) analyses (Fig. 1-15).

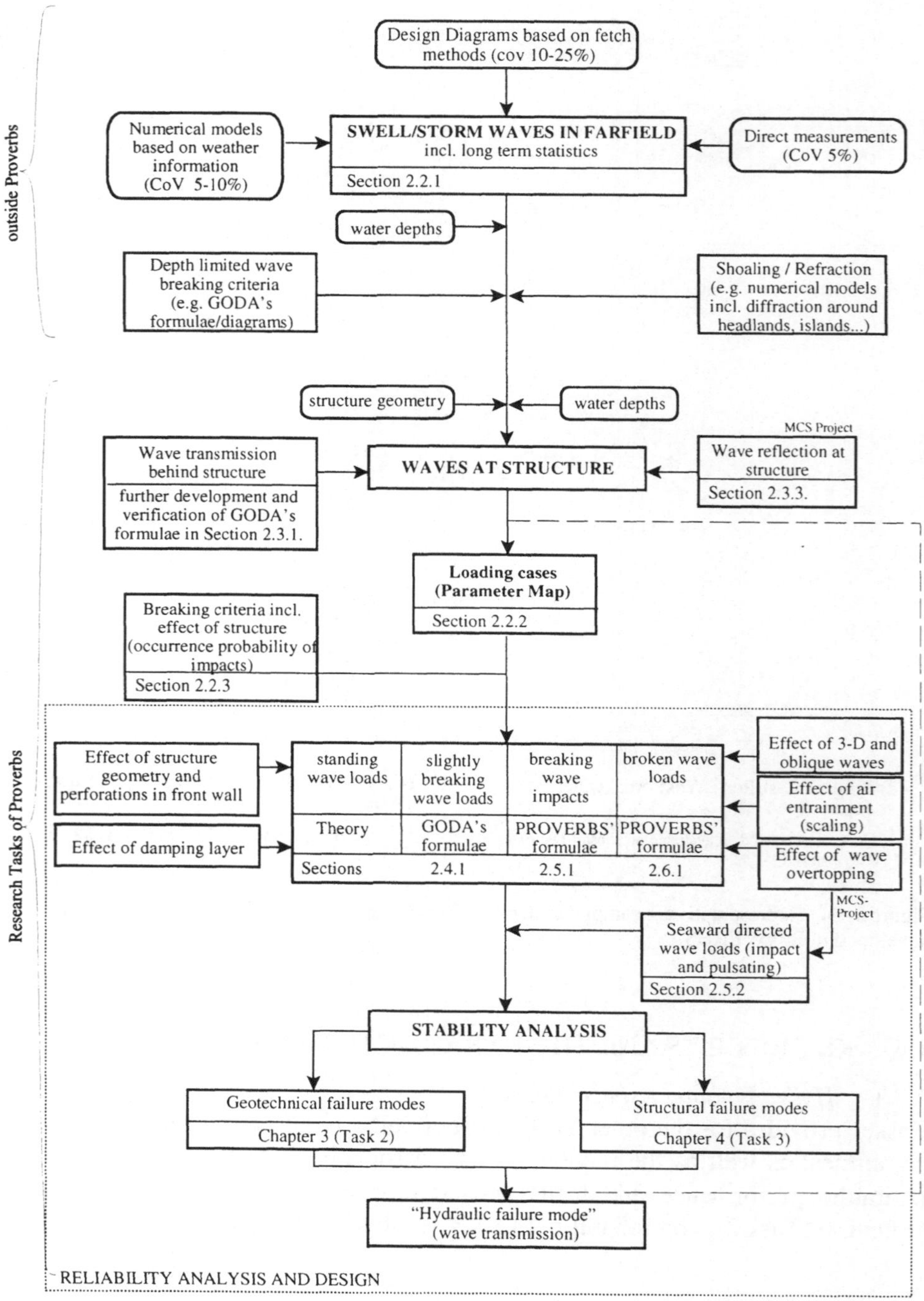

Figure 1-15. Overview of main hydrodynamic aspects of vertical breakwaters in PROVERBS.

The input waves far from the structure, including the various wave transformations in shallower water, have not been dealt with in PROVERBS and were therefore provided by existing methods. The only work of PROVERBS in this respect consisted in selecting the methods/models to be adopted and in assessing the associated uncertainties (e.g. in the form of a coefficient of variation (CoV) and statistical distributions). The actual research work of Task 1 in this respect starts when waves arrive at the structure (Fig. 1-15). The key results which have been achieved in Task 1 are briefly summarised below.

1.3.1.1 *Parameter map for wave load classification*

A practical parameter map for wave load classification has been developed and validated on the basis of laboratory testing at four different institutes (section 2.2.2). It particularly allows to distinguish between (i) impact loads for which the load duration/time history is most relevant for the dynamic response of the structure and which therefore need to be handled with special care and (ii) those "pulsating" wave loads for which the expected response of the structure is such that "quasi-static approaches" might apply (Fig. 1-16).

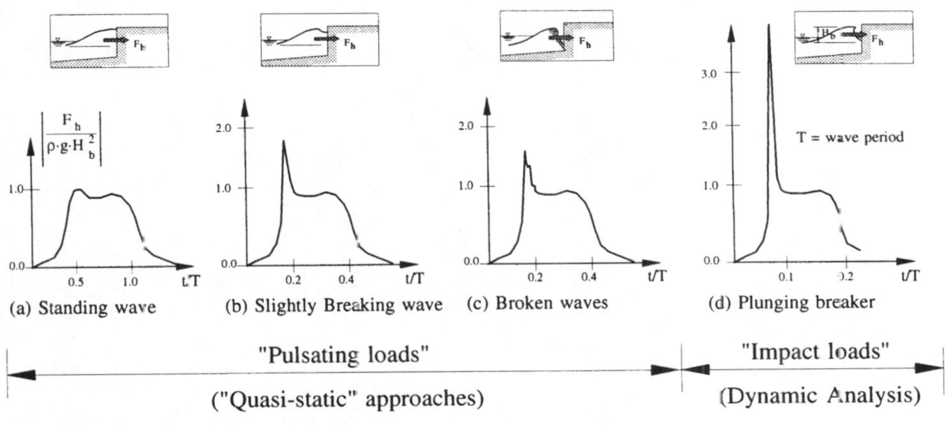

Figure 1-16. Distinction between "impact" and "pulsating" loads.

The parameter map concept as illustrated by Figure 1-17 is based on the use of three simple non-dimensional input parameters related to the structure geometry as well as to the water depth and wave conditions at the structure to decide at a first stage on the type of monolithic structure (relative height h_b^*) and at a second and third decision level on the loading case (relative wave height H_s^* and relative berm width B^*).

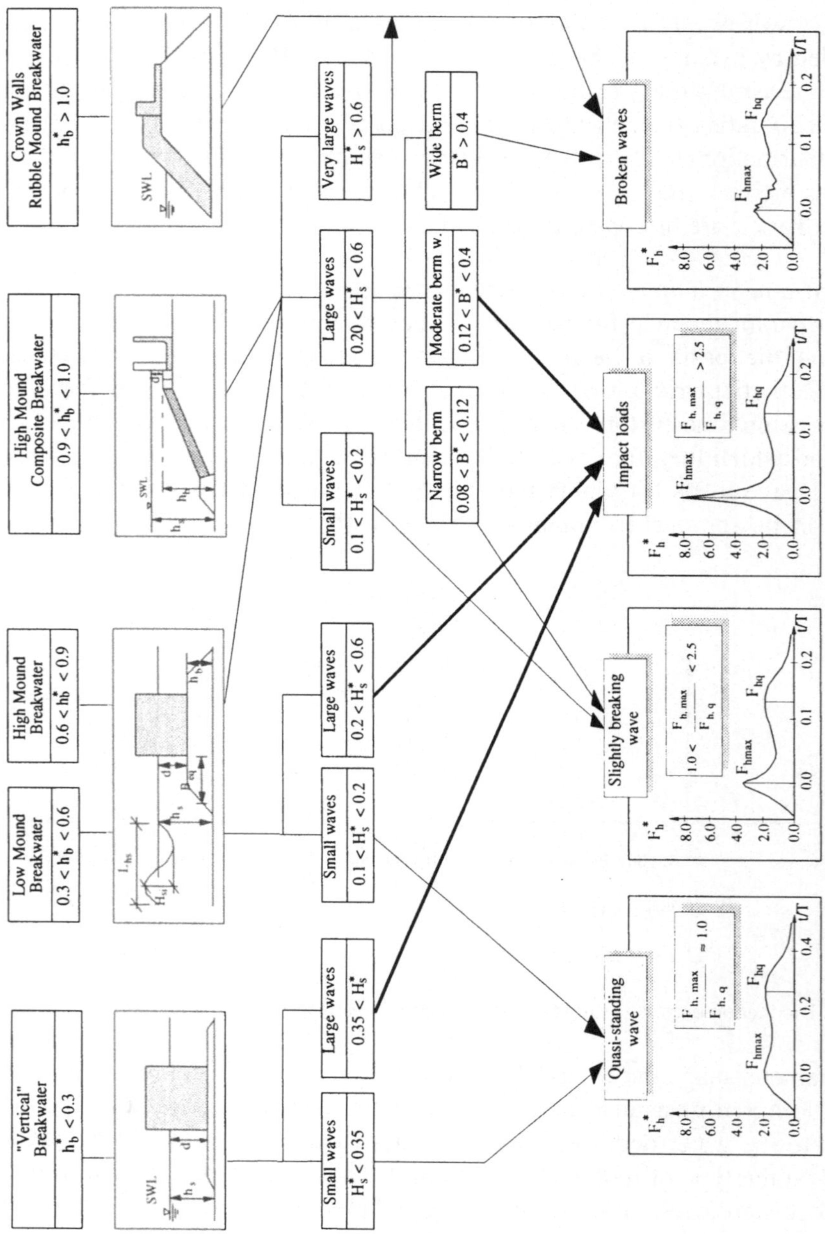

Figure 1-17. Extended parameter map including "high mound composite breakwaters" (Kortenhaus & Oumeraci, 1998).

Using further results from large-scale model tests performed in the Hannover flume (GWK) for two Japanese projects (Oumeraci & Muttray, 1997; Oumeraci et al., 1998) the PROVERBS Parameter map in Figure 1-17 can be extended to include the so-called "High-Mound Composite Breakwater" (HMCB) which is becoming very popular for the protection of artificial islands in Japan (Muttray et al., 1998).

The parameter map in Figure 1-17 has been further supplemented by a simple procedure to predict the occurrence frequency of breaking wave impacts (section 2.2.3 and Eq. 2-46). This procedure is based on two threshold values of the incident wave height $H_{0,4\%}$ at the location of the structure: (a) the breaker height H_{bc} at the structure which is calculated by means of a new breaking criterion taking into account the reflection properties of the structure (Eq. 2-43), and (b) the wave height H_{bs} describing the transition zone at which broken waves occur (Eq. 2-45).

Depending on the purpose and the failure modes for which they are used a further classification of the wave loads is suggested: quasi-static loading (overall failure), cyclic loading (fatigue and stepwise failure), and impact loading (progressive and single overall failure / structural failure modes). This classification (see Fig. 2-5) which may be further refined permits to properly select the wave load type and specify it in more detail according to the failure mode under study (single or progressive, local or overall, structural or geotechnical failure mode).

1.3.1.2 *New formulae to predict impact loading*

New formulae to predict impact loading and associated load duration, including the effect of air content (scale effects) have been developed on the basis of mathematical/numerical analysis, large-scale model testing, small scale model testing with fresh and salt water as well as prototype measurements. The overall research strategy which have been pursued in PROVERBS for the impact loading is schematically illustrated by Figure 1-18. Based on solitary wave theory, impulse-momentum considerations, extensive small-scale laboratory tests, including sophisticated measuring techniques (PIV) and large-scale model testing, a new method has been developed to predict the relative impact forces ($F_{h,max}/\rho g H_b^2$) as a function of the relative rise time $t_r / \sqrt{d_b / g}$ (Fig. 1-18). For further details, including pressure distribution refer to section 2.5.

The new breaking wave impact prediction formulae for horizontal and uplift loading, together with the associated spatial and temporal pressure distributions provide the yet missing and urgently needed loading inputs for any dynamic analysis related to the overall stability of the breakwater, including geotechnical and structural failure modes (see also chapters 3 and 4). These are in fact very important, because the existing conventional design approaches based on static design loads can neither explain the failures experienced by prototype structures, nor do they allow any design optimisation.

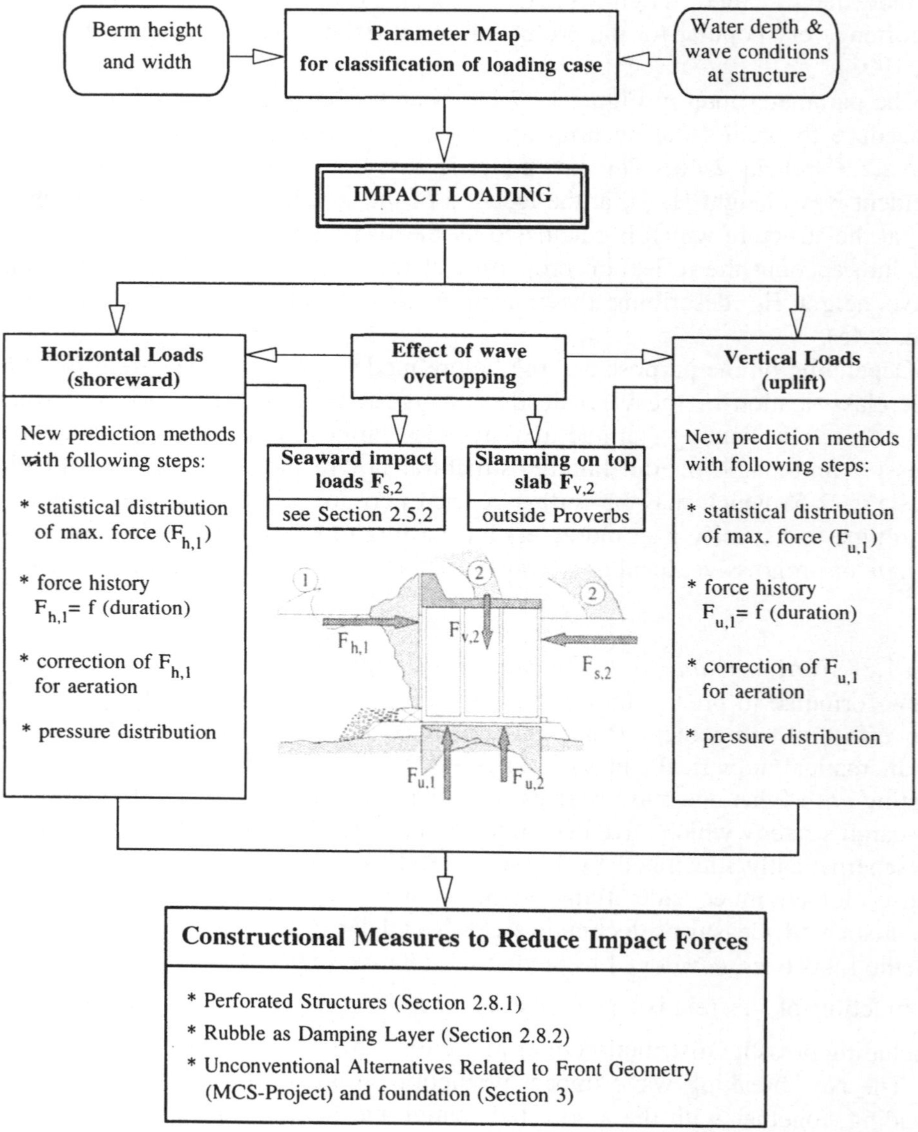

Figure 1-18. Research strategy for wave impact loading of vertical structures in PROVERBS/Task 1.

1.3.1.3 *Effect entrained/entrapped air on scaling impact loads*

Mathematical studies (Peregrine & Thais, 1996) and laboratory testing (Oumeraci & Hewson, 1997) have shown that even a very small fraction of air in water can dramatically reduce the impact pressure by increasing the impact duration. This and the scaling problems illustrated by Figure 2-26 point out towards the necessity of devoting more effort to estimate the volume fraction on entrained/entrapped air during impact. Moreover, comparative impact tests using fresh and sea water have shown that higher impact tests are expected from fresh water tests due to smaller bubble sizes and higher aeration levels in sea water. These efforts have led to the development of a new instrumentation to measure air content which has been deployed in the laboratory and in the field (see Volume IIa, chapter 7).

Moreover, detailed laboratory testing has also led to a much better understanding of the physics of impact loading and the scaling laws associated with each of the physical processes involved within the whole impact history (Fig. 2-26).

In fact, the results have shown that, whether FROUDE or MACH-CAUCHY similarity laws or a combination of both has to be applied for the interpretation of the test results, will strongly depend on the level of aeration and the amount of entrapped air which both determine the compressibility of the impacting fluid mixture. Therefore an improved scaling procedure based on the separation of the different components of the impact load history has been suggested (Fig. 2-27). For further details see Kortenhaus & Oumeraci (1999).

A further more pragmatic approach based on correction factors for the impact load (reduction load factor) and the related rise time (inverse value of reduction load factor) has also been developed (see Eqs. 2-99 and 2-100) which are based on an empirical relationship between the level of aeration (in %) and the number of breaking waves within a given time (Eq. 2-97).

The results which have been achieved in the scaling issue by using theoretical, numerical and experimental studies, together field measurements provide the engineer with much improved tools to scale-up the laboratory results to prototype conditions and to assess the associated uncertainties in modelling wave impact loading. Moreover, they built for the researcher a very good starting basis towards further improvements.

1.3.1.4 *Effect of caisson length, wave obliquity and short-crestedness on impact forces*

In the previous MAST II/MCS-Project (MAS2-CT92-0047) force reduction factors were developed for the pulsating load to account for wave obliquity and shortcrestedness (Franco et al., 1996). In PROVERBS further 3D-tests were performed (i) to extend the previous study for pulsating waves forces by also addressing in more detail the variability of the forces along the breakwater (Volume IIa, section 4.3, Madrigal, 1998) and (ii) to focus more on the occurrence and variability of breaking wave impact forces along the breakwater as a function of the wave obliquity, short-crestedness and berm geometry (Volume IIa, section 5.3).

The new results on the pulsating load case have generally confirmed the previous MCS-conclusions with respect to the reduction of non breaking wave forces (Franco et al., 1996); i.e.

(i) the method of Goda (1974) for long crested waves can be applied to predict the reduction of non breaking (pulsating) wave forces but is too conservative when applied to breaking wave (impulsive) forces,

(ii) short-crestedness may induce a pulsating force reduction (15 - 30% as compared to long crested waves) only for head-on waves ($\beta = 0°$),

(iii) the theoretical force reduction factors suggested by Battjes (1982) for non breaking waves are too conservative for most practical caisson lengths.

The new results on the breaking wave forces and impulsive loading (Volume IIa, section 5.3) have led to the following key results:

(i) while for head-on-waves ($\beta = 0°$) no significant difference occurs between long-crested and short crested sea with respect to the onset of breaking, for oblique waves ($\beta \neq 0$) less breaking waves are observed for short-crested than long-crested sea,

(ii) there are less breaking wave impacts for oblique waves than for head-on waves,

(iii) the onset of breaking as well as the occurrence frequency and the severity of breaking wave impacts are strongly affected by the berm geometry,

(iv) prediction formula has been developed for the impact force reduction as a function of the relative caisson length L_c/L_{op} for various angles of wave obliquity, including long-crested and short-crested sea (Fig. 2-25).

For further details refer to chapter 2 and Volume IIa, section 5.3. These results will allow to quantify spatial variability of the impact force along the breakwater which generally result in a substantial force reduction as compared to the present conservative design approach based on the assumption that the impact force calculated for a one meter long caisson will act unchanged over an infinite length (spatially unlimited impact).

1.3.1.5 *Seaward impact forces induced by wave overtopping*

Overtopping waves plunging into the harbour basin behind the breakwater have been identified by wave flume tests and numerical modelling based on pressure impulse theory as the dominant mechanism generating impact loads directed seawards (Fig. 2-22). The pressure-impulse distribution on the rear face and the bottom slab of the caisson have been determined, showing the governing role of the entrapped air under the overtopping plume in the impact generation.

The improved physical understanding of the generation mechanisms which has been achieved by these results will allow to avoid in future design the seaward tilting failures which have been observed for many prototype breakwaters for excessive wave overtopping conditions during storms (Oumeraci, 1994).

1.3.1.6 *Artificial neural network modelling of wave force*

Taking the opportunity that large data sets from hydraulic model tests performed by several institutes are available, Artificial Neural Network (ANN) Modelling has been performed to improve the prediction of horizontal pulsating wave forces (van Gent & Van den Boogaard, 1998). The ANNs predicted results depict indeed a better agreement with the measurements then the standard Goda method which clearly underpredicts the wave forces (for further details see section 2.4.5).

The results indicate that ANN-Modelling is a promising tool for the interpretation of experimental data, as well as a complementary prediction tool for design purposes. Moreover, a new technique has also been developed to quantify the reliability intervals of the predicted ANN-outputs, thus making ANN-modelling also readily applicable for probabilistic analysis (see also chapter 5).

1.3.1.7 *New prediction formulae for pulsating wave forces on perforated caisson breakwaters*

Due to the better performance of perforated wall structures in terms of wave reflection, overtopping, load reduction and environmental impact, the use of perforated caisson is expected to increase more than that of plain wall caissons structures. Therefore, comprehensive laboratory and field studies have been performed by a group of three laboratories towards an improved physical understanding of the hydraulic performance and a better prediction of the pulsating wave loads. This research has been based on

(i) two case studies, Dieppe West Breakwater (Seine Maritime/France) and Porto Torres breakwater (Sardinia/Italy), including prototype measurements and small-scale model tests for both breakwaters and
(ii) a basic large-scale model study on perforated wave screens and perforated caisson breakwaters with one and multiple wave chambers.

The overall research strategy for perforated structures is schematically illustrated by Figure 1-19.

Different approaches have been used at three laboratories to predict the pulsating wave loads on perforated structures (Fig. 1-20):

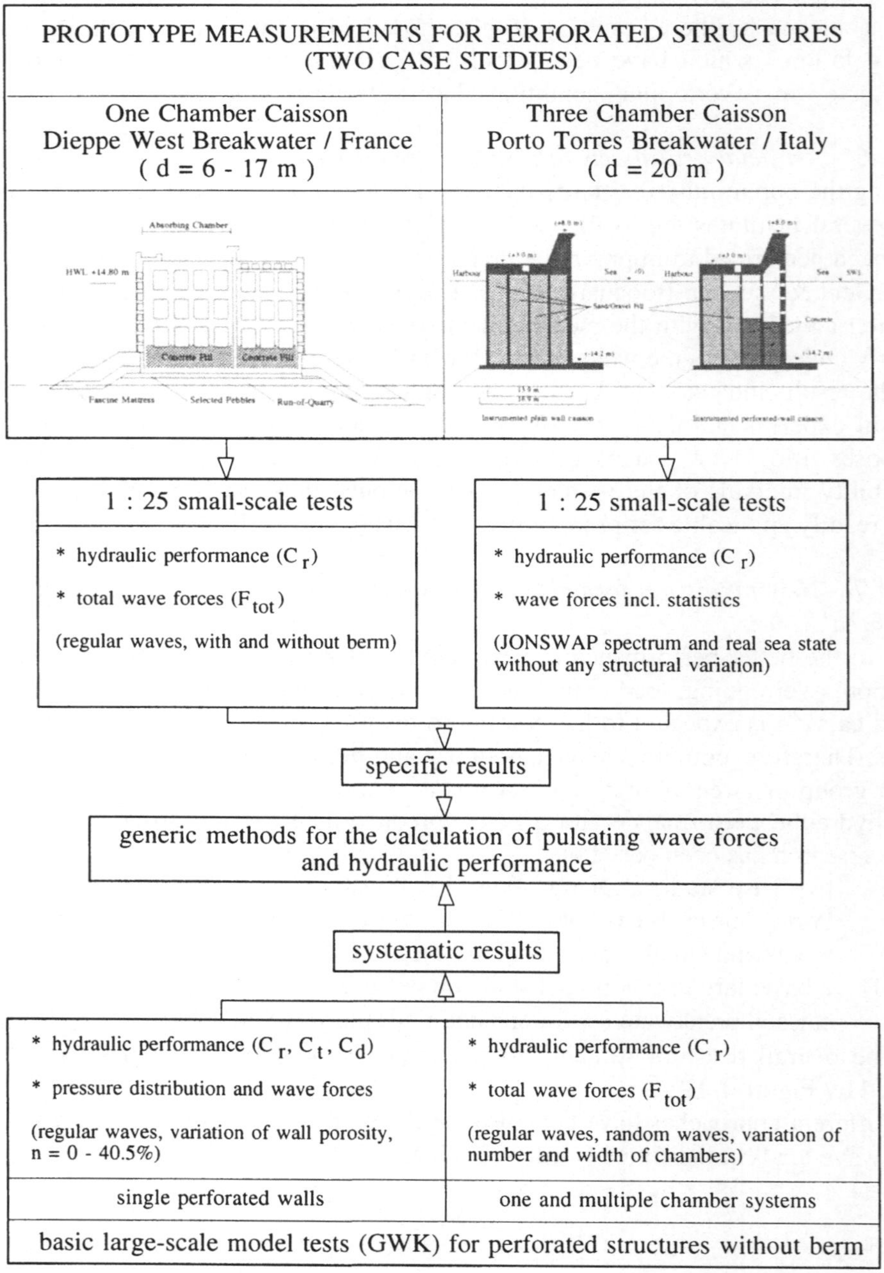

Figure 1-19. Research strategy for perforated structures.

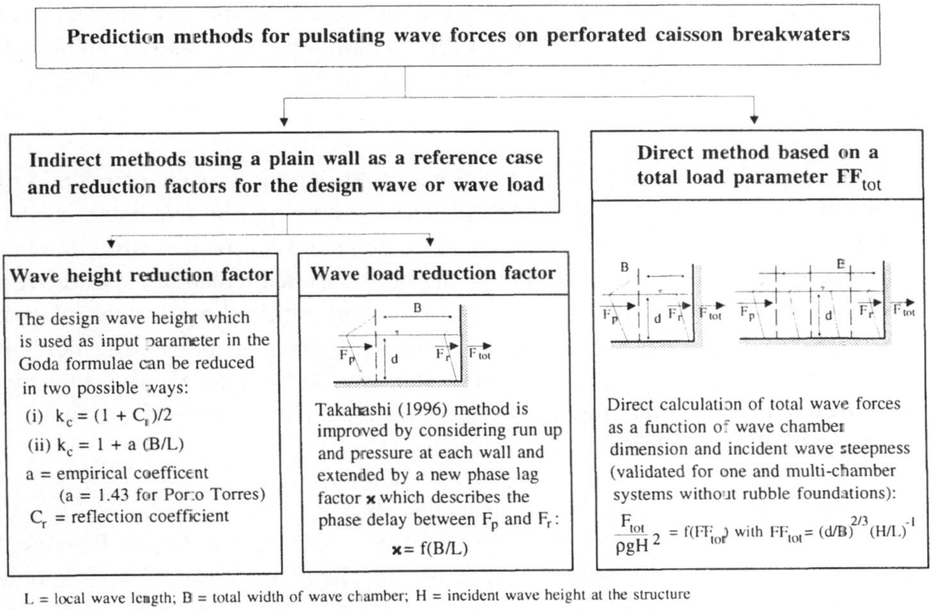

Figure 1-20. Approaches used to develop prediction formulae for pulsating loads on perforated structures.

(i) Indirect method based on reduction factors for the wave height at the structure resulting from lower reflection. The reduction factor K_r which is expressed as a linear function of the relative width of the wave chamber (B/L) is applied to the wave height which then represents the input into the GODA-formulae for the prediction of the pulsating wave loads for plain wall structures. This method has been applied to Porto Torres breakwater (Fig. 2-40) and has been found to be slightly more conservative for $B/L < 0.25$ than the simpler method which consists in calculating the reduction factor K_r directly from the reflection coefficient C_r of the structure ($K_r = (1 + C_r)/2$).

(ii) Indirect method based on reduction factors of the wave loads as calculated by the GODA formulae for plain walls (Takahashi, 1996). A substantial improvement of the Takahashi method has been achieved by the introduction of a new phase lag factor χ which describes the phase delay between the resultant wave forces F_p on the perforated wall and the resultant force F_r on the plain back wall and which allows to calculate directly the total wave force F_{tot} on the caisson breakwater.

(iii) Direct method based on a new total load factor $FF_{tot} = (d/B)^{2/3} (H/L)^{-1}$ which describes the dimensions (depth and total width) of the chamber and the incident wave steepness. The derived empirical formula generally yields results showing that total force reductions in the order of 30% for a single chamber system and in the order of 50% for a three chamber system can be achieved. It is planned to extend this method by accounting for the variation of the berm geometry.

An improved understanding of the wave energy dissipation mechanisms and of the hydraulic performance has also been provided, showing that a substantial improvement of the hydraulic performance can be achieved by using multi-chamber systems instead of the traditional single chamber Jarlan breakwater. Moreover, the hydraulic performance remains at its highest level within the practical relative chamber width (B/L > 0.3), irrespective of the frequency range of the incident waves; i.e.the multi-chamber perforated caissons are much less selective in terms of wave frequency than the single chamber caissons. This represents a considerable improvement for practice.

1.3.1.8 *New wave load formulae for crown walls*

Based on prototype measurements on the crown wall of the Gijon Breakwater/Spain as well as corresponding small-scale and large-scale model tests a new method based on a discrimination between "dynamic" and "reflecting" pressure distribution has been developed for the prediction of wave loads on crown walls of rubble mound breakwaters (Volume IIa, section 6.2, Martín & Losada, 1998) (Fig. 1-21).

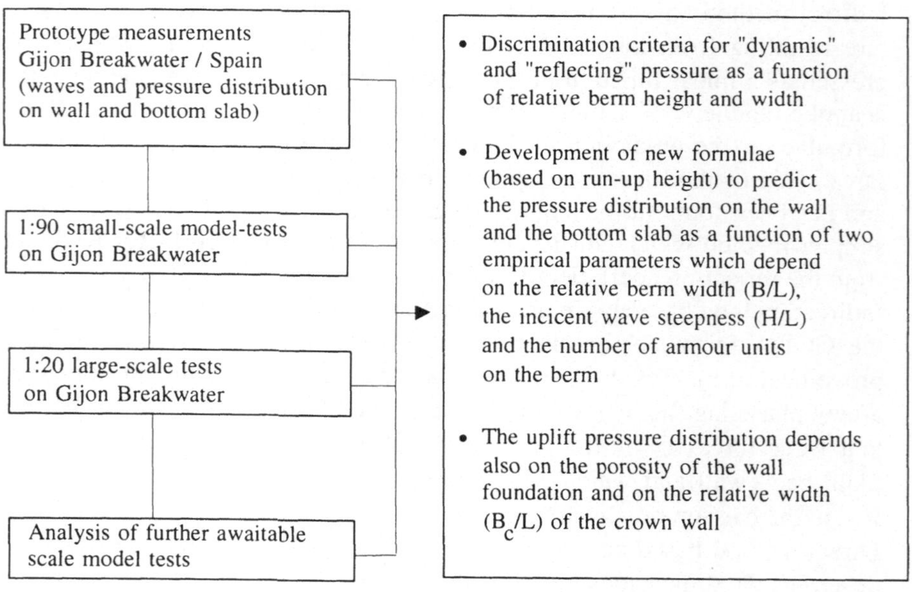

Figure 1-21. Research strategy for wave loads on crown walls.

The new formulae which allows to calculate the pressure distribution on the front wall and the bottom slab are based on the run up height and the Iribarren number as well as on the evaluation of two empirical coefficients, one describing the "dynamic" pressure and the other one the "reflecting" pressure. The first parameter is essentially governed by the relative berm width (B_b/L_p) while the second one is determined by the local incident wave steepness (H_i/L), as well as by the relative berm width. In addition, the uplift pressure distribution depends on the relative width of the crown wall (B_e/L_e) and on the porosity of the berm.

The results of these studies, which will go on even after MAST III, represent the first attempt to develop complete design formulae for crown walls on the basis of prototype measurements, small and large-scale model tests.

1.3.1.9 *Development of wave load formulae for High Mound Composite Breakwaters*

High Mound Composite Breakwaters (HMCB) are rubble mound structures with a concrete superstructure which is smaller than a caisson of a traditional composite breakwater, but larger than a crown wall of a traditional rubble mound breakwater. This old concept has been used in the last century, for instance in Cherbourg and Alderney. Recently, this old concept has been rediscovered and further improved in Japan where it is particularly used for the protection of artificial islands. The improvement consists essentially in providing an innovative slit wall superstructure, thus ensuring a considerable reduction of wave reflection (30%), wave load (horizontal: 50%, uplift: 30%), wave overtopping (30-90%) and splash height (70%) as compared to the older concept using a common plain wall superstructure. Since there are basic differences in the way a HMCB-structure works as compared to traditional composite caisson breakwaters and crown walls, the wave loading is also different.

Therefore, it was necessary to develop new design formulae for the wave loading and the hydraulic performance of these new types of structures.

For this purpose, large-scale model tests were performed within two joint specific projects[1] outside PROVERBS. However, the data were also further analysed for PROVERBS in order to derive generic prediction formulae for this type of structure. Therefore, the research strategy briefly summarised in Figure 1-22 was adopted.

---

1 Jointly supported by Port and Harbour Research Institute (PHRI)/Yokosuka and Leichtweiß-Institut (LWI) for the project on wave loading and structural response and by Civil Engineering Research Institute (CERI)/Hokkaido for the project on wave overtopping and splash reduction (Oumeraci et al., 1997 and Oumeraci et al., 1998)

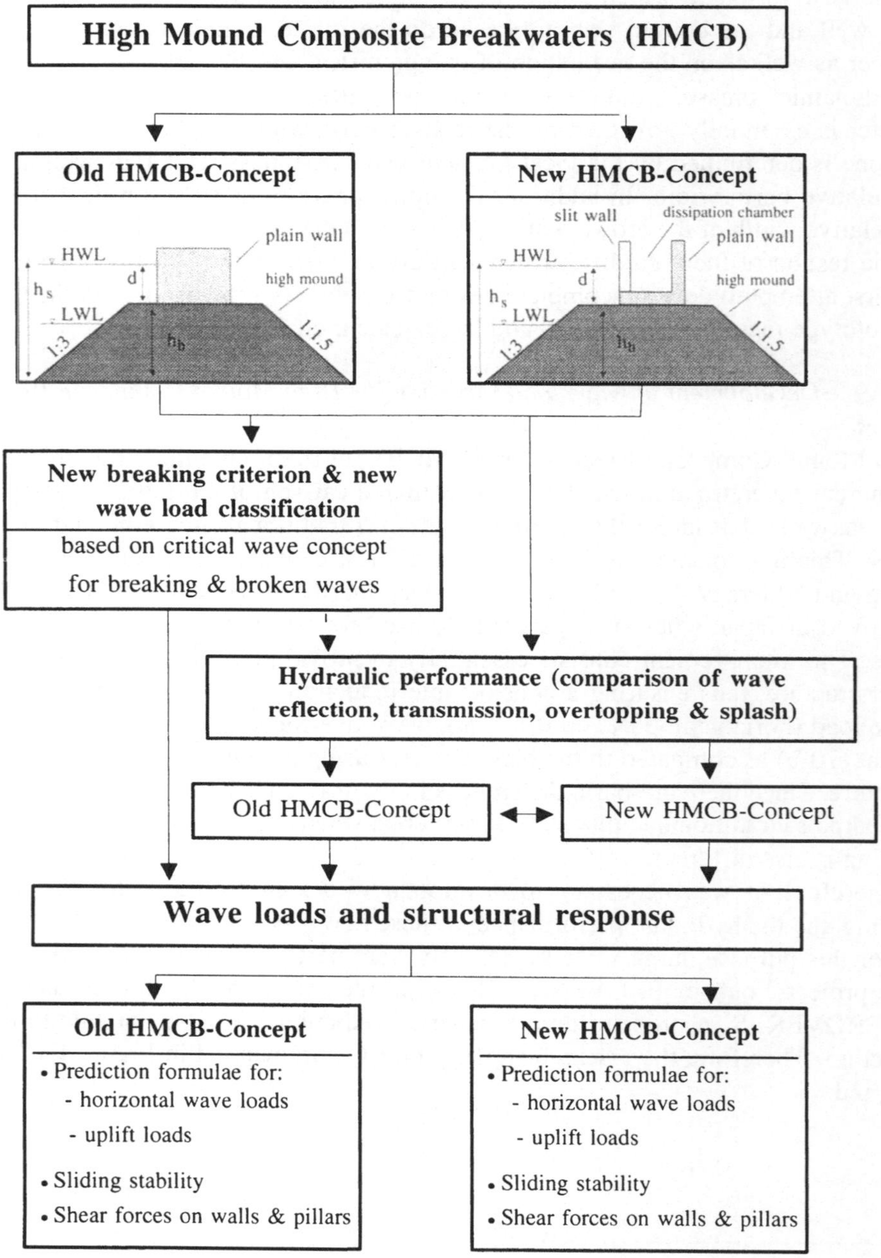

Figure 1-22. Research strategy for High Mound Composite Breakwaters and large-scale model tests in GWK.

Based on the experimental results, new formulae were developed for the hydraulic performance (Muttray et al., 1998) as well as for the horizontal and uplift loads (Volume IIa, section 6.3). These results are very important as they substantially supplement the PROVERBS parameter map (Fig. 1-17) by providing a new class of monolithic structures for which neither the prediction formulae developed for a crown wall nor those developed for a traditional composite caisson breakwater can apply.

1.3.2 *Geotechnical Aspects (Task 2)*

Traditionally, the overall stability analysis of vertical monolithic breakwaters has been performed essentially for (i) sliding over the rubble foundation, and (ii) "overturning" around the heel, while the bearing capacity has only been accounted for by comparing the stresses transmitted into the foundation by the wave load to a threshold value associated with the characteristic of the foundation material (e.g. for rubble ≈ 500 kN/m^2 (Goda, 1985)). This procedure is too simplistic and unrealistic in the sense that it completely ignores the actual physical processes involved in the dynamic structure-foundation interaction, like resonance and amplification, soil degradation due to repetitive loading, instantaneous and residual excess pore pressure effects, as well as further effects which may lead to many forms of bearing capacity failure in the rubble foundation or/and sub-soil, including seaward failure (De Groot et al., 1996; see also Figs. 3-1 to 3-3).

The design sequence from the perspective of a foundation engineer is schematically illustrated in Figure 1-23. The technical basis to implement this design process has been developed in Task 2 of PROVERBS. For further details refer to chapter 3. The key results which have been achieved in Task 2 are summarised in the following sections.

1.3.2.1 *Data base for design soil parameters*

A data base for design soil parameters and a detailed strategy for soil investigations at various design levels (Fig. 1-24) has been developed, including related diagrams for the estimation and interpretation of the required soil parameters as well as the associated uncertainties. Recommendations on how to reduce these uncertainties are also provided.

The developed data base, including a user manual and a procedure for updating and management, represents an unique source for the reliability analysis and foundation design of vertical breakwaters. This is very valuable as the data are obtained by analysing all available results at various laboratories, by performing new experimental and numerical tests and by incorporating the offshore experience with respect to clayey subsoils.

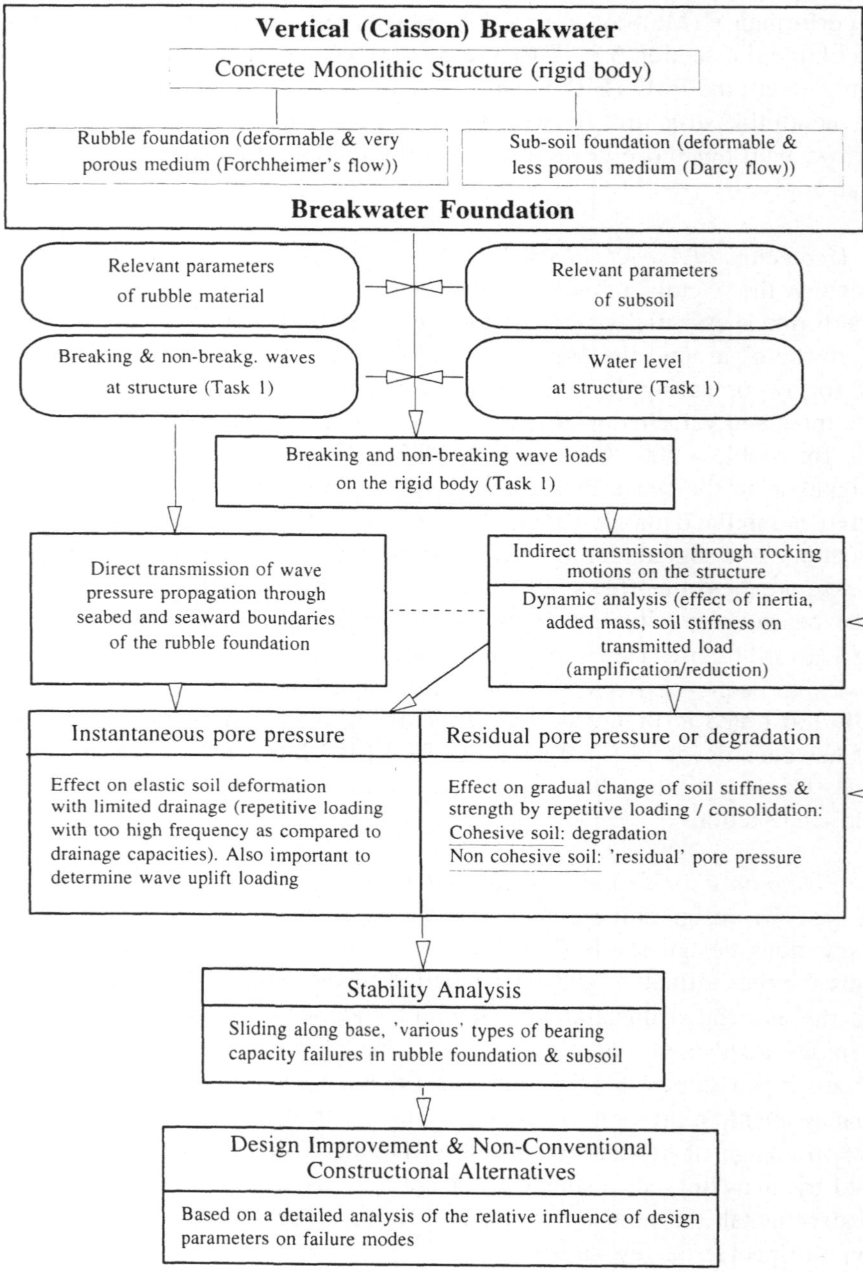

Figure 1-23. Design sequence from the foundation engineer's perspective.

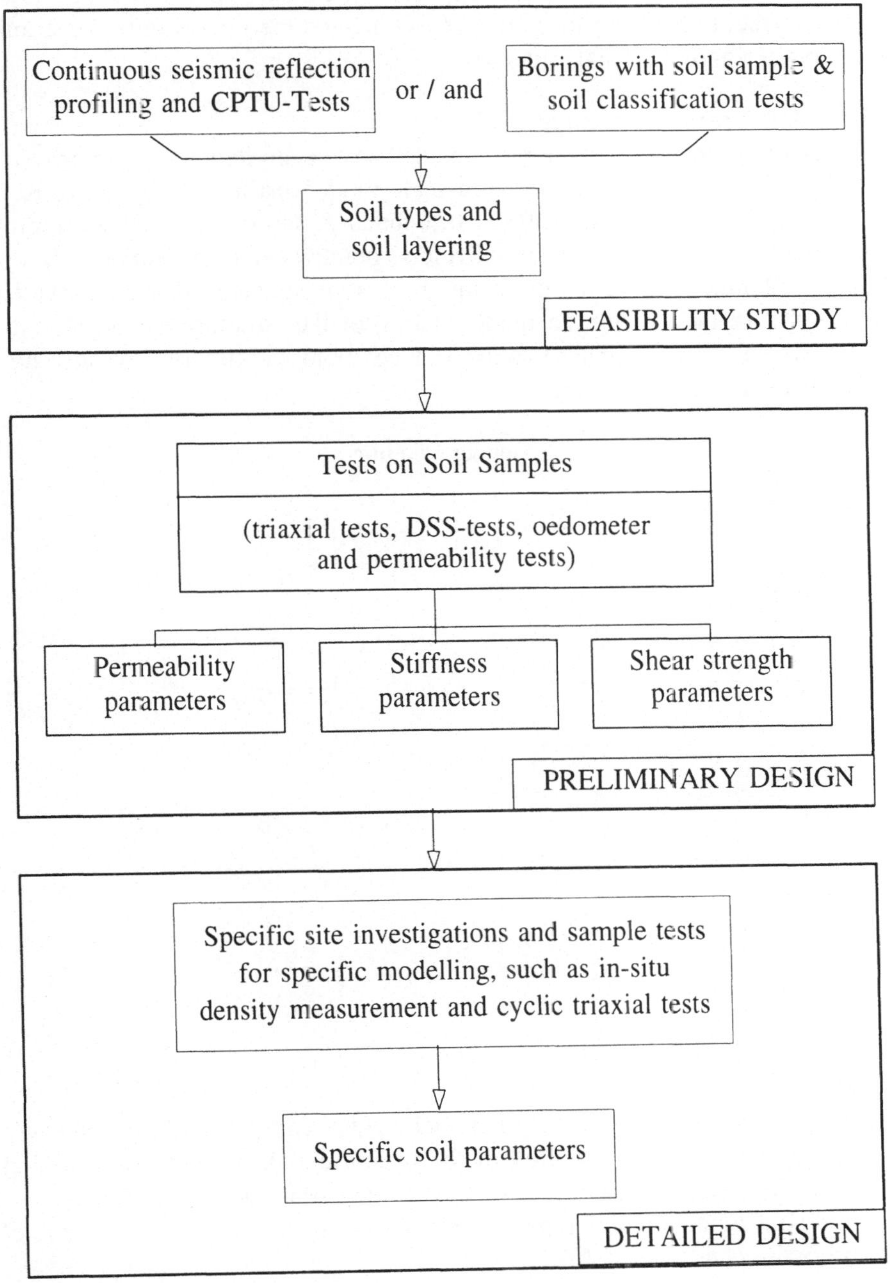

Figure 1-24. Strategy for soil investigations (very simplified representation).

The wide use of this data base is expected to considerably stimulate the feedback from practitioners world-wide and to enhance the opportunity for a unified foundation design for coastal structures.

1.3.2.2 *Engineering "dynamic models"*

A number of engineering "dynamic models" have been developed, improved and implemented which are based on large-scale model testing (Oumeraci & Kortenhaus, 1994; Oumeraci et al, 1995; Kortenhaus & Oumeraci, 1997a), prototype measurements at Genoa Voltri and Brindisi breakwater (Lamberti & Martinelli, 1998a,b) and numerical modelling using more sophisticated FE-Models (Hölscher et al., 1998). In fact, hydraulic model testing of the structure and its foundation have shown that wave impact loading induces both oscillatory motions and permanent displacements (Fig. 1-25).

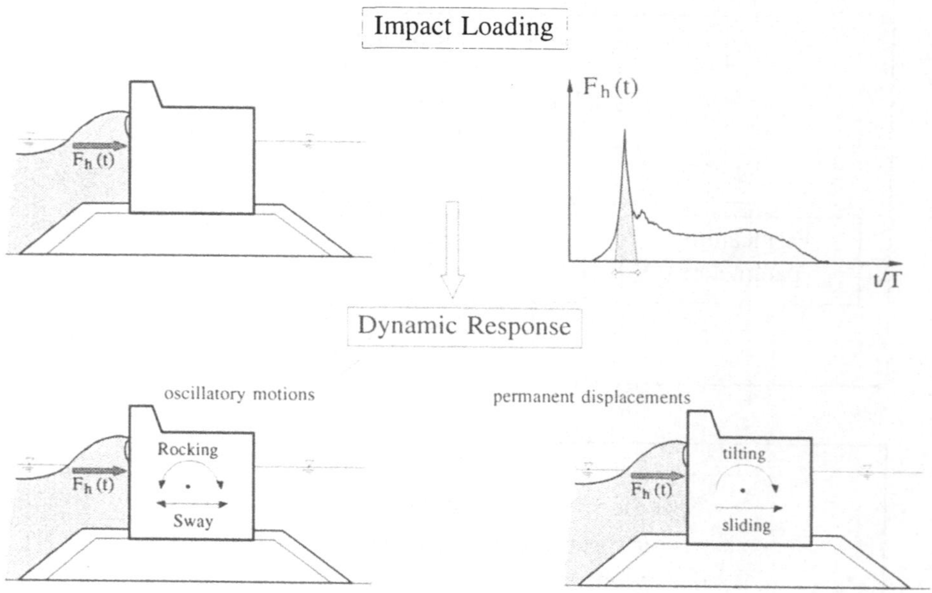

Figure 1-25. Effect of breaking wave impact on structure response.

This requires the development of models which can predict both kinds of motion as both can affect the stability of the structure and its foundation. In fact the repetitive wave impact loading may lead to degradation or/and cumulative permanent displacements which represent stepwise failures before the collapse of the structure will occur (Fig. 1-26).

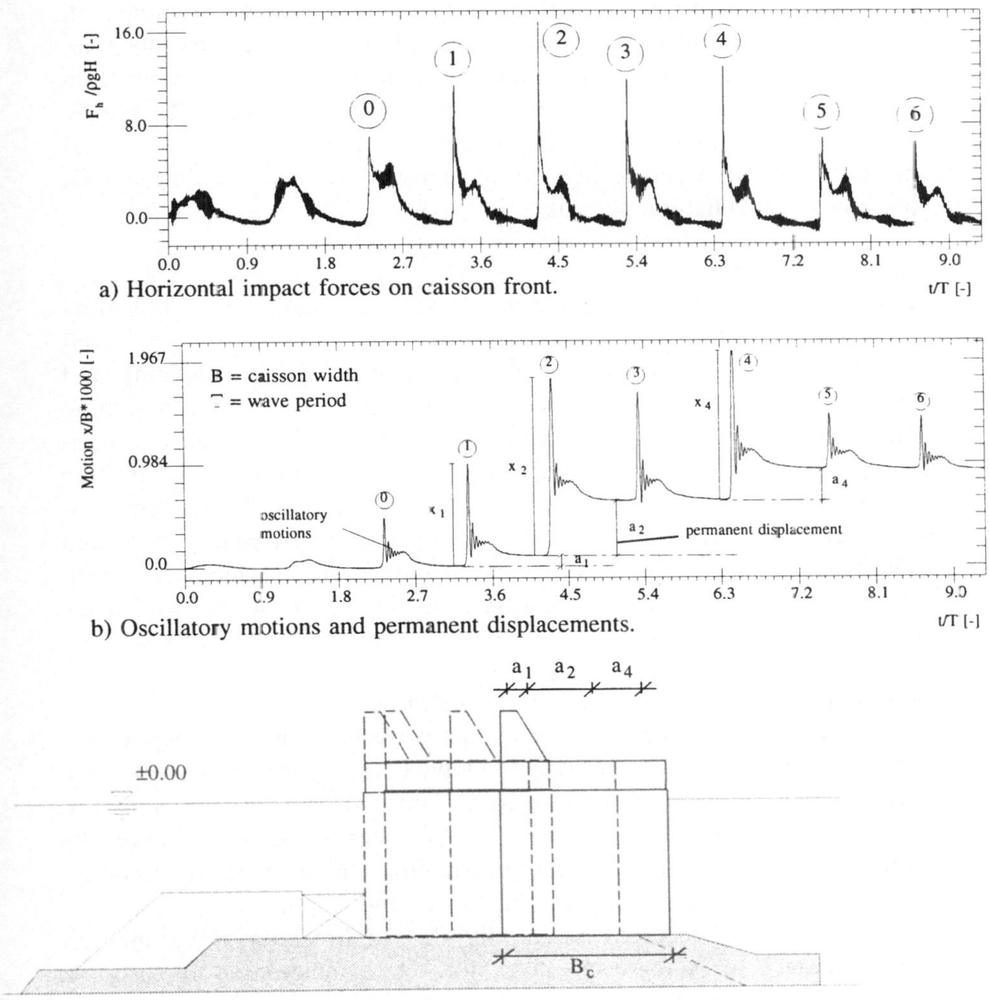

a) Horizontal impact forces on caisson front.

b) Oscillatory motions and permanent displacements.

c) Cumulative permanent displacements of vertical breakwaters.

Figure 1-26. Cumulative effect of repetitive impacts on breakwater stability.

Both elastic and plastic models have already been successfully applied to reproduce some prototype failures (e.g. in Japan, Oumeraci et al., 1995) which otherwise could not be explained by existing standard design formulae and procedures.

This means that the developed plastic models can reproduce the cumulative permanent deformations/displacements related to the sliding failure over the rubble foundation as well as to the bearing capacity failure (rotational motions). In fact, the displacement/deformation induced by a single impact might be too small and non relevant for the overall stability, but the cumulative effect resulting from repetitive impacts may yield the ultimate collapse of the structure. The elastic models are rather used to predict the dynamic amplification effects of load transmission into the foundation and to identify the most relevant oscillation modes, including their associated natural periods.

1.3.2.3 *Instantaneous pore pressures*

The instantaneous pore pressure generated in the rubble foundation and subsoil by direct wave action along the seabed and the rubble foundation boundary, as well as indirectly via the rocking motion of the structure subject to breaking and breaking wave loads, have been investigated by means of large-scale hydraulic model testing (Oumeraci et al., 1994, Kortenhaus, 1996) and an extensive hindcast including theoretical analysis and FEM-calculations with non-stationary two phase flow (Hölscher et al., 1998). Field measurements at Porto Torres breakwater (Franco et al., 1998) have been analysed by means of stationary flow and FEM-simulations. Extremely high positive and negative pore pressures have been observed in the sandy subsoil during centrifuge modelling (Van der Poel & De Groot, 1997).

For the pressure generation in the rubble foundation:

(i) quasi-stationary processes due to non-impact wave loading have been identified and a number of possible phenomena have been suggested to explain the observed deviations from the commonly accepted triangular uplift pressure distribution along the bottom slab of the breakwater. Based on these results the prediction of the wave-induced uplift loading and seepage forces in the rubble foundation can be refined (see Figs. 3-16 and 3-17),

(ii) non-stationary flow processes essentially due to breaking wave impacts which can reduce or increase the uplift force by a magnitude up to 30 % when the caisson is lifted up by breaking wave impact (Fig. 1-27).

For the pressure generation in sandy and silty subsoils undrained conditions (like for clay) are generally assumed so that the effect of wave pressure fluctuations only needs to be considered in a thin top layer of the subsoil. Because of the contribution of negative pore pressure due to dilation, additional strength is mobilised in case of dense sand. For preliminary design it is recommended to use the results of undrained tests and to formulate the strength of sand/silt in terms of undrained shear strength.

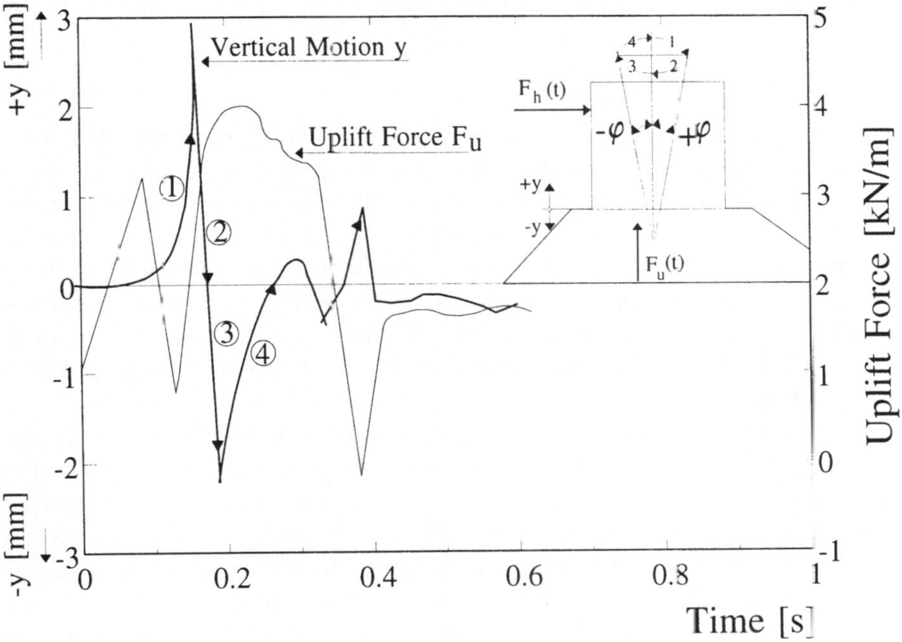

Figure 1-27. Effect of caisson lift up by wave impact on uplift forces (test records for illustration only).

1.3.2.4 *Degradation and residual pore pressures*

The degradation/"residual" pore pressure in the subsoils induced by the cyclic wave loads have been investigated with respect to the relevant physical phenomena and to predict their effect on the reduction of soil stiffness and shear strength. Design diagrams for this purpose have been developed on the basis of centrifuge testing and FE-modelling (Volume IIb, chapter 5, Zwanenburg & De Groot, 1998). The basic differences in the behaviour of clayey subsoils (undrained during a series of storm) and sandy subsoils (partial drainage during a storm) as well as the relative role of "precycling" (loading by smaller storms prior to design storms or by smaller waves prior to design wave) have been examined for both soil types, showing that the beneficial effect of "precycling" should be accounted for.

For sandy subsoils a prediction formula to quantify the shear strength reduction of relatively loose layers is provided (Eq. 3-30).

For clayey subsoil a design diagram for the correction factor which accounts for the cyclic wave loading when using the results of monotonic (static) undrained tests to predict shear strength reduction is provided for various over consolidation ratios and loading conditions (Fig. 3-21). With regards to the shear modulus a reduction in the order of half of the initial value has been found for both types of subsoils.

1.3.2.5 *Limit state equations*

Relatively simple Limit State Equations (LSEs) have been formulated for the main sliding and bearing capacity failure modes during wave crest (shoreward failures) and wave trough (seaward failures). The limitations of these simplified LSEs are discussed; further more refined LSEs based on upper bound theory and optimisation procedures which allow to overcome these limitations are suggested for preliminary and detailed design. Finally, numerical modelling (Bishop slip circle analysis and more sophisticated FE-Modelling) is recommended for preliminary and detailed design, including guidance on the use of these models and their limitations.

The results not only illustrate the comparison of the hierarchy of models developed for various design levels but also show the conservatism of the common assumption of 2D-rupture surfaces in bearing capacity calculations as compared to the actual (3D) rupture surfaces. This effect strongly increases with the depth/length ratio of the rupture surface. Consequently, it is relevant for deep rupture surfaces if the loading out of a limited length as is usual with wave impacts. Then, overestimations of the soil resistance up to 40 % are possible, which strongly increase with the depth/length ratio of the rupture surface. On the other hand, solutions are shown on how to perform stability analysis in the case of cumulative effects and stepwise failure (Volume IIb, chapter 6).

1.3.2.6 *Uncertainties*

The possible uncertainties of soil parameters and of the models are first surveyed by means of a new "operational fault tree" providing a clear picture of the input parameter, models and output parameters. Based on the various steps required to evaluate soil parameters, the associated uncertainties are quantified (Table 3-1) which can be readily used for a feasibility study. Using numerical modelling to hindcast prototype measurements and large-scale model tests, the model uncertainties have also been quantified (Table 3-2) which can be used for a feasibility study. Guidance is then provided for the determination of uncertainties required for higher design levels.

1.3.2.7 *Influence of design parameters on failure modes*

The relative influence of design parameters, load parameters and soil strength parameters on the failure modes has been investigated for structures on a thin bedding layer and coarse grained subsoil under pulsating waves. This has been done in order to identify general tendencies related to the most relevant failure modes and the most relevant parameters which are worth to be varied in the design process. For this purpose, a parameter map has been developed combining most relevant loading case and failure modes as related to the design parameters. It is found that the most relevant parameters which essentially determine the type of failure modes are the relative eccentricity of the dead weight ((e_c/B_c), the angle of internal friction of the subsoil and the relative wave height H_{si}/h_s. The validity of the obtained tendencies has also been examined for further specific cases like high rubble foundation, fine grained subsoils and wave impact loading.

Based on the results on these detailed sensitivity analyses, several possibilities are suggested for design improvement, including a range of "optimal" eccentricities, optimal distribution of structure volume above and beneath still water level, enlarging the rubble foundation, non-conventional constructional alternatives like skirts, etc. (see chapter 3).

1.3.3 *Structural Aspects (Task 3)*

Beside the hydrodynamic and the soil mechanic aspects, the coastal / harbour engineer has also to consider the structural design aspects. Figure 1-28 schematically illustrates the design sequence from a structural engineer's perspective. For further details refer to chapter 4.

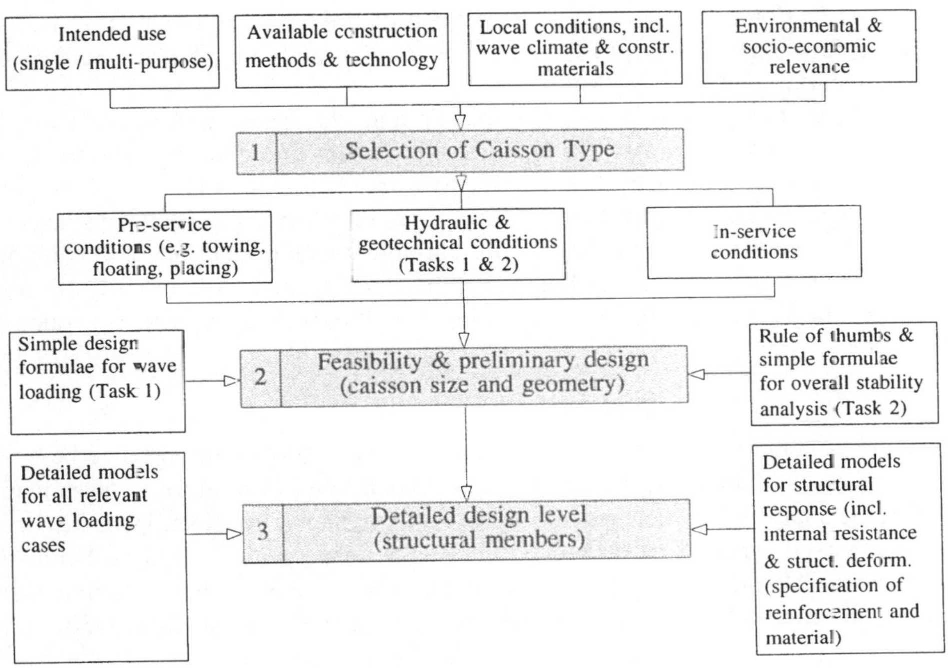

Figure 1-28. Design sequence from a structural engineer's perspective.

In addition to the soil mechanics aspects mentioned above and in chapter 3, the structural engineer has to consider further geomechanical factors like

(i) the unevenness of the foundation to avoid the rupture of the base slab and the misalignment of the caissons,

(ii) the sand fill of the caisson to calculate the internal earth pressure on the caisson walls,

(iii) the gravel filled pocket between caissons (shear keys) to provide load sharing by friction without excessive precision requirements for caisson alignment.

In the following these results are discussed with regards to their importance for practice and further research.

1.3.3.1 *Analysis of existing codes*

The detailed analysis of the existing codes relevant for structural design of reinforced concrete caisson breakwaters has enabled to develop a new approach to include the missing maritime aspects in the existing codes (e.g. Eurocodes) without significantly rewriting these codes. This approach is based on a strategy towards separation of the treatment of uncertainties: the "at source PSF" for actions and materials are applied directly from Eurocodes (materials) and suggested alternatives (actions) without any consideration of limit state function (Fig. 1-29a) while a new approach is developed for PSF (Partial Safety Factor) related to model uncertainties (Fig. 1-29b).

Based on the general Eurocode format and the new approach shown in Figure 1-29 a unified framework for the development of a code for the structural design of caisson breakwaters, including perforated structures has also been suggested (Volume IIc, chapter 2) which accounts for all safety aspects and ensures consistency between the existing codes through the introduction of the model factor concept as illustrated in Figure 1-29b. These results will certainly be very useful to supplement the related Eurocodes as well as the related national codes of practice.

1.3.3.2 *Pre-service failure modes*

Pre-service loading cases and failure modes might in some circumstance be even more relevant than those under in-service conditions. Therefore, a systematic structural analysis for such pre-service conditions had to be developed. This has been achieved for the first time in PROVERBS - even within a full probabilistic framework (Volume IIc, chapter 3), although only the most relevant loading cases and failure modes have been considered in the structural analysis (Fig. 1-30).

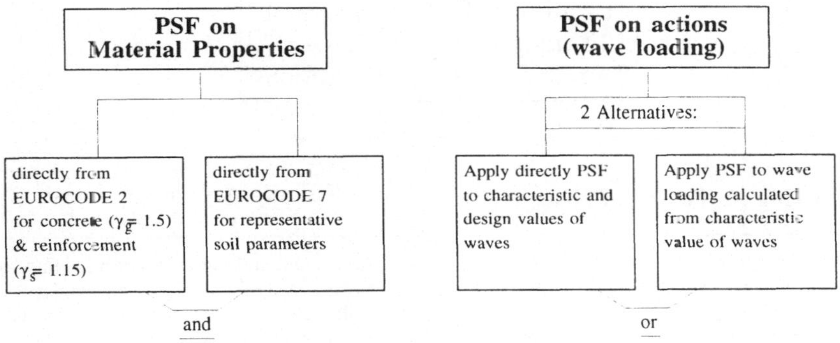

a) Direct introduction of Partial Safety Factors (PSF) related to actions and materials without any consideration of limit state functions ("intrinsic" parameter uncertainties).

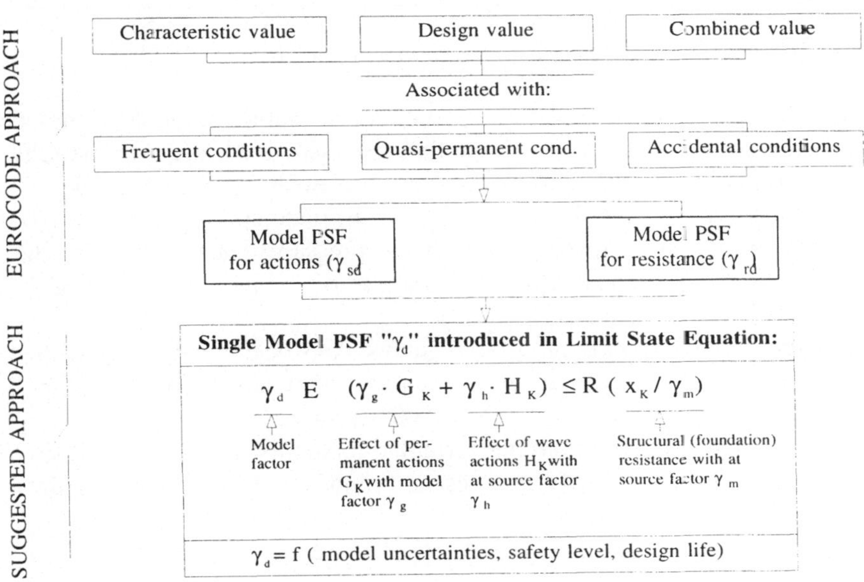

b) Introduction of "Model PSF" at last stage of calibration process using probabilistic analysis (model uncertainties).

Figure 1-29. Suggested approach to include maritime aspects in structural design (Eurocodes).

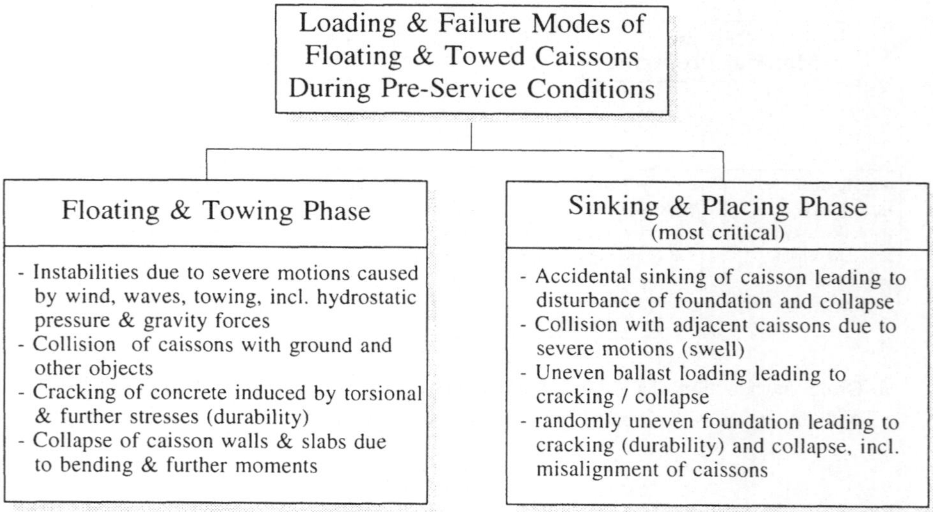

Figure 1-30. Structural analysis of caisson breakwaters during construction.

For the floating & towing phase, a structural probabilistic analysis has been developed for three failure modes, including the associated target probability P_f^t of failure: (i) instabilities due to severe motions with P_f^t = 0,001 (β_t = 3,0), (ii) cracking with and P_f^t = 0.07 (β_t = 1.5) and (iii) plate collapse with P_f^t = 0.001 (β_t = 3.0). In the dynamic load analysis wave pressures, heave motions due to swell as well as roll motions due to wind, waves, towing and manoeuvring operations are considered.

For the sinking and placing phase, the analysis has been conducted for the two failure modes induced by uneven foundation: (i) cracking with P_f^t = 0.07 (β_t = 1.5) and (ii) plate collapse with P_f^t = 0.001 (β_t = 1.5).

The used models vary from simple formulae to sophisticated FE-analysis. The results build a framework for further research and a technical basis for drafting design guidelines.

1.3.3.3 *Loads for in-service conditions*

The loads required for the structural analysis for in-service conditions have been specified, including permanent loads (dead weights & reaction of foundation), variable loads (pulsating and impact loads) as shown in Figure 1-31, and accidental loads (vessel impact during berthing operation, falling masses during cargo handling and seismic loading).

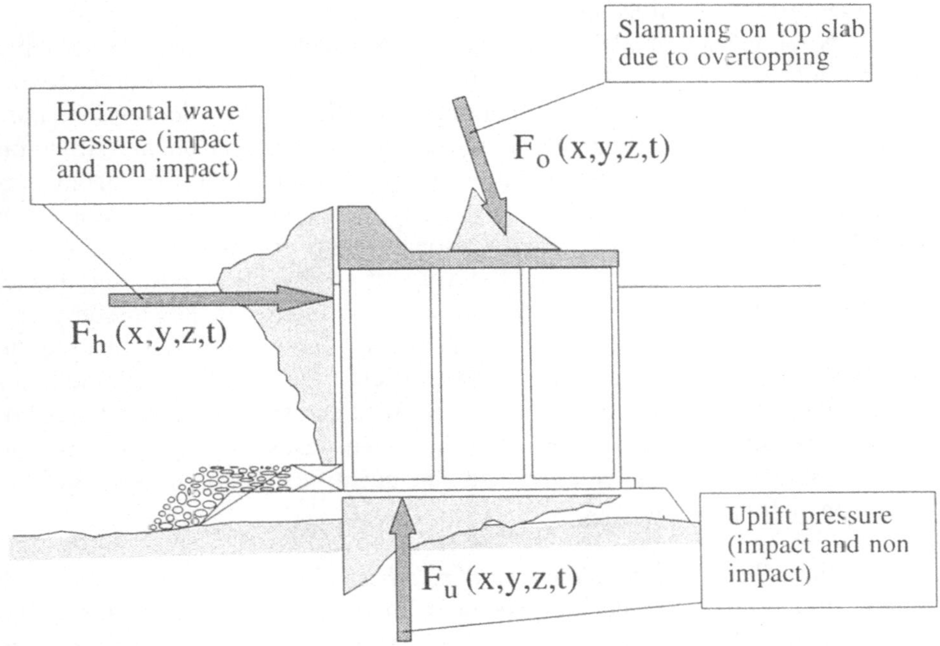

Figure 1-31. Spatial and temporal pressure distribution of pressure relevant for structural analysis of caisson breakwaters during in-service conditions.

The results have highlighted the relative importance of the loading cases and identified the areas where further research is needed (e.g. slamming on top slab and impact loads).

1.3.3.4 *In-service structural failure modes*

The most relevant in-service structural failure modes have been identified (see Fig. 4-4 and section 4.6.2) and the associated simplified limit state equations formulated (section 4.8). Most important is the analysis of the role of each of the eight structural members of a multi-cellular caisson in the load paths (front wall, rear wall, side walls, internal stiffness cross-walls, base slab, top slab, crown wall and shear keys) which is necessary to understand the transmission of the wave forces through all these members to the foundation and to identify the simplest possible idealisation for the four ultimate limit state equations based on beam and slab analogies (flexural failure, shear feature, flexural cracking and chloride penetration/corrosion). These results enables to use simple structural models based on the behaviour of the individual members for preliminary design and to get a first insight into the load transmission and structural behaviour before embarking into more sophisticated FE-models for full 3D dynamic analysis.

1.3.3.5 *Hierarchy of refined models*

A hierarchy of refined models has been developed and used for an example structure (Genoa Voltri Breakwater) subject to an impact loading in order to examine the relative capabilities and limitations of the different models and the possible uncertainties when using the simplified approach based on the structural behaviour of individual members (beam analogy). In addition to the conclusions mentioned in Table 1-3, the results have also shown that an equivalent static structural analysis of the front wall is reasonable only for moderate and longer impact durations ($t_d >> 0.01$ s) and that very often the short duration impact ($t_d < 0.01$ s) are the most critical for the structural response of the caisson members. Furthermore, it is clear from the results that it is now possible to analyse the effect of non-linearity of soil, of the loss of contact of the base slab with the foundation during rocking, of the reduced reinforcement and softened concrete (corrosion!). The results have also highlighted the urgent tasks for future research in modelling fluid-structure interaction and the quantification of uncertainties in structural modelling.

1.3.3.6 *Durability of reinforced concrete members*

The durability of reinforced concrete members in the marine environment has first been put in the related context of understanding the structural consequences in order to avoid the common confusion as to the real significance of cracking of reinforced concrete members. The results have shown that the consequences of concrete degradation is *closely* tied to the location of the structural frame which governs the effect on the failure of the structure (Fig. 1-32).

For the prediction of concrete degradation limit state equations have been formulated to describe chloride penetration/corrosion and cracking during preservice and in-service conditions. This will not only help to predict possible local structural failures, but also to set up a proper strategy for monitoring, inspection and maintenance.

1.3.4 *Probabilistic Design Tools (Task 4)*

In the preparatory phase several detailed reviews and analyses of the available reliability methods have been performed in order to check their applicability/adaptation for vertical breakwaters and thus their involvement and further development in the PROVERBS probabilistic design framework. This review and the subsequent results may also be very useful in the sense that they can considerably help saving research and development efforts when adapting the reliability tools developed in PROVERBS to further classes of coastal structures:

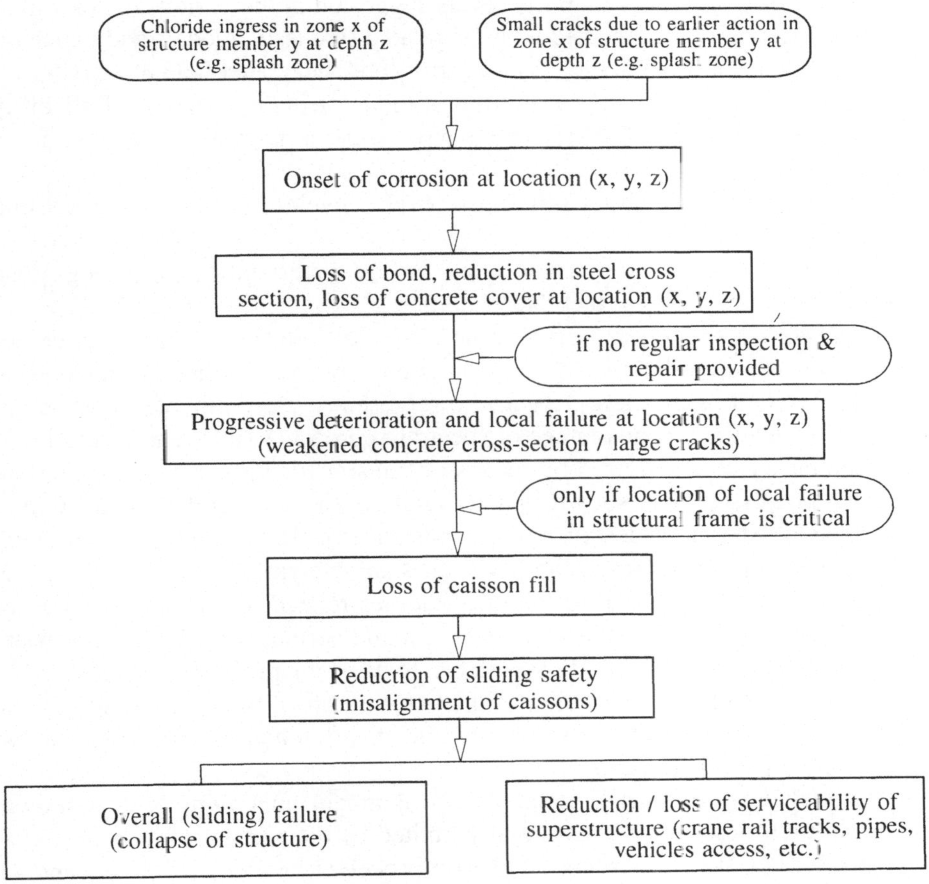

Figure 1-32. Consequences of material degradation for a critical location of local failure in the structural frame and for the case without regular inspection and repair.

- Recommendations have been suggested for the methods to be used for time dependent/independent loads and strengths modelled by various stochastic variables/processes, including software packages (Van Gelder, 1997).
- Guidance has been given on how to select the target reliability levels in close connection with stochastic models describing the uncertain quantities, the safety class, the class and consequences of failure, the type of limit state and further considerations like the basis of acceptable level of risk (social, personal or economic optimisation) (Sørensen, 1997). This guidance also includes a warning notice on

(i) the validity of target reliability levels suggested in the various codes which should always be connected with the related stochastic models and which do not include human and organisation errors and the related field of activities,
(ii) the interpretation of the probabilities of failure using First/Second Reliability Methods (FORM/SORM) which have only a nominal (not absolute!) character,
(iii) the limitations of target reliability levels developed solely on economic optimisation.

➢ Guidance has also been given to evaluate the uncertainties, including some indications on how they affect the reliability:
(i) *Inherent uncertainties:* For periodic stationary processes which can be described by a probability density function a warning notice is included on the limitations of energy spectra to represent a wave climate at a site despite the attractive advantage of spectral approaches (inherent uncertainties can easily be transferred by means of transfer functions). For non stationary processes, it is shown that in addition an autocorrelation-function is needed. The latter describes the persistence of the processes and is therefore important for serviceability limit state analyses. Suggestions are also given on the treatment of the uncertainties related to the joint occurrence of many random variables (wave height and period, wave height and water levels etc.) as well as those related to random processes in space. The limitations of the description of the latter is shown to be essentially due to the lack of practical knowledge and information and illustrated by the example of soil parameters.
(ii) *Model uncertainties*: For the statistical model uncertainties it is shown that their description is essentially limited by the amount of data as well as by further slowly varying stochastic processes like sea level rise and increase of storminess caused by climate changes. For the physical model uncertainties, the limitations strongly depend on the actual source of uncertainties (lack of a sound physical basis for empirical models and too simplified assumptions for theoretical models).
(iii) *Human and Organisation Errors (HOE)*: Although most of experienced failure cases are rather due to HOE, the mechanisms of HOE and their influence on reliability are still not well understood. The most critical HOE might be the omission of a relevant failure mode at the design or/and construction phase. HOE should be considered at each life cycle of the structure (design, construction, operation and maintenance phase).

Finally, the importance of Bayesian approaches is stressed which allows to account for all uncertainties and their updating as more knowledge and more data become available.

The developed probabilistic design tools can be based on the models which describes the hydrodynamic, geotechnical and structural aspects at different levels of detail: feasibility, preliminary and detailed design levels. They have the following main advantages as compared to conventional design methods based on deterministic analysis:

- ➢ They provide a systematic comprehensive framework to exercise engineering judgement in getting scope and priority of further studies and safety improvements. This will also save time and money because the "hot spots" of the detailed design studies can be identified at a very early stage. For instance a preliminary probabilistic analysis conducted at an early stage of PROVERBS on a fictitious breakwater has shown that the contribution of some geotechnical failure modes to the probability of failure of the whole system was determinant and that accounting for wave impact loads considerably increases the probability of failure. This allowed to particularly focus on these critical aspects in the research. These results compared to those of a probabilistic analysis using the models associated with a more detailed design levels have also shown that the probability of failure decreases with increasing level of detail of the design, thus confirming the strategy adopted in PROVERBS which is based on the development of hierarchical models and analyses at different levels of detail: feasibility, preliminary and detailed design level.
- ➢ They allow to better account for inspection and maintenance schedule during the design process, thus leading generally to a less expensive design than conventional deterministic design approaches when the costs are considered over the life time of the structure (Fig. 1-35). In fact, this allows to reduce the excessive costs for rebuilding damaged structures which may amount to more than twice the original construction costs for major failures, including down time costs.
- ➢ They better help to achieve an optimal design over the life time of the structure consistent with Eurocodes and further national codes, i.e. without compromising reliability and without resulting in more costly design. An optimal design can hardly be achieved by conventional design since deterministic analysis can provide only two possible answers: either the structure will fail ($P_f = 1$) or not fail ($P_f = 0$) while probabilistic design can provide much more alternative answers on the structure behaviour, thus making an optimal design more feasible (Fig. 1-33).

Moreover, the probabilistic approach also provides information on the penalty for a possible "underdesign" or "overdesign" in terms of total costs over the life time of the structure (Fig. 1-34) and as function of degree of deviation from the optimal design (target!).

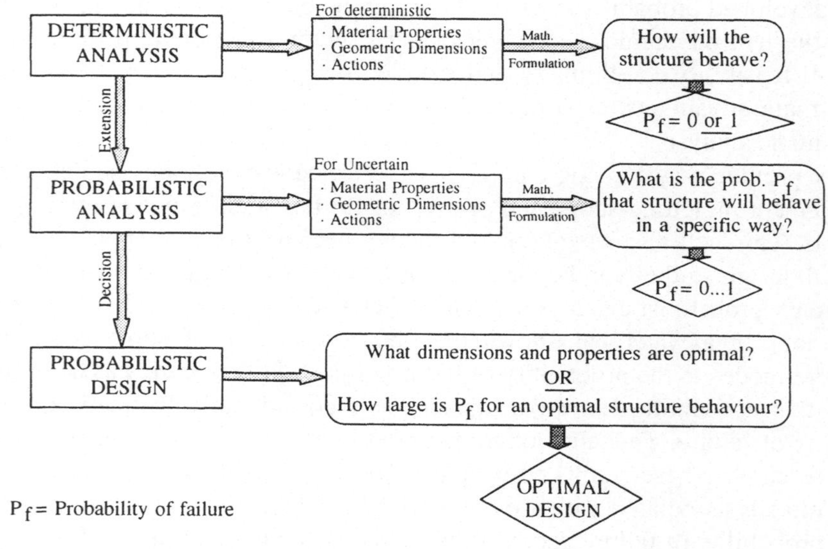

Figure 1-33. Conventional deterministic design vs. probabilistic design considered of structure life time.

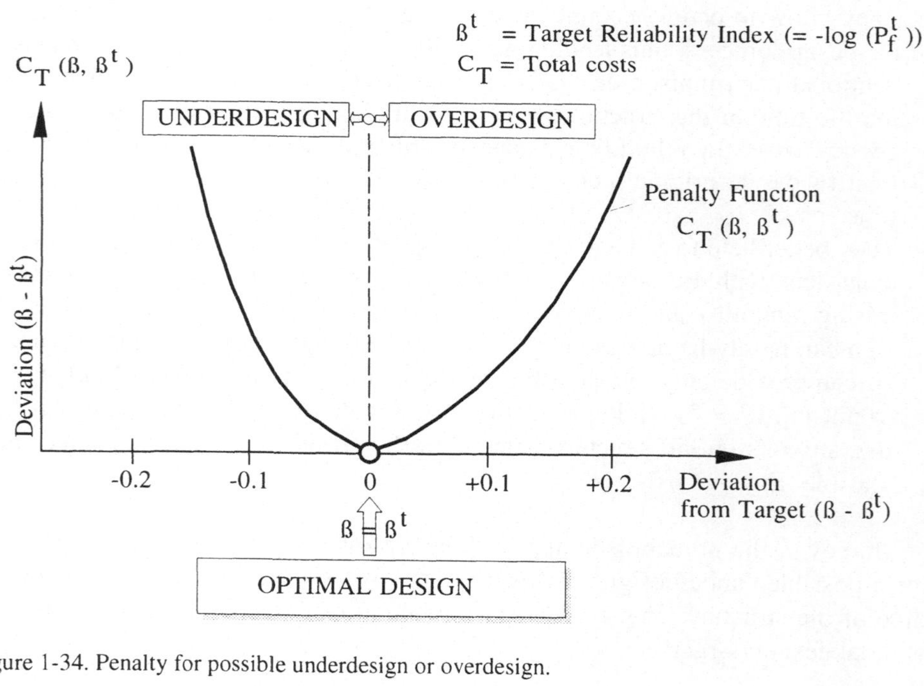

Figure 1-34. Penalty for possible underdesign or overdesign.

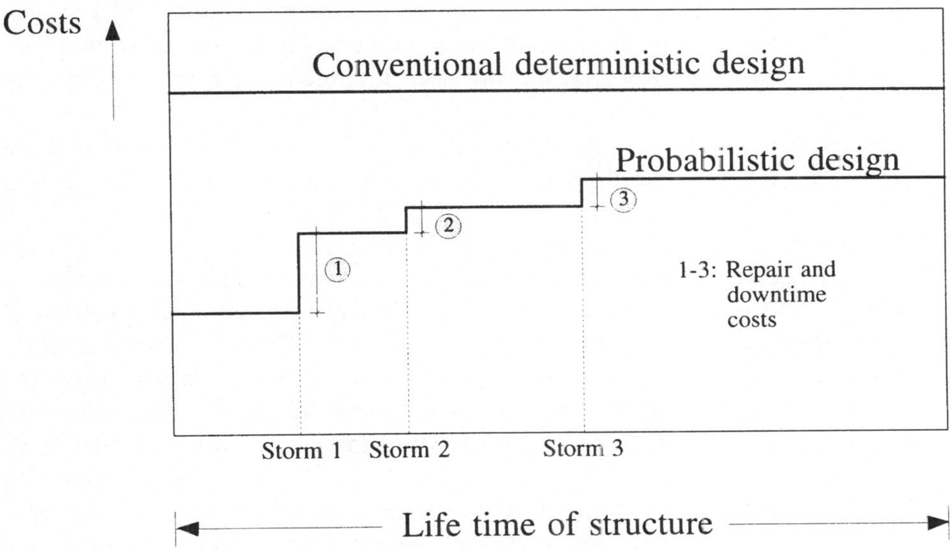

Figure 1-35. Conventional deterministic design and probabilistic design (principle sketch).

The application of the developed design tools to the five representative structures illustrated in Figure 1-14, as well as to other example structures, have brought more light into the following aspects:

(i) The range of failure probabilities P_f for breakwaters which are much higher than those commonly found for other civil engineering structures like bridges, dams etc. which have an annual failure probability of $P_f = 10^{-7}$ to 10^{-4} because human injury and loss of life is generally considered in the later case.
For the three Japanese breakwaters in Figure 1-14 failure probabilities $P_f = 0.037$ to 0.28 have been determined, based on the assumptions of 50 years design life time, strong subsoil and without performance of model tests for wave loading (Volume IId, section 5.3). Even for the Genoa Voltri breakwater (Fig. 1-14) which is considered as a typically rather safe breakwater, failure probabilities of $P_f = 0.10$ to 0.15 have been found for 50 years design life time (Volume IId, section 5.1).

(ii) The relative importance of each failuire modes might differ from the deterministic design approach and according to the design level adopted.
Despite some slight differences there is a general agreement between the results of the five case studies (Fig. 1-14) in the sense that bearing capacity failure modes including subsoil rupture surfaces are more relevant than the failure mode "sliding over the foundation" which has often been observed in prototype (Oumeraci, 1994). This is probably due to the fact that many

bearing capacity failures are associated with large horizontal displacements and only small rotational motions and settlements so that the apparently observed "sliding" in prototype has not been interpreted as a bearing capacity failure.

(iii) The relative influence of the degree of sophistication of the models and the limit state functions used for the failure modes, i.e. the effect of the design level on the final probability.

The results of the exercises on the fictitious Easchel breakwater have shown for instance that using simplified models at feasibility level and more advanced models at preliminary design level could lead to underestimation of the foundation strength in the range of 30% and 10%, respectively as compared to using a more sophisticated finite element model. This confirms the conservatism of the lower design levels (Volume IId, section 5.2). However, this is not always true as has been shown in the case of Genoa Voltri (Volume IId, section 5.1).

(iv) The relative effect of the various load and resistance parameters on the overall result. Most of the results show that the governing parameters are related to the wave loading and the associated uncertainties, and to a lesser extent also to the strength of the subsoil.

Most crucial are the uncertainties related to (a) the calculation of the local design wave height from given deep water wave heights; and (b) the calculation of impact forces (Volume IId, section 5.2). However, also the uncertainties associated with the pulsating wave loads are still important (Volume IId, chapter 5.1).

Using a modified Miche breaking criterion to account for irregular waves instead of the Goda method to calculate the local wave height for the Easchel breakwater has led to failure probabilities which are 3-4 times larger for wave impacts and even an order of magnitude larger for pulsating loads. Considering impacts in addition to pulsating loads generally leads to larger failure probabilities. The differences are strongly dependent upon the model used to calculate the local wave height from given deep water waves and upon the model for impact loading.

(v) The relative effects of the various parameters of the structure on the failure probability which allow to derive the best alternatives for design improvement.

For instance, the caisson width B_c, more specifically the relative eccentricity (e_c/B_c) has been identified as one of the most important parameters which together with the angle of internal friction of the foundation soil and the wave height at the structure essentially decide about the actual failure mode (seaward/shoreward tilt) and about the type of bearing capacity failure. Furthermore, the weight of the superstructure has also been found to greatly influence the failure probability.

1.3.5 *Toward probabilistic risk analysis and management*

Conventional deterministic design methods implicitly recognise uncertainty, but generally react to it with extensive conservatism in the form of large safety factors and by using the calculation upon the worst predictable scenario. As a consequence, seriously overdesigned structures often result (Fig. 1-35).

Further, the existing structures are subject to an ageing process. Conventional deterministic design cannot rationally justify the continued serviceability / maintenance of these ageing structures, nor are they able to account optimally for maintenance in the design process. Global integration, increasing resource scarcity and subsequent increasing competitiveness can no longer permit the overdesign and the lack of transparency in the structure performance over its design life time which both typically characterise conventional deterministic design approaches. PROVERBS has shown that probabilistic analysis represents a more rational and more transparent tool in the sense that it explicitly handles uncertainty and yields a safety improvement and an overall cost reduction by accounting for all relevant failure modes which might occur under pre-service conditions, as well as over the whole life time of the structure (Fig. 1-35).

Probabilistic analysis is a transparent tool in the sense that it provides the best insight into relative contributions of various failure modes to the overall vulnerability and weaknesses of the structure and also allows to obtain the probability for the initiation of repair/maintenance as a function of service time.

However, probabilistic analysis still has serious drawbacks which may be summarised as follows:

(i) Stability analyses in the real engineering practice can vary for instance from quite sophisticated complex numerical modelling through simple analytical / empirical formulae to an almost entirely qualitative and judgement based procedure. The experience made in PROVERBS has shown that there is in fact a tendency to emphasise potential safety deficiencies for which quantitative models exist and underestimate those for which only qualitative knowledge is available.

(ii) Probabilistic analysis explicitly provides just the information on how probable (P_f) or improbable the failure of the structure is. The obtained failure probability P_f is then compared to a target value for decision-making. This technical issue of the safety process may not be sufficient or may not be understood by top decision-makers (finance, harbour authorities and owners) who may be more worried upon the consequences of the failure and who are more familiar with risk concepts rather than with target reliabilities and design points.

Therefore, probabilistic analyses should be extended to probabilistic risk analysis and management (Fig. 1-36) in order to

(i) promote a weighted emphasis on the relative importance of each potential failure modes, no matter whether or not it is amenable to quantitative analysis. This will help to avoid overlooking important issues simply because they may be difficult to address quantitatively,

(ii) arguably and effectively enhance safety more than by any other method without sacrificing traditional safety concern, engineering judgement and qualitative knowledge. This improvement is not only based on technical issues but also explicitly on an economic basis and a societal acceptance, because the consequences of failure can be expressed in monetary terms and the probable net value (NPV) difference between existing conditions and any proposed change can be calculated,

(iii) manage the remaining risk by optimising the monitoring and maintenance strategies – and where personal injury and loss of life is expected also emergency strategies (Fig. 1-36). In fact, planned maintenance is generally much more economical than forced outages caused by failures,

(iv) bridge the gap between engineering and finance, since the expected values defined by risk (defined as the product of failure probability and expected damage costs) is a term already used in the financial community.

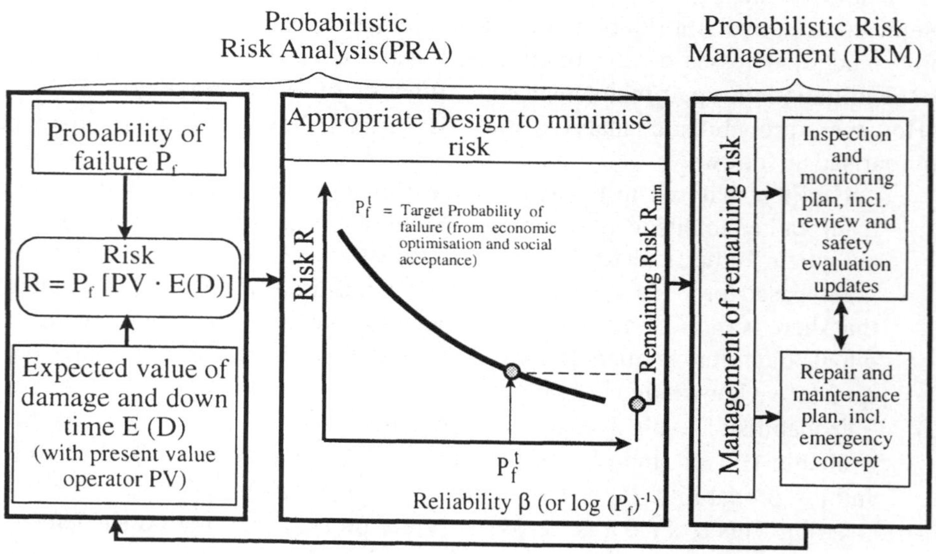

Figure 1-36. Probabilistic risk analysis and management.

ACKNOWLEDGEMENTS

The editor would like to thank all partners of PROVERBS for their valuable contributions and for the enjoyable research atmosphere they created within PROVERBS. Thanks are also due to the European Union for the support of this project. The nice co-operation with the EU representative C. Fragakis is also acknowledged. The support of national research projects by the respective national research foundations and governments which largely contributed to the success of PROVERBS is also gratefully acknowledged. Without the national projects supported by the Deutsche Forschungsgemeinschaft (DFG) and the Bundesministerium für Forschung und Bildung (BMBF) the author would have been unable to efficiently contribute to PROVERBS.

REFERENCES

Battjes, J.A. 1982. Effects of short-crestedness on wave loads on long structures. *Journal of Applied Ocean Research*, vol. 4, no. 3, pp. 165-172.

Bouchet, R.; Cellario, P.; Isnard, J.L. 1994. New types of breakwater: two projects in Monaco. *Proceedings International Conference on Hydro-Technical Engineering for Port and Harbor Construction (Hydro-Port)*, Yokosuka, Japan, vol. 1, Part 1, pp. 581-592.

De Groot, M.B.; Andersen, K.H.; Burcharth, H.F.; Ibsen, L.B.; Kortenhaus, A.; Lundgren, H.; Magda, W. et al. 1996. Foundation design of caisson breakwaters. *Norwegian Geotechnical Institute*, no. 198, 2 volumes, Oslo, Norway, 126 pp; 9 Appendices.

Ditlevsen, O.; Madsen, H.O. 1996. Structural reliability methods. Chichester et al.: John Wiley & Sons, 372 pp.

Franco, C.; Van der Meer, J.W.; Franco, L. 1996. Multidirectional wave loads on vertical breakwaters. *Proceedings International Conference Coastal Engineering (ICCE)*, ASCE, no. 25, Volume 2, Chapter 156, Orlando, Florida, USA, pp. 2008-2022.

Franco, L.; De Gerloni, M.; Colombo, D.; Passoni, G. 1998. Wave forces on solid and perforated caisson breakwaters: comparison of field and laboratory measurements. *Proceedings International Conference Coastal Engineering (ICCE)*, ASCE, Copenhagen, Denmark, vol. 26, 2 pp.

Goda, Y. 1974. New wave pressure formulae for composite breakwaters. *Proceedings International Conference Coastal Engineering (ICCE)*, Copenhagen, Denmark, no. 14, Volume 3, pp. 1702-1720.

Goda, Y. 1985. Random seas and design of maritime structures. Tokyo: University of Tokyo Press, 323 pp.

Hölscher, P.; Zwanenburg, C.; De Groot, M.B. 1998. Hindcast Hannover breakwater. *Research Report, GeoDelft (formerly "Delft Geotechnics")*, CO-364920/127, Delft, The Netherlands, ca. 500 pp.

Kortenhaus, A. 1996. Pore pressures in rubble foundation underneath a caisson breakwater - results of large-scale model tests. *Research Report. MAST III, PROVERBS-Project: Probabilistic Design Tools for Vertical Breakwaters*, Braunschweig, Germany, 25 pp., 3 Annexes.

Kortenhaus, A.; Oumeraci, H. 1997a. First results of spring-dashpot model applied to representative structures. *Proceedings 1st Overall Project Workshop, MAST III, PROVERBS-Project: Probabilistic Design Tools for Vertical Breakwaters*, Las Palmas, Gran Canaria, Annex 2A.

Kortenhaus, A.; Oumeraci, H. 1997b. Selection of representative structures. *Proceedings 1st Overall Project Workshop, MAST III, PROVERBS-Project: Probabilistic Design Tools for Vertical Breakwaters*, Las Palmas, Gran Canaria, Annex C2, 13 pp.; 2 Annexes.

Kortenhaus, A.; Oumeraci, H. 1999. Scale and model effects in breaking waves and impact loading of coastal structures. *Proceedings Hydralab Workshop*, Hannover, Germany. In print.

Lamberti, A.; Archetti, R. 1998. The Punta Riso breakwater at Brindisi. *Research Report. MAST III, PROVERBS-Project: Probabilistic Design Tools for Vertical Breakwaters.*, Final Version, Bologna, Italy, 31 pp.

Lamberti, A.; Martinelli, L. 1998a. Interpretation of dynamic tests of Brindisi and Genoa Voltri breakwaters. *Proceedings 2nd Overall Project Workshop, MAST III, PROVERBS-Project: Probabilistic Design Tools for Vertical Breakwaters.*, Naples, Italy, Class 2 Report, Chapter 2.3a).

Lamberti, A.; Martinelli, L. 1998b. Prototype measurements of the dynamic response of caisson breakwaters. *Proceedings International Conference Coastal Engineering (ICCE)*, ASCE, Copenhagen, Denmark, no. 26, 14 pp.

Lee, D.S.; Hong, G.P. 1994. Korean experience on composite breakwaters. *Proceedings International Workshop Wave Barriers in Deepwaters*, PHRI, Yokosuka, Japan, pp. 172-183.

Madrigal, B.G. 1998. Wave obliquity and multidirectionality effect on composite vertical breakwater. Case of study: Las Palmas, Gran Canaria. *Research Report, MAST III, PROVERBS-project: Probabilistic Design Tools for Vertical Breakwaters*, Madrid, Spain, 17 pp., 1 Annex.

Martín, F.L.; Losada, M.A. 1998. Wave forces on rubble mound breakwaters crown walls. *Proceedings 2nd Overall Project Workshop, MAST III, PROVERBS-Project: Probabilistic Design Tools for Vertical Breakwaters*, Naples, Italy, Class 2 Report, Chapter 1.6a), 31 pp.

Muttray, M.; Oumeraci, H.; Shimosako, K.; Takahashi, S. 1998. Hydraulic performance at a high mound composite type breakwater. *Proceedings International Conference Coastal Engineering (ICCE)*, ASCE, Copenhagen, Denmark, no. 26, 14 pp.

Oumeraci, H.; Partenscky, H.-W.; Tautenhain, E.; Nickels, H. 1991. Large-scale model investigations: a contribution to the revival of vertical breakwaters. *Proceedings of the Conference on Coastal Structures and Breakwaters*, ICE, London, U.K.: Thomas Telford Ltd., pp. 207-220.

Oumeraci, H. 1994. Review and analysis of vertical breakwater failures - lessons learned. *Coastal Engineering, Special Issue on 'Vertical Breakwaters'*, Eds.: Oumeraci, H. et al., Amsterdam, The Netherlands: Elsevier Science Publishers B.V., vol. 22, nos. 1/2, pp. 3-29.

Oumeraci, H.; Kortenhaus, A. 1994. Analysis of dynamic response of caisson breakwaters. *Coastal Engineering, Special Issue on 'Vertical Breakwaters'*, Eds.: Oumeraci, H. et al., Amsterdam, The Netherlands: Elsevier Science Publishers B.V., vol. 22, nos. 1/2, pp. 159-183.

Oumeraci, H.; Kortenhaus, A.; Klammer, P. 1994. Caisson breakwaters: integrated design and wave load specifications. *Proceedings International Conference on Hydro-Technical Engineering for Port and Harbor Construction (Hydro-Port)*, Yokosuka, Japan, vol. 1, Part 1, pp. 453-472.

Oumeraci, H.; Kortenhaus, A.; Klammer, P. 1995. Displacement of caisson breakwaters induced by breaking wave impacts. *Proceedings of the International Conference of the Institution of Civil Engineers - Advances in coastal structures and breakwaters*, Clifford, J.E., London, U.K.: Thomas Telford Ltd., Paper 3, pp. 50-63.

Oumeraci, H.; Muttray, M. 1997. Large-scale model tests on a high mound composite type breakwater. Final Report. *Berichte Leichtweiß-Institut für Wasserbau, Technische Universität Braunschweig*, Nr. 818, Braunschweig, Germany, 115 pp.

Oumeraci, H.; Hewson, P. 1997. Tentative recommendations for scaling wave impact loading. *Proceedings 2nd Task 1 Workshop, MAST III, PROVERBS-Project: Probabilistic Design Tools for Vertical Breakwaters*, Edinburgh, U.K., Discussion note, Annex 1.1.6, 8 pp.

Oumeraci, H.; Kudella, M.; Muttray, M; Kimura, K; Hayakawa, T. 1998. Wave run-up and overtopping on a high mound composite type breakwater. *Internal Report. Leichtweiß-Institut für Wasserbau, Abteilung Hydromechanik und Küsteningenieurwesen*, Nr. 831, Part I: Final Report, Braunschweig, Germany, 88 pp.; 8 Annexes.

Peregrine, D.H.; Thais, L. 1996. The effect of entrained air in violent water wave impacts. *Journal of Fluid Mechanics*, vol. 325, pp. 377-397

Sørensen, J.D. 1997. Target reliability level for various types of structures. *Proceedings 2nd Task 4 Workshop, MAST III, PROVERBS-Project: Probabilistic Design Tools for Vertical Breakwaters*, Delft, The Netherlands, Annex 11, 10 pp.

Takahashi, S. 1996. Design of vertical breakwaters. *Short Course International Conference Coastal Engineering (ICCE)*, Orlando, Florida, 85 pp.

Tanimoto, K.; Takahashi, S. 1994. Japanese experience on composite breakwaters. *Proceedings International Workshop Wave Barriers in Deepwaters*, PHRI, Yokosuka, Japan, pp. 1-24.

Van der Poel, J.T.; De Groot, M.B. 1997. MAST III, PROVERBS results of centrifuge tests on a caisson breakwater. *MAST III, PROVERBS-project, GeoDelft (formerly "Delft Geotechnics")*, CO-364920/89, Delft, The Netherlands, ca. 1000 pp.

Van Gelder, P.H.A.J.M. 1997. Note on reliability methods. *Proceedings 2nd Task 4 Workshop, MAST III, PROVERBS-Project: Probabilistic Design Tools for Vertical Breakwaters*, Delft, The Netherlands, Annex 9, 6 pp.

Van Gent, M.R.A.; Van den Boogaard, H.F.P. 1998. Neural network modelling of forces on vertical structures. *Proceedings International Conference Coastal Engineering (ICCE)*, ASCE, Copenhagen, Denmark, no. 26, 14 pp.

Zwanenburg, C.; De Groot, M.B. 1998. Stability of caisson breakwater tested in centrifuge: prediction and hindcast. *Research Report, GeoDelft (formerly "Delft Geotechnics")*, CO-364920/130, Delft, The Netherlands, ca. 100 pp.

CHAPTER 2

Hydraulic aspects

William Allsop and Andreas Kortenhaus

2.1 INTRODUCTION

2.1.1 *Objectives of Task 1*

Many port or coastal structures, including breakwaters, are formed with vertical, near-vertical or steeply sloping faces. The prediction of wave forces on such structures is complicated, and predictions of forces and pressures often have very high uncertainties. The simple prediction methods for pulsating wave loads by Goda (1985), Ito (1971) and Hiroi (1919) generally predict pressures averaged over the front face up to about $p_{av} = 2\rho g H_s$ where H_s is the incident significant wave height. Studies under PROVERBS have however measured short duration local pressures up to or greater than $p = 40\rho g H_s$, much higher than would be predicted by simple design methods.

Problems due to wave impacts may appear comparatively small. Vertical breakwaters are massive structures and do not seem to respond to very short duration impacts so that the incidence of failure appears to be relatively low. Quays or sea walls subject to wave action are usually backed by fill, and seldom fail in the direction of the principal wave loading. Unfortunately, recent failures of vertical breakwaters in UK, Japan and Italy have demonstrated that design methods are insufficient, and studies in UK and Germany have demonstrated that wave impacts may have considerable influence on loadings.

The present stock of vertical breakwaters, many in Europe constructed after about 1830, are subject to continuing deterioration, so their safety will tend to reduce with time. In some areas, both sea levels and storminess have increased over the last 10-20 years, and show every sign of continuing to reduce safety margins.

Within PROVERBS, the primary objective of Task 1 was to supply essential data and prediction methods to Tasks 2-4 on hydro-dynamic loadings on vertical and composite structures. The second objective was to develop new prediction tools and data to be used more widely by consulting engineers, contractors, and owners in the analysis of performance and safety of such structures. Task 1 was divided into 3 sub-Task areas: Field Measurements, Hydraulic model studies and Numerical Modelling (see chapter 1).

2.1.1.1 *Technical progress*

Wave pressures have been measured on breakwaters at Gijon in Northern Spain, Las Palmas in the Canaries, Dieppe on the Channel coast of France, at Alderney on the Channel Islands and Porto Torres in Sardinia. The measurements in the field reflect the complexities of real seas, provide supplementary information to the laboratory tests, and are intended to provide data for improved scale correction methods. Field measurement are very expensive and need many years of effort, so PROVERBS linked with measurement campaigns supported by national funding.

Hydraulic model tests have been used widely to measure wave forces, and/or wave pressures on sections of simple vertical or composite breakwaters. Most tests have been conducted in narrow 2-dimensional wave flumes with normal wave attack, but some studies have also used large 3-dimensional wave basins with short-crested and/or oblique wave attack. Much of the effort in the first year of the project was devoted to substantially extending the analysis of data from laboratory studies completed under national funding.

This initial analysis confirmed that most wave forces under "pulsating wave" conditions on many structures were much as expected in simple prediction methods. That analysis however also demonstrated that "wave impacts" are also very important in the design of some structures. Previously felt to be only of very short duration, or very rare in occurrence, wave impacts have been shown to have caused failure in a number of breakwaters. The major results of the first part of the project have therefore been:

- new methods to predict conditions that give rise to wave impacts, see example parameter map in Figure 2-6;
- methods to predict the magnitude and duration of impacting wave forces.

Other results are standardised response parameter coverage and measurements; new data for the performance of perforated caissons; identification of effects of steep bed slopes; and new data on wave induced pressures on and in rubble foundations.

2.1.2 *Outline of deterministic design procedure*

Within this section a brief and stepwise procedure is given on how to deterministically design a breakwater or a vertical wall structure. The given steps include the present methods and further developments by the PROVERBS group. The models developed here also build the basis for the limit state equations as used in chapter 5 in which the probabilistic design procedures are explained. The main problem to be addressed is to dimension a vertical breakwater, its structural elements, and its foundation to resist wave action and its effects, and to deliver required hydraulic performance. The main structure design problem is to dimension the caisson or any other type of vertical wall large enough to resist sliding or "overturning" forces, yet small enough to ensure optimum cost for performance. Historically this was achieved by deriving an equivalent sliding load, then configuring the caisson wide enough to generate sufficient resistance to sliding. Early studies in PROVERBS however demonstrated that impacts loads may cause damage or failure, so the effects of these loads must also be considered.

A number of different prediction methods for wave forces on vertical / composite walls have been developed under Task 1 of PROVERBS. For some responses or parameters, it is not yet possible to demonstrate that one particular method is more complete or more reliable than another. For such responses, it is therefore incumbent upon the user to apply the best alternative methods, and then use engineering judgement and experience to decide which gives the most realistic result for the particular application considered. The approach taken here is therefore to identify each main method, and to define the parameters to be derived. In each instance, the parameters that will be derived are shown against the right-hand margin of this note, see example at end of section 2.1.2.1 below.

2.1.2.1 *Step 1: Identification of main geometric and wave parameters*

Identify the following parameter (for notation see Annex 1):

- Water depth and bed slope in front of the structure (h_s and α)
- Width, height and slope of front of berm in front of wall (B_b, h_b and m)
- Crest freeboard above water level, height of caisson face (R_c and h_f)
- Equivalent berm width ($B_{eq} = B_b + (m / 2 \cdot h_b)$)
- Depth of water over the berm for design water level (d)
- Obliquity of structure to (design) wave direction (β)

It should be noted that some of these parameters may take different values for different water levels, for each of which the structure may need to be analysed.

Identify design wave condition(s) given by H_{si} and T_p taking account of wave shoaling and refraction, and of depth-limited breaking. Derive peak period wave length L_{pi} in the water depth of the structure, h_s. Use Goda's simple breaking criteria to calculate $H_{max} = 1.8H_s$ or $H_{max.b}$ (see Eq. 2-16), where the breaking wave depth h_{break} is taken $5H_s$ seaward of the structure.

[h_s, m, α, B_{eq}, d, β, H_{si}, T_m, T_p, L_{pi}, H_{max} or $H_{max.b}$]

2.1.2.2 *Step 2: First estimate of wave force / mean pressure over wall height*
Use Hiroi's formula to estimate an equivalent uniform wave pressure p_{av} on the front face over a wall height h_f up to $1.25H_s$ above still water level, and hence calculate the total horizontal force F_{Hiroi}:

$$p_{av} = 1.5\rho_w g H_s \tag{2-1}$$

$$F_{Hiroi} = 1.5 h_f \, \rho_w \, g \, H_s \tag{2-2}$$

Use F_{Hiroi} to give first estimate of breakwater width B_c to resist sliding assuming no dynamic uplift pressures, but including buoyancy, and friction $\mu = 0.5$.

[p_{av}, F_{Hiroi}, B_c, m_c]

2.1.2.3 *Step 3: Improve calculation of horizontal and up-lift forces*
Use Goda's method (see section 2.4.1) to predict horizontal and up-lift forces at 1/250 level, F_{hGoda} and F_{uGoda}, and related pressure distribution. Wave pressures on the front face are distributed trapezoidally, reducing from p_1 at SWL to p_3 at the caisson base. Above SWL $p = 0$ at the notional run-up point at height η^*. Uplift pressures are distributed triangularly from the seaward edge to zero at the rear heel.

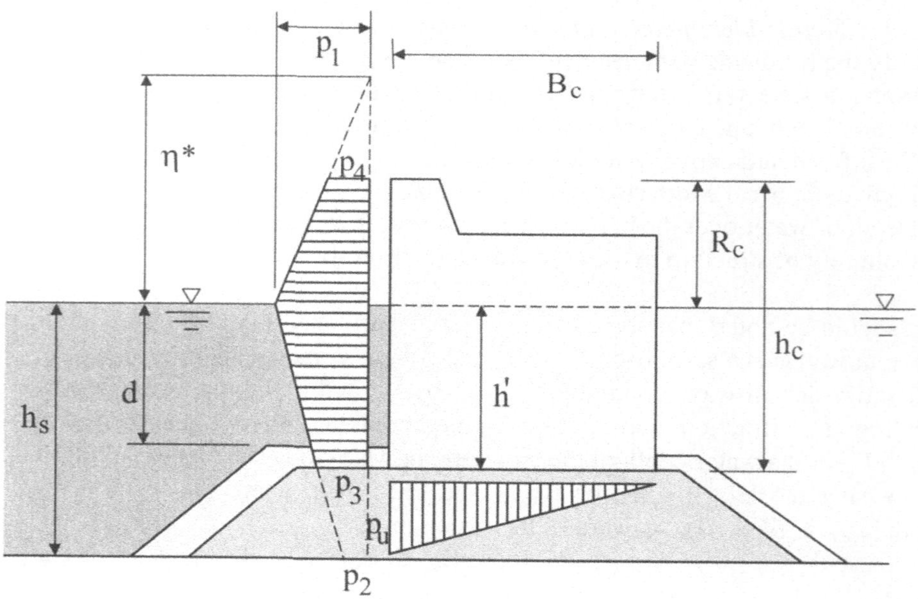

Figure 2-1. Goda's wave pressures.

The total horizontal force F_h (per m length of breakwater) is calculated by integrating pressures p_1, p_3 and p_4 over the front face. The total up-lift force F_u (per m length of breakwater) is given by $F_u = 0.5\ p_u\ B_c$. Check F_u against corrections given in section 3.5.3. All of these values are calculated at 1/250 level when using Goda's method.

Using the 1/250 value, and assuming a Rayleigh distribution, forces at various other exceedance levels may be estimated from the following ratios of $F_{i\%}/F_{1/250}$:

Exceedance level	$F_{i\%}/F_{1/250}$
50%	0.33
90%	0.59
98%	0.77
99%	0.84
99.5%	0.90
99.8%	0.97
99.9%	1.03

[p_1, p_2, p_3, p_u, F_{hGoda}, F_{uGoda}, $F_{i\%}$]

2.1.2.4 *Step 4: Revise estimates of caisson size*

Use both horizontal and up-lift forces F_{hGoda}, F_{uGoda}, to revise estimate of caisson width, assuming friction $\mu = 0.6$ or other given value. [B_c, revised]

Use simple overtopping methods (see sections 2.3.1 to 2.3.2) to check crest elevation against required wave transmission or overtopping limits, and confirm or revise crest freeboard, R_c. [R_c, revised]

2.1.2.5 *Step 5: Identify loading case using parameter map*

Calculate key decision parameters:

- relative berm height ($h_b^* = h_b/h_s$)
- relative wave height ($H_s^* = H_{si}/h_s$)
- relative berm width, ($B^* = B_{eq}/L_p$.)

Use these in parameter map (section 2.2.2, Fig. 2-6, Volume IIa, section 2.2) to determine loading case type.

[Quasi-standing, slightly breaking, impact or broken waves]

2.1.2.6 *Step 6: Initial calculation of impact force*

If parameter map indicates "Slightly breaking waves" or "Impact loads", then use Allsop & Vicinanza's method to calculate a first estimation of the impact force, $F_{h.A\&V}$, again at 1/250 level:

$$F_{h.A\&V} = 15\ \rho_w\ g\ d^2\ (H_{si}/d)^{3.134} \tag{2-3}$$

Use this simple estimate of impact force as a reference value (for design see chapter 2.5.1) if $F_{h.A\&V}/F_{hGoda} > 1.2$

[Impact force, $F_{h.A\&V}$]

2.1.2.7 *Step 7: Estimate percentage of breaking waves leading to impacts Pi%*
Use the new method described in section 2.2.3 to determine the percentage of waves that give impact breaking onto the wall, P_i. Calculate a breaking wave height, H_{bc}, when waves start to break and derive estimate of $P_{i\%}$.

Note that H_{bc} is a fictional rather than measured parameter, and may differ significantly from breaking significant wave heights determined by other methods, see particularly Volume IIa, section 2.1, Weggel (1972), Owen (1980), Durand & Allsop (1997).

Use $P_{i\%}$ to decide loading case

$P_{i\%} < 2\%$	Little breaking, wave loads are primarily non-breaking
$2 < P_{i\%} < 10$	Breaking waves give impacts
$P_{i\%} > 10\%$	Heavy breaking may give impacts or broken loads

[Pi%, confirmation of loading case type]

2.1.2.8 *Step 8: Estimate impact force using Oumeraci & Kortenhaus' method*
If $P_{i\%} > 2\%$, use PROVERBS impact method (section 2.5.1 or Volume IIa, section 5.1) to calculate $F_{hO\&K}$.

[$F_{hO\&K}$]

Compare $F_{hO\&K}$ against $F_{hA\&V}$. If the difference is large, check that case is in range of the test data. If this is the case use $F_{hO\&K}$ otherwise take most conservative approach to estimate $F_{hImpact}$.

[$F_{hImpact}$]

2.1.2.9 *Step 9: Estimate impact rise time and duration*
Use PROVERBS impact method (section 2.5.1 or Volume IIa, section 5.1) to estimate limiting impact rise times, t_r and impact duration t_d.

[t_r, t_d]

2.1.2.10 *Step 10: Estimate uplift forces under impacts*
If step 7 gives impacts, use PROVERBS impact method (section 2.5.1 or Volume IIa, section 5.1) to calculate uplift force, $F_{uK\&O}$ and pressure. $p_{uK\&O}$. Check against section 3.5 in chapter 3 and apply corrections if necessary.

[$F_{uK\&O}$, $p_{uK\&O}$]

2.1.2.11 *Step 11: Scale corrections*

If condition in step 6 and/or 7 is pulsating, scale F_{hGoda} and F_{uGoda} by Froude, i.e. scale correction factor of unity is applicable.

If forces are impacts in steps 6 and 7, then use PROVERBS scaling method (section 2.5.1.5, section 2.5.4.2 or Volume IIa, section 5.4) to:

a) estimate aeration from level of $P_{i\%}$;
b) estimate attenuation of $F_{hImpact}$ from level of aeration;
c) apply scale correction to $F_{hImpact}$ based on aeration-induced attenuation.
d) scale rise time and impact duration, t_r & t_d by duration correction.

[Scale corrections to $F_{hImpact}$, t_r and t_d for impact conditions]

2.1.2.12 *Step 12: Pressure distributions*

If condition in step 6 and/or 7 is pulsating, plot pressures calculated in step 3.

If forces are impact in steps 6 and 7, then use PROVERBS method as indicated in section 2.5.1 or Volume IIa, section 5.1.

[Pressure distributions under impact conditions]

For structures where major elements may respond to short duration loadings, a dynamic approach may be taken where a more complete description of the loading is used togehter with a description (model) of the dynamic characteristics of the structure. A full description of this approach is given in section 3.4 and Volume IIb, chapter 3, a number of simplified methods are described by Torrini et al. (2000).

2.2 WAVES AT THE STRUCTURE

2.2.1 *Wave conditions at the structure*

For the design of breakwaters, the main interest is related to sea state conditions giving large hydro-dynamic effects on the structures in terms of wave loads and wave overtopping. These conditions are most often caused by severe storms, although it should be noted that swell from distant storms may cause worse responses for some processes.

Waves are generated by wind fields over the sea offshore from the coast of interest. In most cases the ratio of water depth to wave length in the generating area allows the approximation of deep water waves, i.e. the sea bed has no or marginal influence on the waves. The waves can then be defined solely by their surface elevations and the speed and direction of propagation of these elevations. In engineering practice, it is more convenient to use wave heights, defined trough to crest. The history of surface elevations may then be divided using zero-downcrossing to define individual waves given by height H and period T.

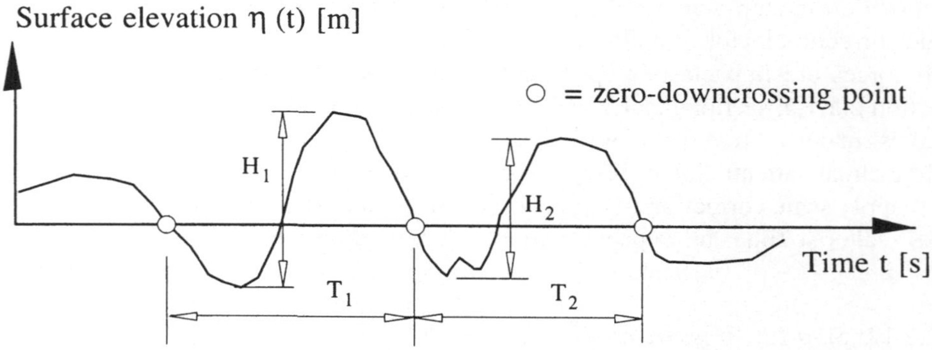

Figure 2-2. Zero-downcrossing analysis of surface elevation record.

Analyses of measurements as well as theory show that wave heights in deep water follow closely a Rayleigh distribution. There is however no generally applicable statistical distribution for wave periods. Nor is there a general correlation between wave period and wave height as wave steepness (s = H/L) again depends on both the stage of wave generation during the storm history, and the location of interest relative to the wave field. Analysis of data from several locations all over the world indicates that H_s and s_{op} can be considered uncorrelated in extreme storm conditions (Repko et al, 2000).

The distribution of individual wave heights in a wave field in deep water can generally be described by a Rayleigh distribution. The only parameter in the Rayleigh distribution is the significant wave height H_s which is therefore suitable to describe the long-term wave statistics for a site. Any loading which may be related directly to wave height may be conveniently described to the same exceedance probabilities, but for some responses it may be more useful to use lower exceedance probabilities, such as:

H_s or $H_{1/3}$	significant wave height given by the average of the largest 1/3 of wave heights
$H_{1/250}$	mean of the 4‰ largest wave heights
$H_{2\%}$ or $H_{98\%}$	height exceeded by 2% of the waves, or not exceeded by 98%
$H_{1\%}$ or $H_{99\%}$	height exceeded by 1% of the waves, or not exceeded by 99%
$H_{1‰}$ or $H_{99.9\%}$	height exceeded by 1‰ of the waves, or not exceeded by 99.9%

A single surface elevation record does not provide information on direction of wave propagation. This has to be estimated from knowledge about the wind direction if not measured by other techniques.

By Fourier analysis of the surface elevation record a spectral presentation of the waves can be given.

From the spectral moments any statistical value of the wave height and estimates of wave periods can be derived. Denoting the moments of the spectrum by m_n, where sub-index n refers to the order (m_0 represents the area of the spectrum for n = 0), we obtain for example for the significant wave height

$$H_{m0} = 4.0 \cdot \sqrt{m_0} \tag{2-4}$$

and for the mean wave period, T_{m0}:

$$T_{m0} = \frac{m_0}{m_1} \quad \text{or} \quad T_{mo} = \sqrt{\frac{m_0}{m_2}} \tag{2-5}$$

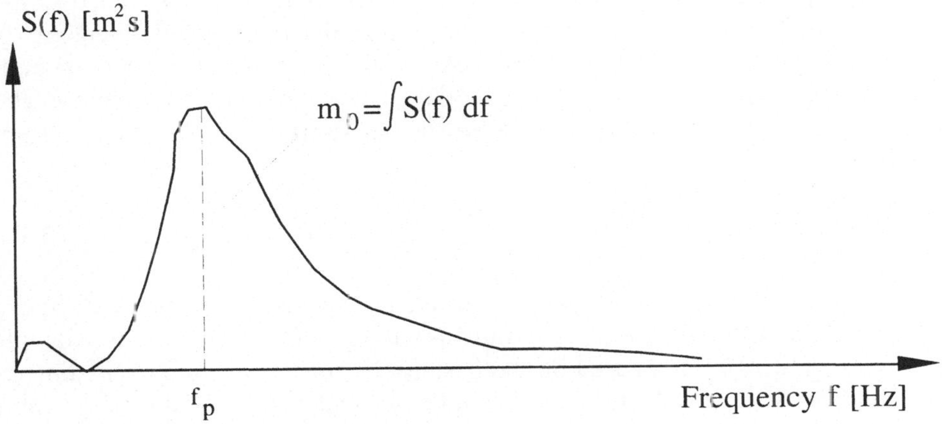

Figure 2-3. Spectral representation of surface elevation (one- dimensional).

where m_1 and m_2 are the spectral moments (first and second order) of the energy spectrum defined as follows:

$$m_1 = \int S(f)\,df \tag{2-6}$$

$$m_2 = \int S(f)^2\,df \tag{2-7}$$

The shapes of deep water wave spectra have been standardised into main types. For growing seas, i.e. wave heights are limited by the fetch, the JONSWAP spectrum is generally applied. For fully arisen seas, i.e. where a balance between energy input from the wind and energy dissipation has been reached, the Pierson-Moscowitz spectrum is generally applied.

The ratio between the spectral peak period, T_p, and the spectral mean period, T_m, depends on the shape of the spectrum. For wider spectra such as the Pierson-Moskowitz, it may be initially assumed that:

$$T_m / T_p \approx 0.7 \tag{2-8}$$

And for narrow spectra such as JONSWAP, it may be initially assumed that:

$$T_m / T_p \approx 0.8\text{-}0.87 \tag{2-9}$$

A spectrum as exemplified in Figure 2-3 assumes that the waves are long-crested, i.e. 2-dimensional with energy only propagating in one direction. In practice however, most wind generated waves are three-dimensional, so called short-crested, with horizontal spread of energy around the mean wind direction. A spreading function $D(f,\theta)$ is then introduced. $D(f,\theta)$ is assumed dependent on both the direction of energy transport given by the angle θ, and on the frequency. The latter dependency is often disregarded. For the short-crested waves the spectrum is given by :

$$S(f;\theta) = S(f) \cdot D(f;\theta) \tag{2-10}$$

where $S(f)$ is the one-dimensional spectrum. The spreading function $D(f, \theta)$ can be defined in different ways. A commonly used definition (see for example Mitsuyasu et al., 1975) is based on the cosinus-2s-distribution and depending on the angle of wave attack, the Gamma function and the spreading parameter s where various s values correspond to various degrees of energy spread as given in Table 2-1. A summary and further details of the spreading function can also be found in Goda (1985).

Table 2–1. Relationship between spreading parameter s and standard deviation of θ.

Standard deviation of θ (σ_θ)	15°	30°	45°
spreading parameter (s)	28	7	3

A typical 3-dimensional wave spectrum is depicted in Figure 2-4.

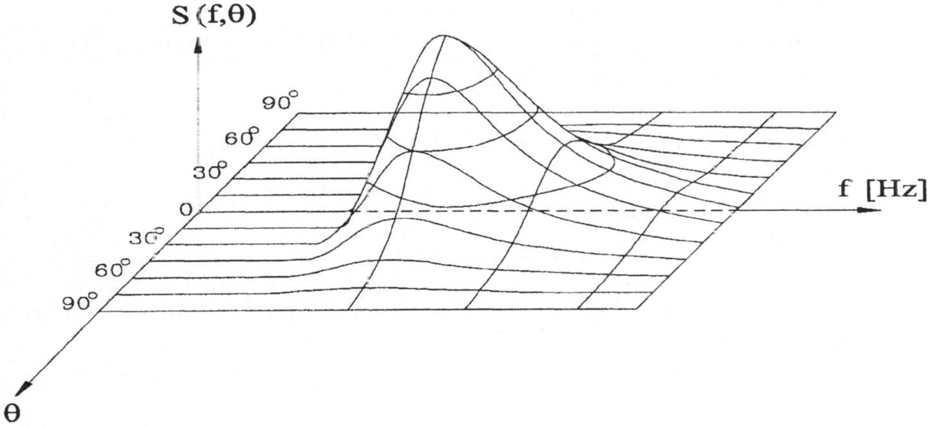

Figure 2-4. Example of 3-dimensional wave spectrum.

For design purpose the long-term statistics of either the max. H_s-values in each storm or the individual wave heights are used. In order to establish long-term extreme statistics for storms characterised by maximum values of H_s, it is necessary to estimate the largest H_s-values over a span of years. This can be done in various ways depending on the available information. If long-term wave recordings are available, e.g. by accelerometer buoys, then uncertainty on single H_s-values will be small. If however only long-term wind statistics given as accumulated occurrence of wind speeds for interval directions are available, then estimates of H_s-values must be based on fetch diagrams or on parametric formulations of the wave spectra in which wind speed and fetch length are included. Interpretation of durations of various wind speeds within single storms will be necessary, and estimates of H_s-values will be rather less certain.

If historical weather maps for meteorological conditions covering a longer span of years are available, then numerical models for wave prediction can be applied. Using modern advanced numerical models and reliable weather maps the uncertainty on values of H_s will be reduced.

The uncertainty on H_s can be implemented as a normal distributed multiplicative stochastic variable F_{Hs} with mean value of unity. Typical values of the coefficient of variation COV_{Fs} for F_{Hs} are given in Table 2-2, dependent on the method of estimation.

Table 2-2. Uncertainty on H_s - estimates (after Burcharth 1991).

Method of estimating H_s	**Coefficient of variation, COV_{Fs}**
Accelerometer buoy, pressure or similar measurements	5%
Advanced numerical modelling based on weather maps	5-10%
Simplified prediction methods based on fetch diagrams	10-25%

Theoretical extreme distributions such as Gumbel and Weibull distributions may be fitted to the data.

2.2.1.1 *Near-shore wave transformation*

The methods of calculation of wave impacts on the structures discussed in this report are based on the incident waves at the location of the structure, i.e. wave conditions as they will appear if the structure were not there.

As the wave propagates from offshore into shallow water, different wave transformations will take place. When the water depth to wave length ratio becomes small, the sea bed influences the waves (usually when $h/L < 0.5$), then the waves start to shoal, reducing the wave length, but increasing the wave height. For waves that are oblique to the depth contours, they start to refract the direction of wave propagation to reduce obliquity between the wave fronts and the bottom contours.

Shoaling occurs when waves travel into water of decreasing depth and slow down. If wave energy is not dissipated, then the wave energy flux remains constant, so the energy density of the waves, and hence the wave height, must increase. Based on linear wave theory the shoaling coefficient K_s may be predicted as follows:

$$K_s = 1 / \sqrt{\{ [1 + (2 k h/\sinh(2 k h)] \tanh(k h)) \}} \qquad (2\text{-}11)$$

where the wave number is $k = 2\pi/L$.

Refraction may also be expected, where waves approach the coast at oblique incidence and therefore change their direction as they travel into shallower depths, and wave crests tend to align more closely with the sea bed contours. On a parallel contour coast with waves approaching at obliquity β_o, refraction will alter the wave direction to β_h in local water depth, h.

As well as altering the wave direction, refraction will change the energy density (and hence the wave height) of the incoming waves. Depending on the foreshore geometry an alteration in wave height, given by the refraction coefficient K_r will be obtained:

$$K_r = \sqrt{\{\cos \beta_o / \cos \beta_h\}} \tag{2-12}$$

Further details on refraction and shoaling are given in Goda (1985). In most practical cases, these two effects should be considered together, so the design wave height. H_n, relative to the offshore wave height, H_o, is given by:

$$H_n / H_o = K_r K_s \tag{2-13}$$

Graphical methods are available for the estimation of refraction and diffraction for regular waves, but for realistic sea bed topographies and wave conditions, the processes of shoaling and refraction become more complicated, and must be modelled by numerical models able to accommodate realistic bed forms and wave conditions. Such models should also include the process of wave diffraction around obstacles, particularly islands, headlands, or breakwaters.

Where water depths become shallower, continuing shoaling of the waves, perhaps assisted by refraction, will lead some waves to approach limiting steepness. Thereafter, any further increase will lead to wave breaking, and the attendant energy losses.

2.2.1.2 *Depth-limited breaking*

Wave breaking remains one of the phenomena which is difficult to describe fully in any mathematical way, in part because the processes are very complex, and because some are not yet completely understood. The primary cause for breaking is that the wave steepness has reached the fundamental limit:

$$(H / L)_{max} = 0.142 \tag{2-14}$$

Wave breaking involves significant loss of energy and is not reversible. In shallow depths h_s, wave heights may be strongly influenced by the degree of wave breaking, depending on the relative depth, h_s/gT_m^2, and the seabed slope, tan α. For very shallow bed slopes, usually taken as flatter than 1:100 or 1:50, it is often assumed that a simple limit to the individual (maximum) wave height H_{maxb} relative to the local water depth h_s in shallow water may be given by:

$$H_{maxb} / h_s = 0.78 \qquad \text{for } \tan \alpha < 1/50 \tag{2-15}$$

Goda (1985) has recommended the following simple formula to estimate the limiting wave height of individual waves:

$$\frac{H_b}{L_0} = A\left\{1 - \exp\left[-1.5\frac{\pi h}{L_0}\left(1 + 15\tan^{4/3}\alpha\right)\right]\right\} \qquad (2\text{-}16)$$

where A is an empirical parameter which is 0.17 for regular waves and in between 0.12 and 0.18 for random waves. Goda's method for wave heights in the surf zone (Goda, 1975) used Shuto's method (Shuto, 1974) instead of simple linear wave methods to give the shoaling coefficient K_s. For wave breaking where $h/L_o \geq 0.2$ and for cases where H_{max} is needed the following method is recommended:

$$H_{1/250} = 1.8\ K_s\ H_{so} \qquad (2\text{-}17)$$

For $h/L_o < 0.2$:

$$H_{1/250} = \min\ \{\ (\beta_0{*}\ H_{so} + \beta_1{*}\ h),\ \beta_{max}{*}\ H_{so},\ 1.8\ K_s\ H_{so}\ \} \qquad (2\text{-}18)$$

where:

$$\beta_0{*} = 0.052\ (H_{so}/L_o)^{-0.38}\ \exp\ (20\ \tan\ \alpha^{1.5}) \qquad (2\text{-}19)$$

$$\beta_1{*} = 0.63\ \exp\ (3.8\ \tan\ \alpha) \qquad (2\text{-}20)$$

and

$$\beta_{max}{*} = \max\ \{\ 1.65,\ 0.53(H_{so}/L_o)^{-0.29}\ \exp\ (2.4\ \tan\ \alpha)\ \} \qquad (2\text{-}21)$$

For wave steepnesses H_{so}/L_o larger than 0.04 Goda recommends the use of design diagrams rather than using equations 2-17 to 2-21 (Goda, 1985). The factor of 1.8 in Equation 2-17 and 2-18 in fact depends on the distribution of individual wave heights in a wave field (usually Rayleigh) and the number of individual wave heights in a storm. The value of 1.8 is valid for a storm field with 250 independent waves. Section 2.4.4.1 provides more details on the dependence of the multiplication factor to the number of waves. In section 5.4.2 an example application is given where the randomness of the multiplication factor is explicitly considered.

For cases where $H_{1/3}$ rather than H_{max} is needed, Goda suggested a similar method for $H_{1/3}$.

For $h/L_o \geq 0.2$:

$$H_{1/3} = K_s\ H_{so} \qquad (2\text{-}22)$$

For $h/L_o < 0.2$:

$$H_{1/3} = \min\ \{\ (\beta_0 H_{so} + \beta_1 h),\ \beta_{max} H_{so},\ K_s H_{so}\ \} \qquad (2\text{-}23)$$

where:

$$\beta_0 = 0.028\ (H_{so}/L_o)^{-0.38} \exp(20 \tan \alpha^{1.5}) \qquad (2\text{-}24)$$

$$\beta_1 = 0.52 \exp(4.2 \tan \alpha) \qquad (2\text{-}25)$$

and

$$\beta_{max} = \max \{\ 0.92, 0.32(H_{so}/L_o)^{-0.29} \exp(2.4 \tan \alpha)\ \} \qquad (2\text{-}26)$$

Noting that for steep bed slopes, waves may shoal substantially before breaking starts, Owen (1980) developed a simple method to provide first-estimates of the upper limit to the (significant) wave height H_{bs} (equivalent post-breaking wave height) in any water depth h_s for five different bed slopes. H_{bs} is not necessarily the wave height which would be obtained by direct measurement, but is designed to give the correct overtopping discharge as confirmed from physical model tests where significant wave breaking took place. The original method was graphical, but polynomial equations were later fitted to the curves. They are recommended for use where overtopping discharges are relevant for design.

Bed slope	breaking limit, H_{bs}/h_s	
1/100	$H_{bs}/h_s = 0.58 - 2\ (h_s/\ gT_m^2)$	(2-27)
1/50	$H_{bs}/h_s = 0.66 - 10.58\ (h_s/\ gT_m^2). + 229.2\ (h_s/\ gT_m^2)^2$	(2-28)
1/30	$H_{bs}/h_s = 0.75 - 20.08\ (h_s/\ gT_m^2) + 479.2\ (h_s/\ gT_m^2)^2$	(2-29)
1/20	$H_{bs}/h_s = 0.95 - 38.42\ (h_s/\ gT_m^2) + 895.8\ (h_s/\ gT_m^2)^2$	(2-30)
1/10	$H_{bs}/h_s = 1.54 - 97.83\ (h_s/\ gT_m^2) + 2542\ (h_s/\ gT_m^2)^2$	(2-31)

Further breaker index results from flume model tests with straight beds of slope 1:10, 1:20 and 1:30 including comparison with various breaker index formulae are given in Durand & Allsop (1997).

2.2.2 *Use of parameter map*

A classification of wave loads induced by different types of breaking waves at the structure is depicted in Figure 2-5. Loads can be classified into (a) quasi-static loading (which may lead to an overall failure of the structure), (b) cyclic loading (causing fatigue and stepwise failure), and (c) impact loading possibly causing local overall failures and structural failure modes. These types of loadings may be further classified and need to be predicted by new methods.

A parameter response map for prediction of the type of wave loading on vertical and vertically composite breakwaters based on structure geometry and wave conditions has therefore been developed, Figure 2-6. This has been validated against a number of model data sets from HR, LWI, UE and UoN (see Annex 2 for abbreviations).

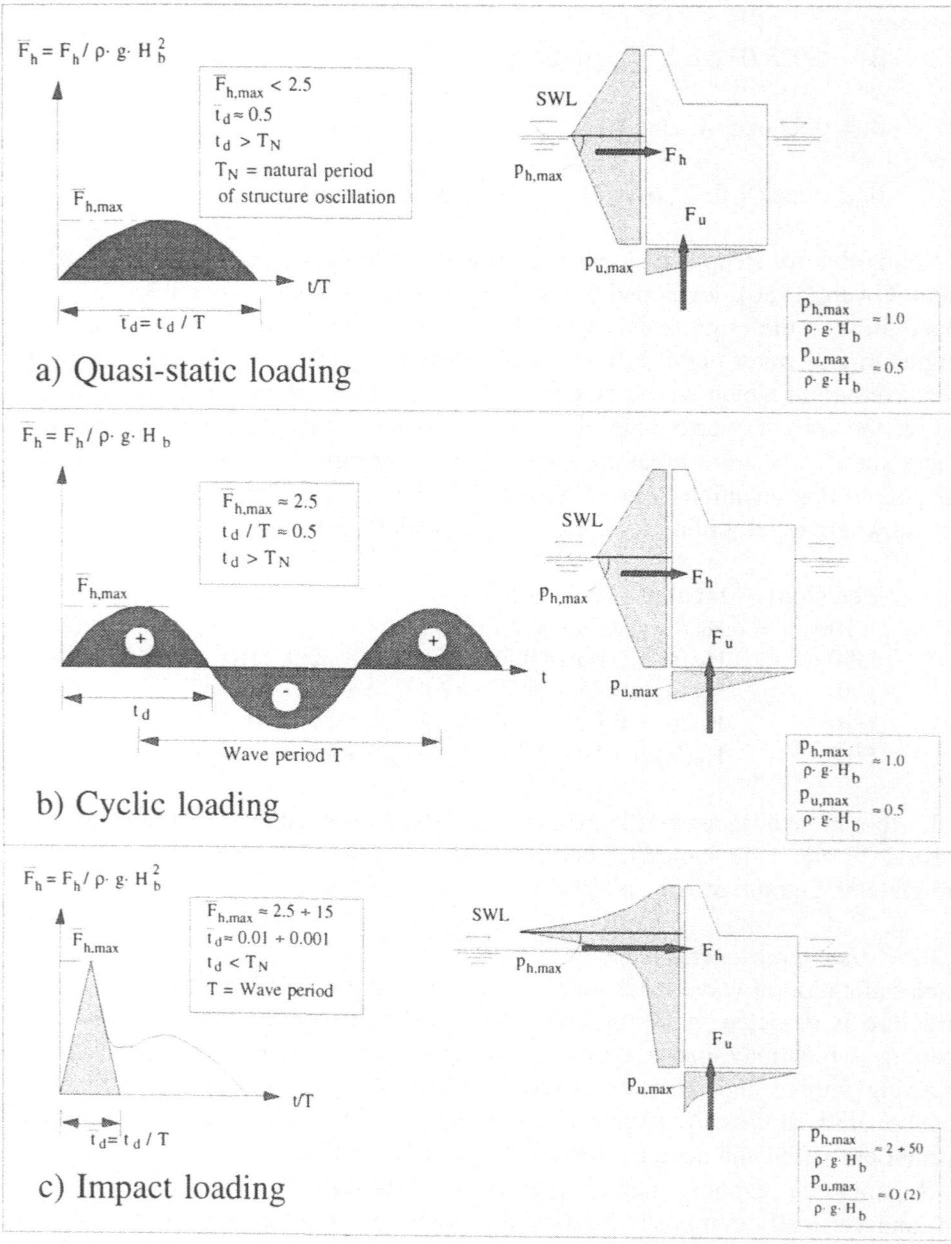

Figure 2-5. Specification of wave loading for monolithic structures (Oumeraci et al., 1995).

The main parameters used in the parameter map for determination of the wave loading on the structure are the relative berm height, $h_b^* = h_b/h_s$, the relative wave height, $H_s^* = H_{si}/h_s$, and the relative berm width, $B^* = B_{eq}/L_{pi}$. The wave parameters H_{si} and L_{pi} are determined in the water depth h_s, and L_{pi} is determined by linear wave theory. The geometric parameters h_b, h_s and d are defined in the standard PROVERBS notation (Annex 1). The equivalent berm width is defined halfway up the berm. $B_{eq} = B_b + m \cdot h_b/2$.

The relative berm height, h_b/h_s determines the type of structure: a simple vertical wall, a composite structure with a low mound, a composite structure with a high mound or a rubble mound with a crown wall. The high mound composite breakwater as described in chapter 2.6.3 was included in the parameter map at a later stage of the project and is described in Kortenhaus & Oumeraci (1998).

The relative wave height, H_{si}/h_s, then determines whether wave impacts will occur, based on the depth of water at the toe of the caisson, which in the case of a composite structure will be the water depth on the berm.

For high mound breakwaters ($0.6<h_b/h_s<0.9$) exposed to large waves ($0.25 < H_{si}/h_s < 0.60$) a further sub-division is made, based on the relative width of the berm, B_{eq}/L_{pi}.

The parameter map indicates that wave impacts will occur for three categories of conditions:

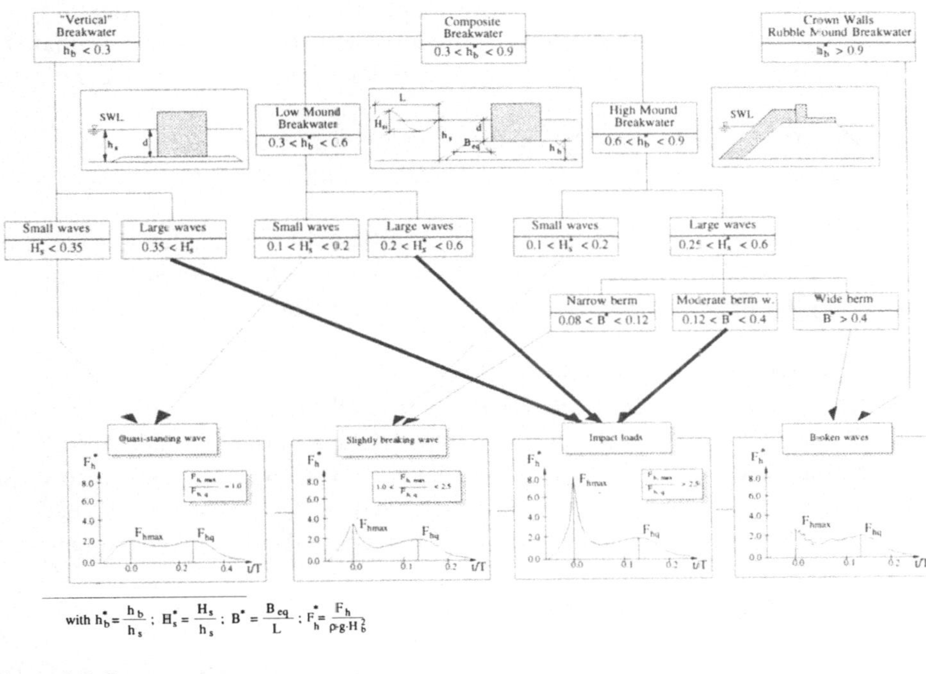

Figure 2-6. Parameter map.

- vertical walls with large waves ($H_{si}/h_s > 0.35$)
- low mound breakwaters with large waves ($0.25 < H_{si}/h_s < 0.60$)
- high mound breakwaters with moderate berms ($0.12 < B_{eq}/L_{pi} < 0.4$) and large waves ($0.25 < H_{si}/h_s < 0.60$).

In a probabilistic approach, inputs to the parameter map are considered as random variables. By application of simulation techniques, the probability of ending up in impact conditions may be derived. This procedure can be used to derive the distribution of design forces on the caisson (see section 5.4.2 for an example).

It is important to emphasise that if the structure is in relatively shallow water on a bed slope (tan α) shallower than 1:50, it is quite possible that many of the larger waves can break before the structure. The sea in front of the structure will be highly aerated and wave forces will be reduced. In this case the parameter map needs a procedure to identify broken or non-broken conditions at the structure.

Using shoaling and a modified Miche (1944) breaking criterion (Vrijling and Bruinsma, 1980) it is possible to iterate a breaking depth (h_{br}):

$$H_{bs} = K_s(h_{br}) \cdot H_{so} = 0.093 \cdot L_p \cdot \tanh\left(\frac{2 \cdot \pi \cdot h_{br}}{L_p}\right) \quad (2\text{-}32)$$

If the ratio between the breaking wave depth (h_{br}) and the water depth at the structure (h_s) is greater than a "decision coefficient" α_{br}, it is possible that the structure is in the surf zone, i.e. many waves have broken before reaching the structure:

- $h_{br} / h_s > \alpha_{br}$ broken waves at the structure
- $h_{br} / h_s < \alpha_{br}$ back to the parameter map for the hydrodynamic condition at the structure

The ideal value for the decision coefficient is $\alpha_{br} = 1.0$, but analysis of breaking methods suggested that a safety factor for deterministic design should be applied. Initial analysis suggests that $\alpha_{br} = 1.2$ may give reasonably safe results.

2.2.3 *Estimation of proportion of impacts*

A simple procedure has been developed within PROVERBS to give first estimates of the likely percentage of wave impacts on a vertical or composite wall (Volume IIa, section 2.3). In this procedure, wave breaking is assumed to occur when, at the location of the structure, the incident wave height with exceedance probability of 0.4%, ($H_{99.6}$) is higher than a breaker height H_{bc} calculated below:

a) Identify geometric parameters as follows:

- Water depth in front of the structure for design case(s) h_s
- Width, height and slope of front face of berm in front of wall B_b, h_b and m
- Depth of water over the berm for design case(s) d

➢ Effective or equivalent berm width $B_{eq} = B_b + (m \cdot h_b / 2)$

b) Identify design wave condition(s) given by H_{si} and T_p taking account of effects on the local wave heights caused by refraction, and shoaling. Derive the peak period wave length L_{pi} in the local water depth h_s, solving:

$$L_{pi} = (gT_p^2/2\pi) \tanh (2\pi\, h_s/L_{pi}) \qquad (2\text{-}33)$$

c) Calculate the wave height at breaking H_{bc}:

$$H_{bc} = (0.1025 + 0.0217\, C^*)\, L_{pi} \tanh (2\pi\, k_b\, h_s/L_{pi}) \qquad (2\text{-}34)$$

where

$$C^* = (1 - C_r) / (1 + C_r) \qquad (2\text{-}35)$$

where C_r is the overall reflection coefficient from the structure so that Equation (2-34) includes wave reflection from the structure which has not been considered in earlier modifications of Miche's formula.

The empirical factor $k_b = 0.0076\, (B_{eq}/d)^2 - 0.1402\, (B_{eq}/d) + 1$ is illustrated in Figure 2–7 against the relative berm width B_{eq}/d.

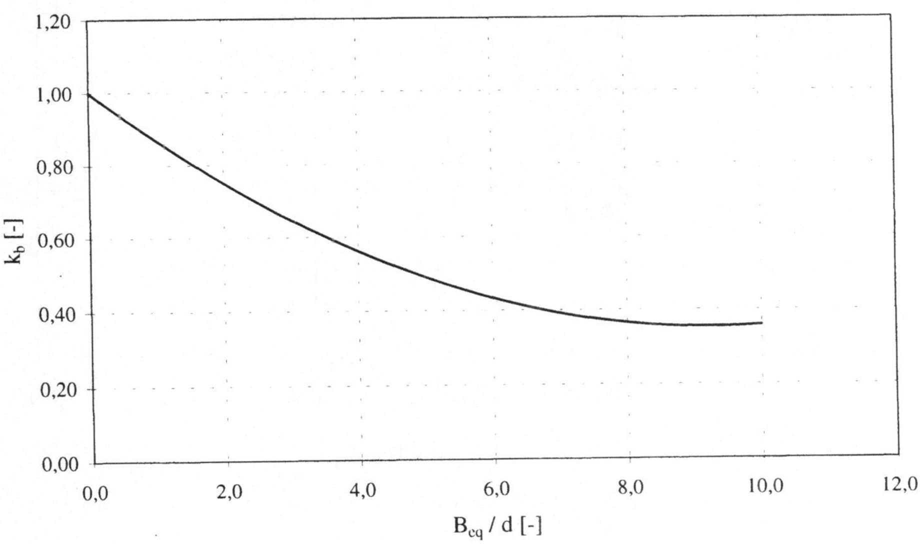

Figure 2-7. Influence of B_{eq}/d on the mound parameter k_b

d) Values of C_r may be estimated for the particular structure combination:
For simple vertical walls and small mounds, high crest $C_r = 0.95$
For low-crest walls, $(0.5 < R_c/H_{si} < 1.0)$ $C_r = 0.8 + 0.1\,R_c/H_{si}$
For composite walls, large mounds, and heavy breaking $C_r = 0.5$ to 0.7

It will be noted that there are still many uncertainties in the prediction of onset of breaking in front of vertical / composite breakwaters. For these structures with high reflective walls, it may be assumed that $C_r = 1$, in which case Equation (2-34) will reduce to:

$$H_{bc} = 0.1025\,L_{pi} \tanh(2\pi\,k_b\,h_s/L_{pi}) \qquad (2\text{-}36)$$

As shown in Figure 2-8, Equation 2-36 succeeds in distinguishing impact conditions from pulsating conditions, although this method does overestimate the proportion of impacts.

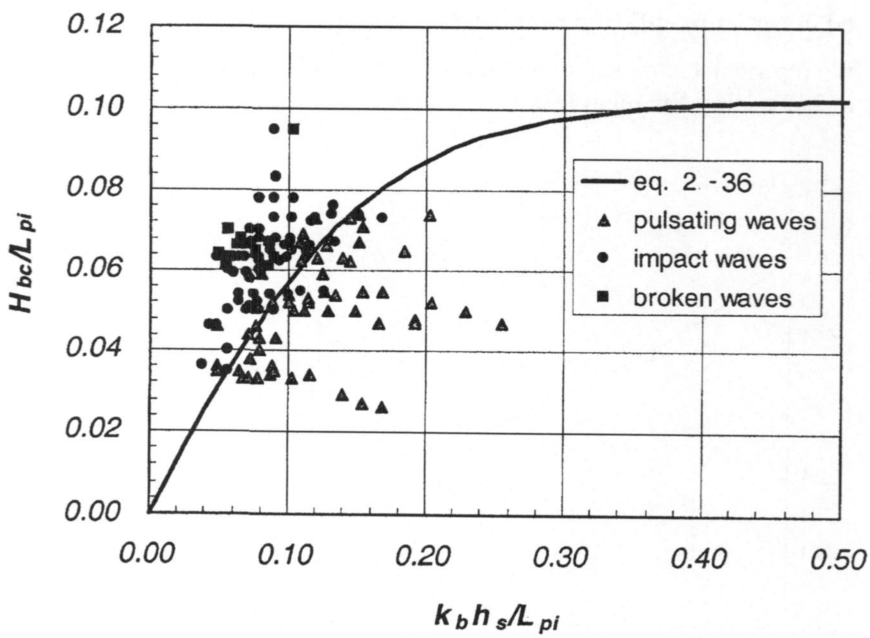

Figure 2–8. Breaking curve (Eq. 2-36) and measured values of H_{bc}/L_{pi} (Vol. IIa, section 2.3).

e) Compare the incident wave height, H_{si}, with the breaking height, H_{bc}, calculated in (Eq. 2-36):

- $H_{si} / H_{bc} \leq 0.6$ — No evident breaking occurs and wave load is non-breaking
- $0.6 < H_{si} / H_{bc} < 1.2$ — Wave breaking occurs and waves may give impacts
- $H_{si} / H_{bc} \geq 1.2$ — Heavy breaking or waves may give broken loads

f) Estimate the proportion (or percentage) of breaking waves P_b using H_{bc} and the incident wave height H_{si} :

$$P_b = \exp\left[-2 \left(H_{bc} / H_{si}\right)^2 \right] \tag{2-37}$$

The predicted value of P_b (shown as a percentage in Figure 2-9) should be considered as an upper limit for the percentage of impact loadings because it expresses the fraction of both impact and broken waves.

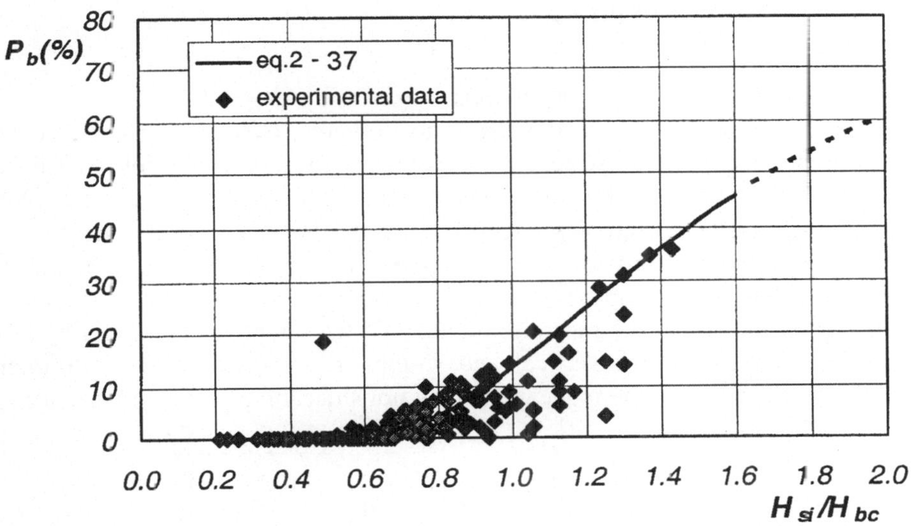

Figure 2-9. Occurrence frequency of breaking waves (Vol. IIa, section 2.3).

Equations from 2-33 to 2-37 are valid for the following conditions:

- $0.08 \leq h_s/L_{pi} \leq 0.2$;
- $0 \leq B_{eq}/d < 10$;
- $H_{si}/d \leq 1.3$.

When the structure is within the surf zone some waves arrive already broken at the structure. For this case the portion of broken waves should be subtracted from P_b given by equation 2-37.

Experimental data seems to confirm that the maximum wave height which describe the transition from impact to broken mode may be estimated assuming $B_{eq}/d=0$ and $C_r = 0$ in equation (2-36) which will reduce to:

$$H_{bs} = 0.1242\ L_{pi} \tanh\left(2\pi\ h_s/L_{pi}\right) \tag{2-38}$$

Therefore the proportion (or percentage) of waves that may actually break directly onto the structure causing impacts, P_i can be calculated as follows (Volume IIa, section 2.3):

$$P_i = \{\exp\left[-2\left(H_{bc}/H_{si}\right)^2\right] - \exp\left[-2\left(H_{bs}/H_{si}\right)^2\right]\} \times 100\% \tag{2-39}$$

2.3 HYDRAULIC RESPONSES

2.3.1 *Wave transmission over caissons*

The main function of any breakwater is to prevent excessive wave attack in a certain area. Assuming the structure itself survives, it is still possible that it fails to fulfil its function due to wave energy passing over the breakwater. Wave transmission is most commonly expressed by the transmission coefficient $C_t = H_{st} / H_{si}$ where the transmitted wave is expressed as a significant wave height, H_{st}, as is the incident wave height, H_{si}.

An expression for wave transmission adapted from a previous model by Goda (1969) was developed for vertical and sloping faced breakwaters, introducing three new parameters in the model. The new transmission model is written as:

$$C_t = \begin{cases} \sqrt{0.25\left(1-\sin\dfrac{\pi}{2\left(\alpha_1+\alpha_X\right)}\left(\dfrac{R_c}{H_1}+\left(\beta+\beta_X\right)\right)\right)^2 + 0.01\left(1-\left(\dfrac{d+d_c}{h_s}+\gamma_X\right)\right)^2} \\ \qquad \text{for } \beta+\beta_X-\left(\alpha_1+\alpha_X\right) < \dfrac{R_c}{H_1} < \alpha_1+\alpha_X-\left(\beta+\beta_X\right) \\ 0.1\left(1-\left(\dfrac{d+d_c}{h_s}+\gamma_X\right)\right) \qquad \text{for } \dfrac{R_c}{H_1} \geq \alpha_1+\alpha_X-\left(\beta+\beta_X\right) \end{cases} \tag{2-40}$$

The parameters α_l and β are derived in the same way as for the original Goda model. Values of the parameters α_x, β_x, and γ_x are shown in Table 2-3 for several caisson types illustrated in Figure 2-10. More caisson types and some additional background information on the derivation of the model are given in Volume IIa, section 3.1.

Table 2-3. Summary of dimensionless parameters for several types of breakwaters (for further types of caisson structures see Volume IIa, section 3.1).

Type of breakwater	α_x	β_x	γ_x	**Region of validity, R_c/H_s**
Conventional breakwater	-0.9	-0.34	0	0.30 – 1.66
Parapet breakwater	0.2	0.13	0	0.61 - 1.19
Perforated breakwater	-0.3	-0.22	0	0.95 – 1.55
Sloping top breakwater	3.1	1.05	0	0.66 - 1.80
Horiz. composite breakwater	-0.4	-0.03	-0.3	0.00 - 0.82

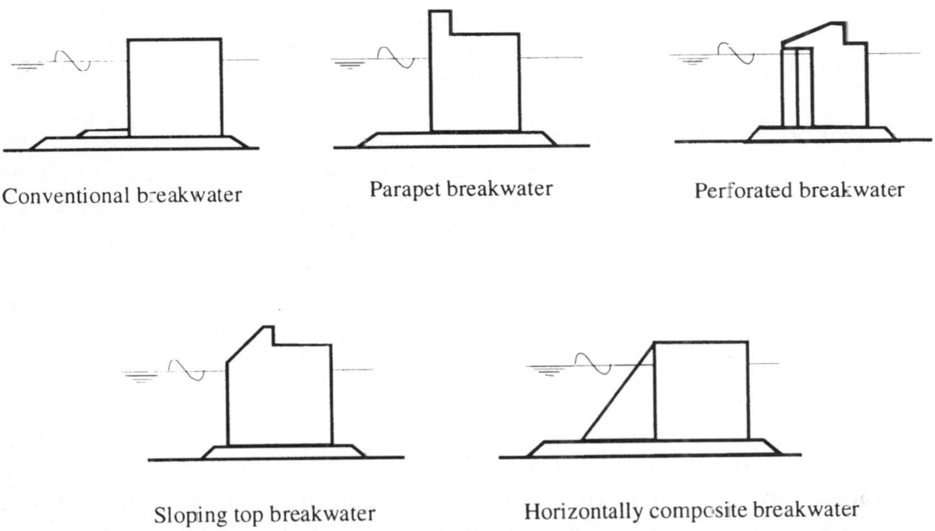

Figure 2-10. Types of caissons.

Comparison of the new model with several measurements in Figure 2-11 shows that the model is in good agreement with the data.

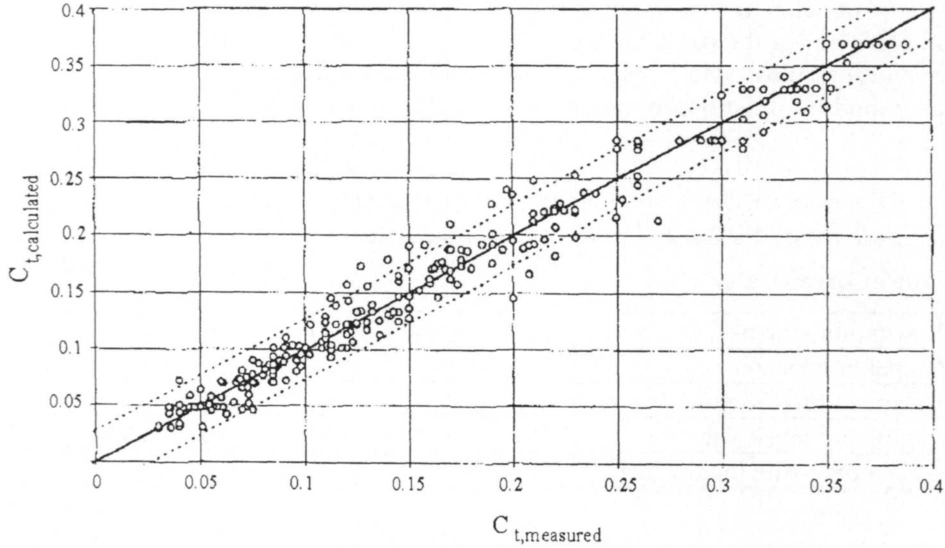

Figure 2-11. Comparison of transmission model with data (Vol. IIa, section 3.1).

2.3.2 *Wave overtopping discharges*

Wave overtopping discharges have been investigated within the previous MAST II / MCS project. Several reports were published by the project partners, and some were also used in PROVERBS. It is recommended here to use the prediction method(s) by Franco & Franco (1999) for mean and peak overtopping discharges for structures in deep water, and by Besley et al (1998) for those influenced by wave breaking. More details can be found in Volume IIa, section 3.2.

2.3.3 *Wave reflections*

2.3.3.1 *Vertical breakwaters and seawalls*

Prediction of wave reflections from simple and composite walls, as well as effects of low-crest levels or oblique / short-crested wave attack have been investigated under the previous MAST II/MCS-project and results have been published in reports and papers. Within PROVERBS these results (Canel, 1995, Allsop & McBride, 1996) have been summarised and further discussed in Volume IIa, section 3.3.

2.3.3.2 *Perforated structures*

The reflection performance of perforated structures have been studied within PROVERBS by means of experimental and mathematical approaches: the former were conducted by LWI with single and multi-screen tests in the GWK and ENEL with random wave tests on the Porto Torres caisson model, the latter by PM trying validation of a mathematical model proposed by Fugazza & Natale (1992).

Although the mathematical approach compared against some GWK data showed a promising prediction capability, it cannot be proposed as the validation was done for experimental conditions with regular waves and for a single chamber only.

The experiments conducted at GWK with single screen showed that the reflection performance of such structures is a function of a reflection parameter RP for regular and random waves defined according to the following formula (Bergmann & Oumeraci, 1999):

$$RP = (H_i/d)^{0.65}/n^{0.95} \tag{2-41}$$

where H_i is the incident wave height, d is the water depth and n is the screen porosity.

The tests on single chamber systems (OCS) and multi-chamber systems (MCS) have also shown that the well known dependence of the reflection parameter by the relative chamber width (B/L) has the typical parabolic shape which is much flatter for multi-chamber systems compared to the OCS (Fig. 2-12). However, these tests have been conducted with regular waves and wave spectra on a vertical structure without a berm which impels a generalisation of the findings for breakwaters on a rubble foundation.

For all the random wave tests performed on the Porto Torres caisson 1:20 model, the reflection coefficients were calculated with the three probes method. The reflection performance shows the well known (Allsop et al., 1994) parabolic shape of the relationship between C_r and B/L, with a minimum around B/L = 0.2.

Joining the data from partial standing waves for Dieppe caisson model tests and random wave tests from 3D model tests performed by PM at Delft Hydraulics an interpolating equation is proposed to calculate the reflection performance as a function of B/L (Fig. 2-13) which at present research stage is the most reliable (or less unreliable) available:

$$C_r = 18.6\left(\frac{B}{L}\right)^2 - 7.3\left(\frac{B}{L}\right) + 0.98 \tag{2-42}$$

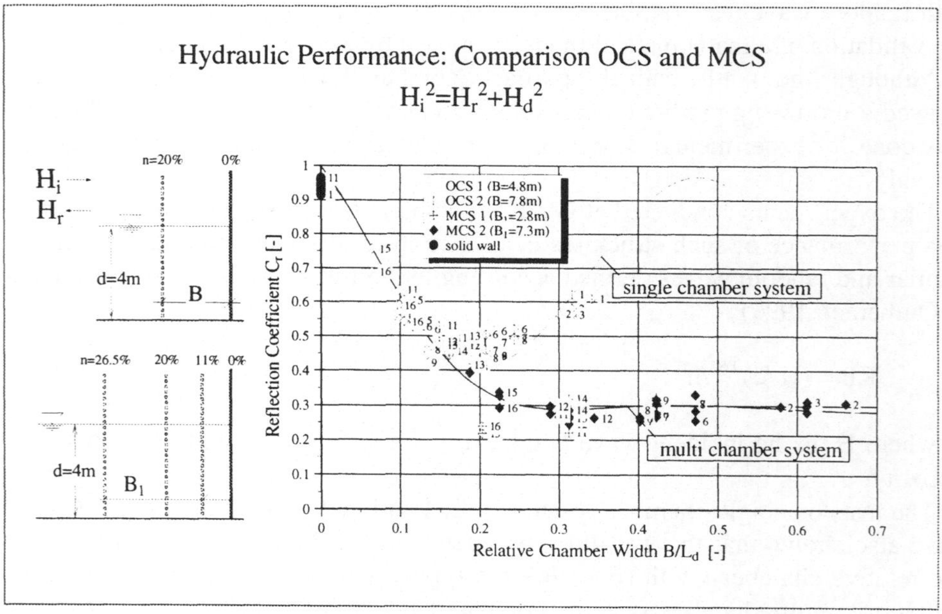

Figure 2-12. Comparison of single- and multi-chamber systems for a simple vertical structure (wave spectra, GWK) (Bergmann & Oumeraci, 1999).

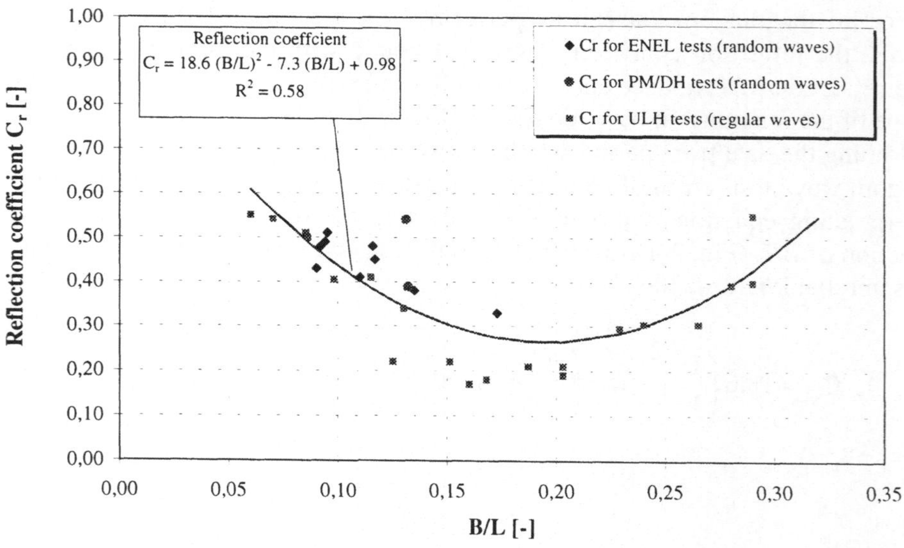

Figure 2-13. Reflection performance as a function of the relative chamber width.

2.4 PULSATING WAVE LOADS

2.4.1 *Horizontal and vertical forces / pressures*

The basis of the Goda model (Goda, 1985) is assumed pressure distributions over the height and width of the caisson which, when integrated over the front face and underside give equivalent sliding forces. The Goda formulae are written as (see Fig. 2-1):

$$\eta^* = 0.75\,(1+\cos\beta)\lambda_1 H \tag{2-43}$$

$$p_1 = 0.5\,(1+\cos\beta)(\lambda_1\alpha_1 + \lambda_2\alpha_2\cos^2\beta)\,\rho g H \tag{2-44}$$

$$p_3 = \alpha_3 p_1 \tag{2-45}$$

$$p_4 = \alpha_4 p_1 \tag{2-46}$$

$$p_u = 0.5\,(1+\cos\beta)\,\lambda_3\alpha_1\alpha_3\,\rho g H \tag{2-47}$$

where η^* is the water elevation above the still water level, H is the incident wave height in front of the structure; β is the obliquity of wave attack relative to normal to the structure; λ_1, λ_2, λ_3 are multiplication factors dependent on the geometry of the structure, and α_1, α_2, α_3, α_4 are multiplication factors dependent on the wave conditions and the water depth which are given by:

$$\alpha_1 = 0.6 + 0.5\left(\frac{4\pi h_s/L_p}{\sinh(4\pi h_s/L_p)}\right)^2 \tag{2-48}$$

$$\alpha_2 = \min\left(\frac{\left(1-\frac{d}{h}\right)\left(\frac{H}{d}\right)^2}{3}, \frac{2d}{H}\right) \tag{2-49}$$

$$\alpha_3 = 1 - \left(\frac{d+d_c}{h}\right)\left(1 - \frac{1}{\cosh(2\pi h/L_p)}\right) \tag{2-50}$$

$$\alpha_4 = 1 - \frac{R_c^*}{\eta^*} \tag{2-51}$$

where h_s is the water depth in front of the structure; L_p is the wave length; d is the depth in front of the caisson; d_c is the height over which the caisson protrudes in the rubble foundation; and R_c^* is the minimum of the freeboard R_c and the notional run-up elevation η^*.

When the wave pressures are known, the wave forces are given by:

$$F_{h;\,Goda} = \frac{1}{2}(p_1 + p_4)R_c^* + \frac{1}{2}(p_1 + p_3)(d + d_c) \tag{2-52}$$

$$F_{u;\,Goda} = \frac{1}{2}p_u B_c \tag{2-53}$$

in which B_c denotes the width of the caisson bottom. The uplift force may need some corrections depending on the geometry and the grain sizes of the rubble foundation as proposed and discussed in section 3.5 of chapter 3.

The lever arms of the wave forces with respect to the centre of the caisson bottom are given by:

$$l_{h;Goda} = d + d_c + \frac{R_c^{*2}(p_1 + 2p_4) - (d + d_c)^2 (p_1 + 2p_3)}{3R_c^*(p_1 + p_4) + 3(d + d_c)(p_1 + p_3)} \tag{2-54}$$

$$l_{v;Goda} = \frac{1}{6}B_c \tag{2-55}$$

With respect to the heel of the caisson, half of the caisson width B_c has to be added to $l_{v;Goda}$ resulting in $l_{v;Goda} = 2/3 \cdot B_c$. Using the expressions for the wave forces and the lever arms, the total moment due to the wave forces can be calculated by:

$$M_{Goda} = l_{h;Goda} \cdot F_{h;Goda} + l_{v;Goda} \cdot F_{v;Goda} \tag{2-56}$$

The calculated forces and moments serve as input in several limit state equations describing the stability of the breakwater. The extension of the Goda model for waves breaking at the structure as described by Takahashi (1996) is not implemented here as PROVERBS developed its own model for impact waves at the structure (see section 2.5.1). Forces on perforated caissons have been investigated under PROVERBS and results are discussed in section 2.8.1 or in more detail in Volume IIa, section 8.1.

2.4.2 *Seaward or negative forces*

Most design methods for caisson and other vertical breakwaters concentrate on forces that act landward, usually termed positive forces. It has however been shown that a significant proportion of caisson or blockwork breakwaters have failed by sliding or rotation seaward, indicating that net seaward or negative forces may indeed be greater than positive forces. To some extent, this may be explained by the low crests of some of these breakwaters causing a reduction of shoreward forces by overtopping. But even with higher crests the seaward force may be larger than the shoreward force, as will be explained below.

Previous prediction methods have been ascribed to Sainflou (1928) and to Goda (1967). Both are based on (relatively) deep water, and non-breaking or pulsating waves. The new methods described here are being discussed by McConnell et al. (2000).

2.4.2.1 *Sainflou's prediction method*

Sainflou (1928) introduced prediction method derived from trochoidal theory. "Simplified Sainflou" formulae provide a pressure distribution at wave crest and trough are:

$$p_1' = \rho g\,(H_{max} - h_0) \qquad (2\text{-}57)$$

$$p_2' = \rho g H_{max} / \cosh(2\pi h/L_{pi}) \qquad (2\text{-}58)$$

$$h_0 = (\pi H_{max}^2/L)\coth(2\pi h/L_{pi}) \qquad (2\text{-}59)$$

where h is the water depth in front of the structure. The parameter h_0 takes into consideration the asymmetry of waves in front of a structure. The wave height used in Sainflou's formulae is assumed to be H_{max}. The main parameter definitions are given in Figure 2-14.

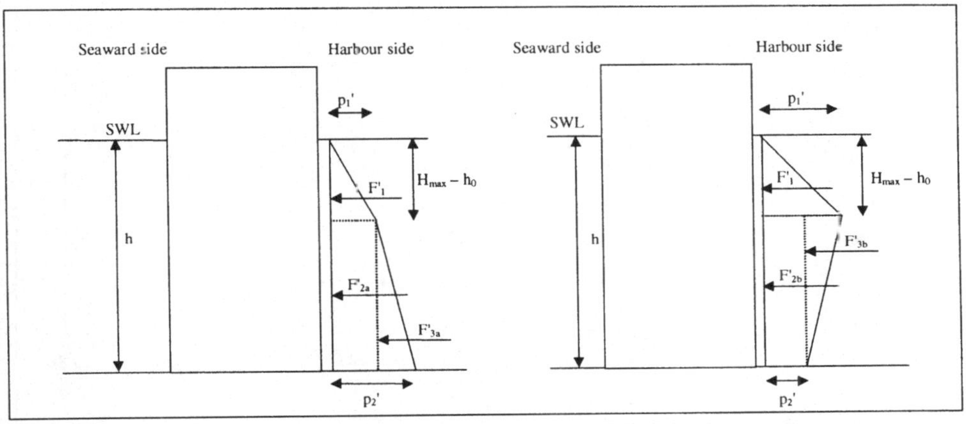

Figure 2-14. Net negative pressure distribution according to Sainflou (on the left: $p_2' > p_1'$; on the right: $p_2' < p_1'$).

From these formulae the net negative horizontal force F_{hs} relative to static water conditions using Sainflou's method may be calculated as follows:

$$F_{hS} = (H_{max}-h_0)\,p_1'/2 + (p_1' + p_2')\,(h-H_{max} + h_0)/2 \qquad (2\text{-}60)$$

Goda's (1967) simple method indicates that, for relative depths h/L < 0.25, net negative forces may be higher than positive forces for most wave steepnesses. This implies that negative forces may govern primary design responses for these cases. This has been tested by examining data from tests within PROVERBS. Values of $F_{hmin(1/250)}/F_{hmax(1/250)}$ are plotted against H_{si}/h_s in Figure 2-15, and show that there is some risk that negative forces exceed positive forces for small relative wave heights, $H_{si}/h_s<0.3$. For deeper water conditions, $H_{si}/h_s < 0.2$, most test data give measured suction forces that are greater than the conventional landward forces ($F_{hmin(1/250)}/F_{hmax(1/250)} > 1$).

2.4.2.2 *Probabilistic design approach for negative forces*

In a probabilistic design approach, any conservatism in the models applied should be avoided. Instead, a model should be used that describes the mean behaviour and an estimate of the spread around the model. The spread (model uncertainty) is then taken explicitly into account in the calculation (see chapter 5 for a discussion of the probabilistic approach and the kinds of uncertainty that may be considered).

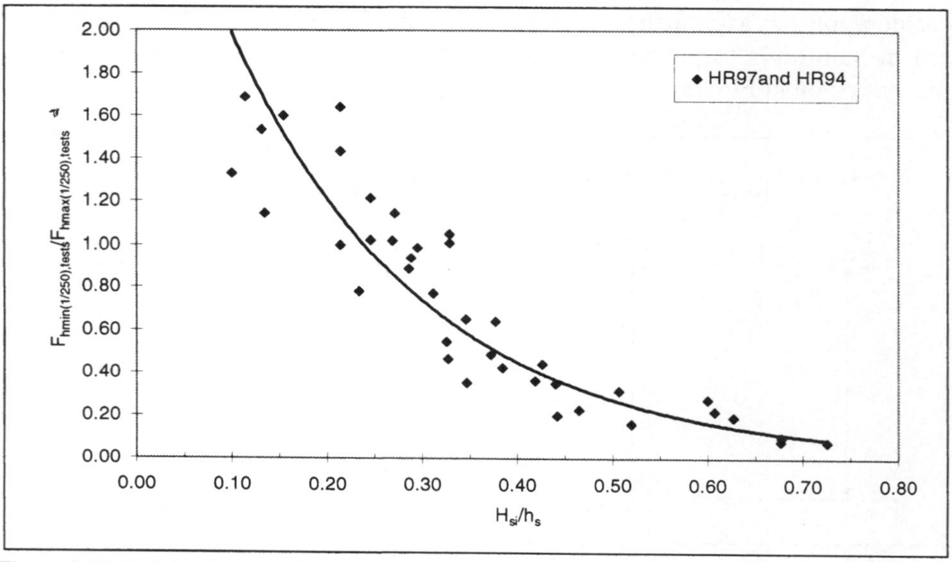

Figure 2-15. Relation between F_{hmin} and F_{hmax} at 1/250 level.

In probabilistic calculations of $F_{hmin1/250}$ use can be made of the mean and standard deviations calculated from the HR94 and HR97 data. For these tests, the mean value of the ratio of measured to Sainflou predictions was 1.126 and the standard deviation $\sigma_\gamma = 0.1508$, giving a coefficient of variation of 13%. The resulting formula for negative forces on simple vertical walls with a 1:50 approach slope, and for $H_{si}/h_s < 0.6$. is:

$$F_{hmin} = 1.126\, F_{hS} \pm 13\% \tag{2-61}$$

2.4.2.3 *Deterministic design approach for negative forces*

Deterministic calculations of $F_{hmin1/250}$ use the following for negative forces on simple vertical walls with 1:50 approach, and for $H_{si}/h_s < 0.6$, illustrated in Figure 2-17.

$$F_{hmin} = 1.27 \cdot F_{hS} \tag{2-62}$$

In all calculations of net negative force described here, it is assumed that the same static water level acts on both sides of the breakwater, and that there are no additional wave-induced forces acting on the harbour side of the structure. Wave action within the harbour, or overtopping impacts behind the breakwater, may act on the caisson to give forces additional to those discussed here.

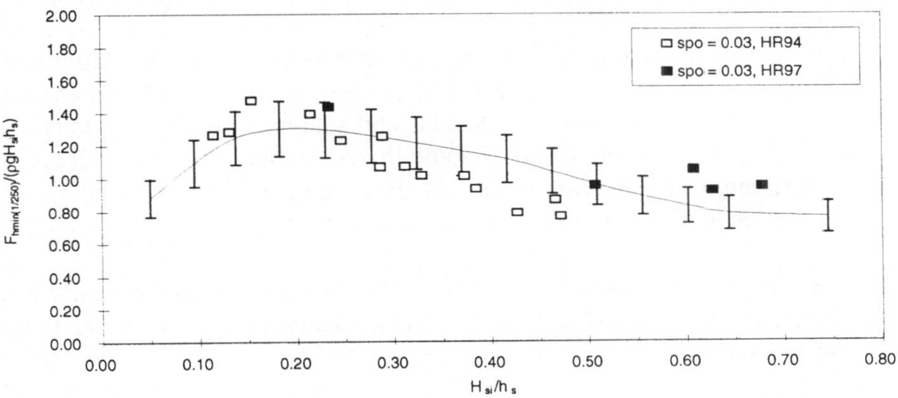

Figure 2-16. Improved Sainflou prediction line (probabilistic design approach) for $s_{po} = 0.03$.

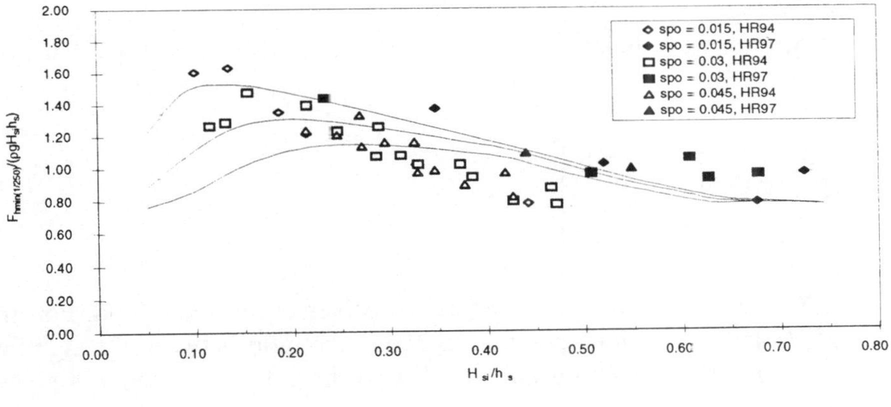

Figure 2-17. Improved Sainflou prediction lines (deterministic design approach).

2.4.3 *Effects of 3-d wave attack on pulsating loads*

Principal investigations on effects of 3-d wave attack on pulsating loads were performed under the MAST II/MCS-project. Results of these tests have been published in internal reports and in journals and conference papers. Within PROVERBS new tests have been performed at AU and HR which are reported and summarised in Volume IIa, section 4.3 (see particularly Allsop & Calabrese, 1998, 1999 which include discussion of the method by Battjes, 1982). It is recommended to either use these methods or use the results from the MCS-project as e.g. reported in Franco et al. (1996).

2.4.4 *Uncertainties and scale corrections*

2.4.4.1 *Uncertainties*

In this section the uncertainty analysis of the wave height of non-breaking waves and the resulting quasi-static forces and moments on vertical structures is described. A more detailed description of kinds and sources of uncertainties is given in chapter 5. Uncertainties on the quasi-static load originate from two sources:

- The uncertainty of the maximum individual wave height in a wave field;
- The model uncertainty in the Goda wave force model.

The distribution of individual wave heights in a wave field can generally be assumed to follow a Rayleigh distribution, in which there is only one parameter, the significant wave height:

$$F_{\underline{H}}(H) = 1 - e^{-2\left(\frac{H}{H_s}\right)^2} \qquad (2\text{-}63)$$

The maximum wave height in a wave field is given by the following distribution:

$$F_{\underline{H_{max}}}(H) = \left(1 - e^{-2\left(\frac{H}{H_s}\right)^2}\right)^N \qquad (2\text{-}64)$$

where *N* denotes the number of individual waves in the wave field. For given values of H_s and N it is now possible to derive the ratio between the significant wave height and the maximum individual wave height. An example is given in Table 2-4.

Table 2-4. Multiplication factors (Short-term Rayleigh).

Number of waves in wave field	**H_s [m]**	**H_{max} [m]**	**Factor**
N=250	6.50	11.90	1.83
N=1000	6.50	12.98	1.99
N=3000	6.50	13.80	2.12

The multiplication factor of 1.8 that is present in the breaking model of Goda (1985) appears to correspond to a wave field with 250 individual waves (see also section 2.2.1.2).

The model uncertainties of the Goda wave force model are treated separately from the uncertainty on the maximum individual wave height. Comparison with model tests showed that the Goda model provides an overestimation of the wave forces (Van der Meer et al, 1994; Volume IIa, section 4.1). Further, an estimate of the spread around the model was derived. In a probabilistic approach, the bias of the model should be taken out and the spread should be explicitly dealt with. This can be done by introducing a model factor f_m to the force components originating from the Goda wave force model, according to:

$$F_{design} = f_m \cdot F_{Goda} \tag{2-65}$$

The model factors are all Normal distributed. The parameters of the distribution vary for different force components (summarised in Table 2-5).

Table 2-5. Distribution properties of model factors of Goda wave force model.

	Mean	**Standard deviation**
Horizontal Force	0.9	0.2
Horizontal Moment	0.81	0.37
Vertical Force	0.77	0.2
Vertical Moment	0.72	0.34

2.4.4.2 *Scaling*

Froude scaling relates the relative influence of inertial and gravity forces where Froude's number is equal both in model and prototype so that the following scaling relations can be derived:

Table 2-6. Principal scaling parameters for Froude's and Cauchy's scaling laws.

Parameter	**Froude**
Length	N_L
Area	$N_A = N_L^2$
Volume	$N_V = N_L^3$
Time	$N_t = \sqrt{N_L}$
Velocity	$N_u = \sqrt{N_L}$
Acceleration	$N_a = 1$
Mass	$N_m = N_\rho N_L^3$
Pressure	$N_p = N_\rho \cdot N_L$
Force	$N_F = N_\rho \cdot N_L^3$
Force per m	$N_{F/m} = N_\rho \cdot N_L^2$

In Table 2-6 N_L is the length scale of the model (length in prototype divided by length in model). It is recommended to use Froude's law for scaling results from hydraulic model tests as long as non breaking (quasi-static) waves occur at the structure.

2.4.5 *Use of numerical models*

As discussed in section 2.4.1 the Goda-method can predict horizontal forces on vertical structures and uplift forces underneath the caisson. Although this method is a valuable design method, no detailed information on flows and pressures under wave conditions can be obtained, nor can the influence of certain parameters such as the rock size of the rubble mound foundation be quantified. Numerical models that can simulate the flow pattern in front of the structure and flows/pressures inside the rubble mound foundation might be able to overcome such problems.

Since the flow pattern in front of vertical structures can be very complex due to breaking waves, wave impacts, wave interaction with complex shapes of the structure and wave interaction with the flow inside the rubble mound foundation, the development of a numerical model covering all relevant aspects in detail is extremely complex. Several types of numerical models can however contribute to modelling and understanding of relevant processes. In addition to empirical and mathematical methods, depth-averaged non-linear and weakly non-linear shallow-water wave equations are widely used in coastal- and breakwater engineering practise. Other types of numerical models solve the more fundamental Navier-Stokes equations or make use of the potential flow theory. In PROVERBS these two types of models are applied for simulating relevant processes for vertical breakwaters. Of this two-track approach, one is based on the Navier-Stokes equations solved by the so-called Volume-of-Fluid method (VOF). This wave-model

Skylla, includes breaking waves, porous media flow and allows for very complex shapes of the structure but cannot deal with wave impacts. Reference is made to Volume IIa, chapter 4, section 4.5, where examples are shown of wave interaction with two of the selected vertical breakwaters. For the second type of models pressure-impulse modelling is used for impact waves. This includes techniques to predict pressure-impulse along a berm and pioneering work for studying impulsive seaward forces due to overtopping events. The methods to predict wave impact loads are discussed in section 2.5.5. In this section some of the results for pulsating loads are discussed.

Figure 2-18 shows results of a computation with a non-impact wave on a vertical breakwater. VOF modelling can not only produce detailed profiles of the wave evolution, but also wave force histories at the breakwater (see Volume IIa, chapter 4, section 4.5). Figure 2-19 shows an example of the computational results, for Genoa Breakwater (one of the PROVERBS selected cases) for the total horizontal force at the front side of the structure.

In addition to generic tools such as numerical models other methods exist to predict wave forces. An innovative type of modelling is neural network modelling which makes direct use of results from physical model tests. Neural network modelling is a technique capable of generalising information if a large data-set is available. Since the partners from PROVERBS made their test-results available a large data-set resulted which formed the basis for neural network modelling. Not only was a neural network developed but also a method to obtain reliability intervals of the predictions of the neural network. This is of special interest for probabilistic design methods which require not only a prediction but also a quantification of the reliability (Van Gent & Van den Boogaard, 1998).

2.4.6 *Pressures on berms*

Wave induced pressures on the rubble berm in front of a breakwater wall are needed as input for numerical models simulating pore pressures in the rubble foundation under the breakwater. A method to predict these pressures has been developed for pulsating wave conditions using data from three different models with permeable berms (large-scale model tests in the GWK Hannover, mid-scale model tests at Franzius-Institut (WKS) Hannover and at HR Wallingford).

Inspection of the data with appropriate visualisation software showed that pressure distributions on the berm are almost triangular. Peak pressures decreases with increasing distance from the wall.

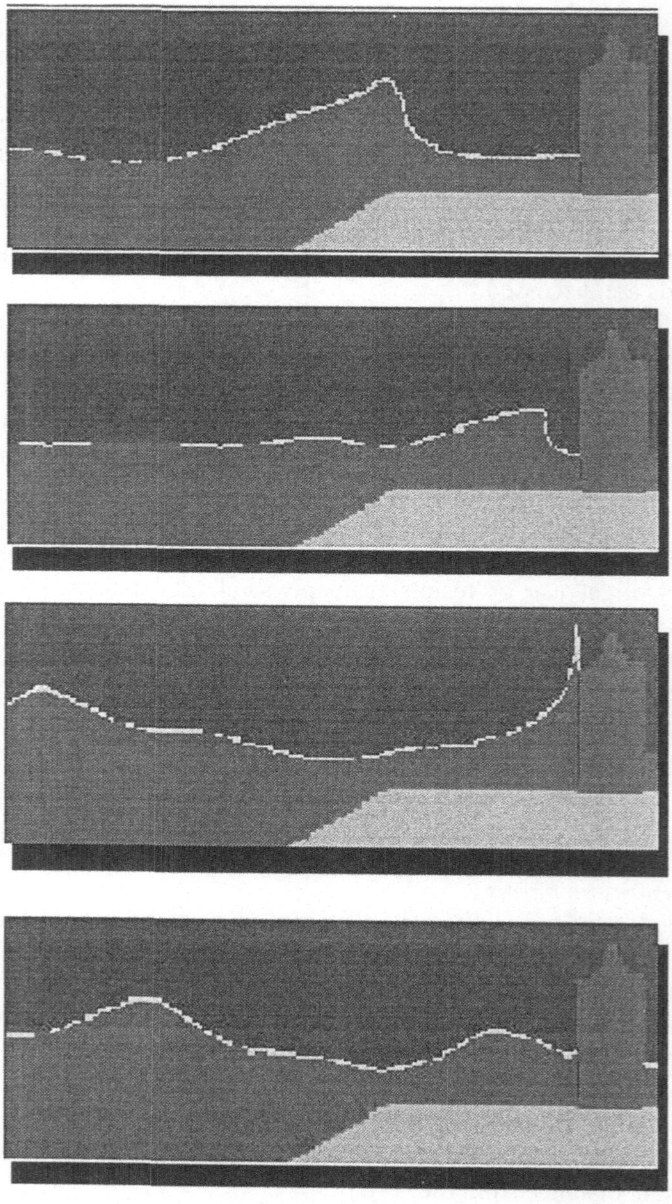

Figure 2-18. SKYLLA: Simulation of wave motion in front of a vertical structure on top of a permeable foundation (Vol. IIa, section 4.5).

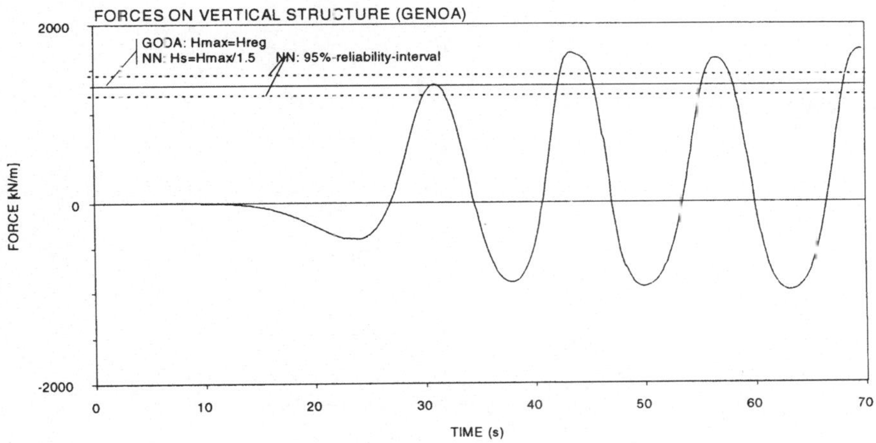

Figure 2-19. Numerical model results for regular waves in comparison to Goda's-method and neural network (NN) results (Vol. IIa, section 4.5).

The problem has therefore been reduced to predicting the pressure at the wall at the height of the berm, p_A, and the extent of the pressure distribution l_{pb} from the wall in the seaward direction where the pressure is positive (Fig. 2-20). Since pressures are wave induced it can be assumed that the distance l_{pb} where the pressure becomes zero, corresponds to the distance where the water level elevation is zero. A visual analysis of many tests confirms a good agreement between these two distances.

Goda's method may be used to calculate wave forces for non breaking waves, so parameters calculated by this method (subscript 'G') are also used to calculate p_A (Eq. 2-66). In order to calculate the pressure $p_{A,G}$ at the height of the berm it is necessary to interpolate between the pressure at the still water level ($p_{1,G}$) and the pressure at the base of the structure ($p_{3,G}$) as follows:

$$p_{A,G} = p_{3,G} + [d_c/(d_c + d)]\,(p_{1,G} - p_{3,G}) \qquad (2\text{-}66)$$

An empirical correction formula for $p_{A,G}$ is proposed based on the measurements as described above resulting in a better prediction of the pressure in the height of the berm $p_{A,cal}$:

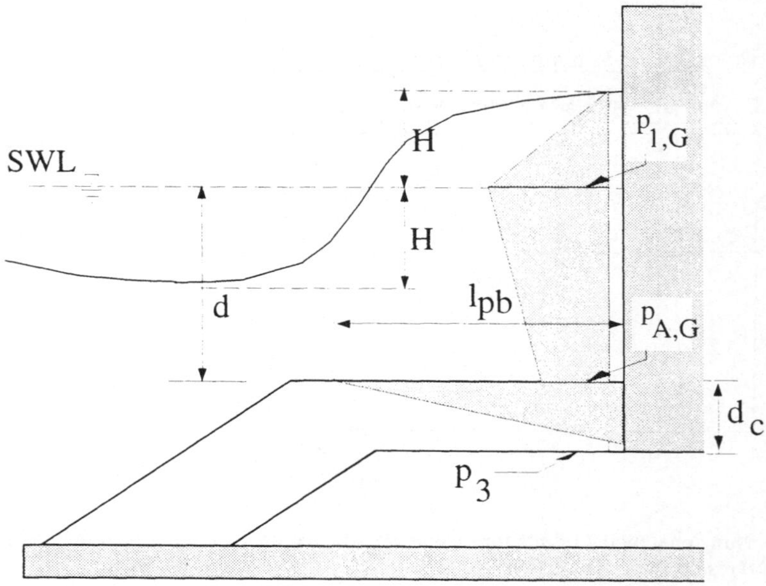

Figure 2-20. Definition of parameters.

$$\frac{p_{A,cal}}{\rho g d} = 0.7 \cdot \left[\frac{p_{A,G}}{\rho g d} \right]^{0.861} \qquad (2\text{-}67)$$

The length of the pressure distribution over the berm can then be expressed by:

$$l_{pb} = 0.26\, L_{pi}^{\;0.87} \qquad (2\text{-}68)$$

For calculation of L_{pi} see Equation (2-33). Equation (2-68) is not dimensionally correct but empirically gives the best correlation coefficients.

2.5 WAVE IMPACT LOADS

2.5.1 *Horizontal and vertical forces / pressures*

Wave impacts are among the most severe and dangerous loads on vertical breakwaters. Whilst many design procedures for these structures are well established, recent research in PROVERBS and elsewhere has shown that some methods are limited in their application and may over- or under-predict loadings under important conditions. This will then lead to over-designed and expensive structures or, even more dangerous, to under-design and consequently to danger to personnel and properties.

There are few design formulae for wave impact loads. Within PROVERBS a new procedure is proposed based on statistical distribution of forces and theoretical considerations derived from solitary wave theory (Oumeraci & Kortenhaus, 1997). This procedure is summarised in the following sections.

2.5.1.1 *Horizontal force and rise time*

The relative wave force F_h^* may be defined as the ratio of the horizontal wave force F_h over $\rho g H_b^2$ where H_b is the individual or maximum wave height at breaking calculated using Eq. 2-34. A wave impact in case of a stationary wave field with $P_i > 1\%$ or 2% (section 2.2.3) can then be calculated using the following Generalised Extreme Value (GEV) distribution:

$$F_h^* = \frac{\alpha}{\gamma}\left\{1 - \left[-\ln P\left(F_h^*\right)\right]^{\gamma}\right\} + \beta \qquad (2\text{-}69)$$

where $P(F^*_h)$ is the probability of non exceedance of the impact force and α, ß, and γ are the statistical parameters for the GEV distribution which can be taken from model tests, or alternatively estimated as $\alpha = 3.97$, $\beta = 7.86$, and $\gamma = -0.32$. These values are based on large-scale model tests which were believed to be closest to prototype conditions. However, Task 1 has also delivered more detailed analyses of small and mid-scale model tests performed and the following values were derived (Tab. 2-7).

Table 2-7. Values of α, β, and γ for GEV distribution of relative horizontal force.

Bed slope	**no. test waves**	α	β	γ
1:7	116	2.896	6.976	-0.526
1:10	159	10.209	12.761	-0.063
1:20	538	3.745	7.604	-0.295
1:50	3321	1.910	3.268	-0.232

The number of waves for the 1:7 and 1:10 bed slope is not really sufficient so that results have to be compared to the initially calculated impact force (Eq. 2-3) and the Goda force $F_{h,Goda}$ (Eq. 2-52). A reasonable estimate of the maximum value, $F^*_{h,max}$, during the design storm can be found by introducing a $P(F^*_h)$ value larger than 90%, depending on the number of wave impacts to be expected during the design storm. The maximum horizontal force can then be calculated by

$$F_{h,\,max} = F^{*}_{h,\,max} \cdot \rho\, g\, H_b^2 \qquad (2\text{-}70)$$

where H_b is the wave height at breaking (Eq. 2-34) and ρ is the density of the water. The measured rise time of the impact force $t_{r,Fh}$ can then be calculated from:

$$t_{rFh} = k \cdot 8.94 \cdot \frac{\sqrt{d_{eff} \, / \, g}}{F^{*}_{h,\,max}} \qquad (2\text{-}71)$$

where k is the mass proportion involved in the impact process which may be estimated to $k = 0.25$ (upper bound value) for deterministic calculations. For probabilistic calculations $k \cdot 8.94$ on the right side of Equation 2-71 may be summarised as k' and the measured rise time t_{rFh} on the left side is substituted by the 'triangular' rise time t_r assuming a constant relationship between both of them:

$$t_r = k' \cdot \frac{\sqrt{d_{eff} \, / \, g}}{F^{*}_{h,\,max}} \qquad (2\text{-}72)$$

The factor k' can then be described by a Log-Normal distribution with a mean value of 0.086 and a standard deviation of 0.084. More details on these relations can be found in Volume IIa, section 5.1.

The effective water depth in front of the structure d_{eff} in Equation 2-72 can be assumed to be identical to the water depth in which the wave breaks and may be calculated as follows (see Annex 1 for notations):

$$d_{eff} = d + B_{rel} \cdot m_{rel} \cdot (h_s - d) \qquad (2\text{-}73)$$

where B_{rel} is the part of the berm width which influences the effective water depth (B_{rel} equals 1 for no berm):

$$B_{rel} = \begin{cases} 1 & \text{for smaller berm widths } (\, B_b \, / \, L \leq 1 \,) \\ 1 - 0.5 \cdot \dfrac{B_b}{L} & \text{for larger berm widths } (\, B_b \, / \, L > 1 \,) \end{cases} \qquad (2\text{-}74)$$

and m_{rel} is the part of the berm slope influencing the effective water depth (m_{rel} equals zero for simple vertical walls):

$$m_{rel} = \begin{cases} 1 & \text{for steeper slopes } (\, m < 1 \,) \\ \dfrac{1}{\sqrt{m}} & \text{for flatter slopes } (\, m \geq 1 \,) \end{cases} \qquad (2\text{-}75)$$

The total duration can then be calculated from a statistically derived relationship as follows:

$$t_d = - c / \ln (t_r) \qquad (2\text{-}76)$$

where c is an empirically derived parameter (dimension: [-s*ln(s)]) normally distributed with a mean value of 2.17 and a standard deviation of 1.08.

The effective impact force being transmitted to the foundation is dependent on the dynamic response characteristics of the structure and its foundation. If quasi-static calculations are to be performed, the applied forces must be corrected by these dynamic response characteristics. A detailed description of this approach is given in chapter 3, section 3.4.4.

2.5.1.2 *Vertical pressure distribution*

Based on the analysis of almost 1000 breakers of different types hitting a vertical wall, the simplified distribution of impact pressure just at the time where the maximum impact force occurs, can tentatively be determined according to Figure 2-21. Three or four parameters need to be calculated in order to describe the pressure distribution:

(a) the elevation of the pressure distribution η_* above design water level;
(b) the bottom pressure p_3;
(c) the maximum impact pressure p_1 which is considered to occur at the design water level;
(d) the pressure at the crest of the structure p_4 if overtopping occurs.

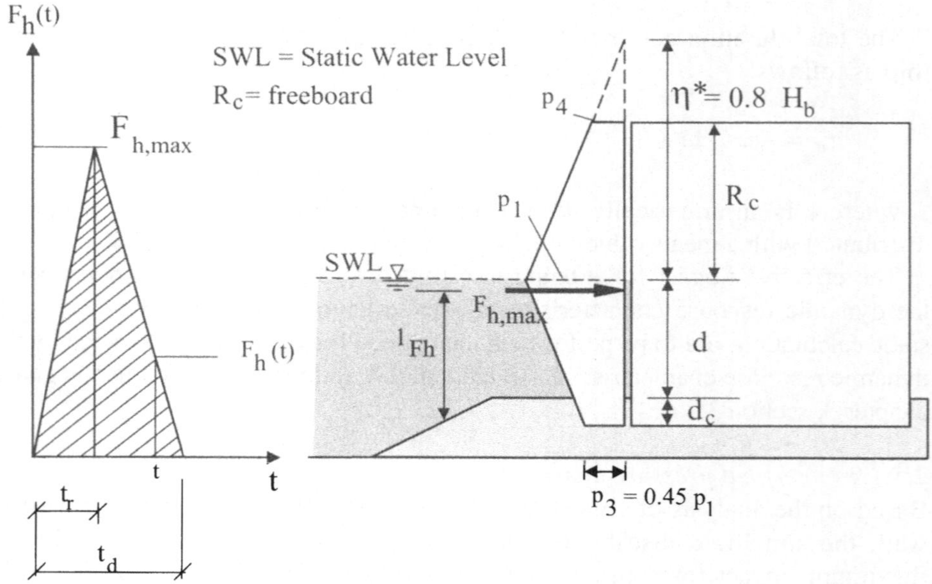

Figure 2-21. Simplified vertical pressure distribution at the caisson front wall (Vol. IIa, section 5.1).

Notional run-up level

The notional run-up level η_* may be calculated from the following tentative formula (see Fig. 2-21):

$$\eta^* = 0.8\, H_b \tag{2-77}$$

Bottom pressure p_3

The bottom pressure p_3 may be derived as a function of the maximum pressure at the height of the still water level as follows (see Fig. 2-21):

$$p_3 = 0.45\, p_1 \tag{2-78}$$

Maximum pressure p_1

The maximum impact pressure p_1 can be calculated directly from the equivalent force history, see example in Figure 2-21, since $F_h(t)$ represents the area of the pressure figure at any time of the history (assuming an infinitely high wall):

$$F_h(t) = \frac{1}{2} p_1(t) \cdot 0.8\, H_b + (d + d_c) p_3 + \frac{1}{2}(d + d_c)(p_1(t) - p_3) \tag{2-79}$$

Substituting Equation 2-78 in Equation 2-79 yields after some rearrangement:

$$p_1(t) = \frac{F_h(t)}{0.4 \cdot H_b + 0.7 \cdot (d + d_c)} \tag{2-80}$$

Pressure at the crest of the structure p_4

If the waves in front of the structure are high enough, overtopping is expected to occur. This will reduce the total impact force as parts of the wave momentum will be lost. This effect can be taken into account by cutting off the pressure distribution at the top of the structure (Fig. 2-21) so that the pressure at the crest of the structure can be described as follows:

$$p4 = \begin{cases} 0 & \text{for } \eta^* < R_c \\ \dfrac{\eta^* - R_c}{\eta^*} \cdot p_1 & \text{for } \eta^* \geq R_c \end{cases} \tag{2-81}$$

The horizontal impact force[2] is then reduced to:

$$F_{h,max,ov} = F_{h,max} - \frac{1}{2} \cdot (\eta_* - R_c) \cdot p_4 \tag{2-82}$$

A more detailed pressure distribution is proposed by Hull et al., 1998 and is described in Volume IIa, section 5.1.

Lever arm of horizontal force

Finally, the lever arm for the horizontal impact force can be calculated from the pressure distribution at the wall as follows:

$$l_{Fh}(t) = \frac{p_1 \eta_{ov}^2 + 3 p_1 d' \eta_{ov} + 3 p_4 \eta_{ov} d' + 2 p_4 \eta_{ov}^2 + 2 p_1 d'^2 + p_3 d'^2}{6 \; F_h(t)} \tag{2-83}$$

In Equation 2-83 η_{ov} is defined in relation to the height of the wave crest to the wall as follows:

$$\eta_{ov} = \min\{\eta_*; R_c\} \tag{2-84}$$

[2] Consequently, the statistical distribution parameters given in section 2.5.1.1 are no longer valid for this reduced force as the statistical distribution was determined for relatively high caisson structures and almost no overtopping.

and d' is defined as:

$$d' = d + d_c \tag{2-85}$$

2.5.1.3 *Uplift force*

The uplift force acting on the underside of the caisson can be calculated for impact conditions in the same way as the impact force only considering some small parameter differences as follows:

- calculate the relative uplift force by using Equation 2-69 but different statistical parameters $\alpha = 2.17$; $\beta = 4.384$; $\gamma = -0.11$ (derived for a bed slope of 1:50 from large-scale tests);
- calculate the rise time of uplift forces using Equation 2-72 and a mean k'-factor of 0.16 (standard deviation of 0.17);
- calculate the total duration using the same formula as for impact forces (Eq. 2-76) but assuming a mean of 1.88 and a standard distribution of 0.99.

The uplift force calculated in this way may be considered as an upper bound. Another upper bound may be found by assuming $p_u = p_3 = 0.45\, p_1$ (p_1 is taken from Eq. 2-80), $p_{ru} = 0.2\, p_u$ and a linear pressure distribution underneath the caisson where the least of both should be taken.

2.5.1.4 *Uplift pressure distribution*

In general, the first estimate of wave uplift pressures underneath vertical breakwaters should be calculated by Goda's method. For some impact conditions however greater uplift forces or pressures may be produced. These pressures can be calculated using the approach described in chapter 3, section 3.5.3 where the instantaneous pore pressure development underneath the breakwater is described.

As soon as the maximum uplift force $F_{u,max}$ is known a simple approach was derived empirically to estimate the pressures underneath the breakwater. It is based on hydraulic model test data using 'upper bound' envelopes which may lead to conservative estimates. Therefore, all results should be compared to the pressures derived by Goda (see section 2.4.1) which should be taken as the 'lower bound'. The form of the distribution is assumed to remain constant over time where a triangular distribution is assumed with the maximum pressure at the time of the maximum uplift force. The pressure underneath the shoreward edge of the breakwater p_{ru} can be calculated as follows:

$$p_{ru} = \rho\, g\, H_b \cdot \left(\frac{H_b}{h_s} - 0.1 \right) \tag{2-86}$$

Knowing the pressure at the shoreward side of the structure, the pressure underneath the seaward edge p_u can be calculated as follows:

$$p_u = \frac{2\ F_{u,max}}{B_c} - p_{ru} \tag{2-87}$$

where B_c is the structure width, F_{umax} is the maximum uplift force and p_{ru} is the pressure at the shoreward side of the structure. The pressure p_u calculated in this way must be considered as an upper bound. Another upper bound is $p_u = p_3$. The pressure distribution may deviate from the linear one and the value of p_{ru} may be different as well, as can be found from the approach described in section 4.5.

Finally, the lever arm for the uplift force can be calculated from the pressure distribution underneath the structure for each time step as follows:

$$l_{Fu} = \frac{B_c^2 \cdot (p_{ru} + 2\ p_u)}{6\ F_{u,max}} \tag{2-88}$$

2.5.1.5 *Effect of aeration*

Much of the data used to derive the prediction methods for wave impact forces/pressures were themselves measured in small or mid-scale tests in fresh water. There may however be significant differences in the aeration of salt water at full scale, and these will influence the magnitude and duration of impact pressures/forces.

A simple scaling correction for impact forces was developed by Allsop et al. (1998) based on field measurements of wave impacts then reproduced in fresh water wave flume at 1:32 scale. This suggested that impact forces are to be corrected by a factor of 0.40-0.45. A more complete approach requires an estimate of aeration The aeration in impact waves at full scale in seawater can be estimated in first instance from:

$$P_a = 2.0 + 5.3 \cdot N_i \tag{2-89}$$

where N_i is the number of breaking waves per minute and P_a is the percentage of aeration in the breaking wave (given in percent). The number of breaking waves per minute may not be known in advance, but may be estimated using the percentage of breaking waves P_i (Eq. 2-39):

$$N_i = P_i \cdot \frac{N_W}{t_{tot}} \tag{2-90}$$

where N_W is the number of waves in a test, and t_{tot} is the total length of the test given in minutes. Under prototype conditions N_W may be replaced by the number of waves in a storm and t_{tot} by the duration of the design storm.

From the aeration percentage obtained by Equation 2-89 a force reduction factor k_{fa} according to Hewson et al. (1998) can be calculated as follows:

$$k_{fa} = \frac{4.0}{[(2.5 + p_a)(97.5 - p_a)]^{0.252}} \tag{2-91}$$

Aeration in fresh water of small-scale models is generally small (p_a in the range of 0 to 0.5%). Hence, the value of k_{fa} found for full-scale can be considered as a reduction factor with respect to the force found in a small-scale model and then scaled according to Froude's law. If the force, however, was found from large-scale tests, the reduction is less because k_{fa}(model) < 1.0, especially if seawater is used in the tests.

The total force impulse seemed to remain independent from the aeration level of the breaking wave. Assuming this impulse to be more or less equal to the triangular impulse longer rise times due to aeration can be calculated as the inverse of the force reduction factor:

$$k_{ta} = \frac{1}{k_{fa}} \tag{2-92}$$

2.5.2 *Seaward impact forces*

Forces generated by wave action on the seaward side of a caisson breakwater are usually critical for its stability (including the quasi-static suction forces described in section 2.4.2). Many breakwaters may experience significant static and dynamic loading on both seaward and rear faces, and indeed there are a number of examples of breakwater failure which have involved a seaward motion confirming that consideration may need to be given to forces that can be generated on the harbour side (Fig. 2-22).

The most likely mechanism for the generation of large dynamic loads from the harbour side is the plunge into the harbour of an overtopping wave - a mechanism suggested by van der Meer (private communication).

This type of loading may be of particular relevance where economic pressures for reduction in caisson size have led to use of low crest elevations and/or sloping roofs which reduce landward forces. Such modifications may allow greater overtopping, but it has not been previously appreciated that this in turn may result in significant seaward impulsive loading. These impulsive forces, the reduced caisson mass and the quasi-static seaward forces (section 2.4.2) increase the risk of seaward failure.

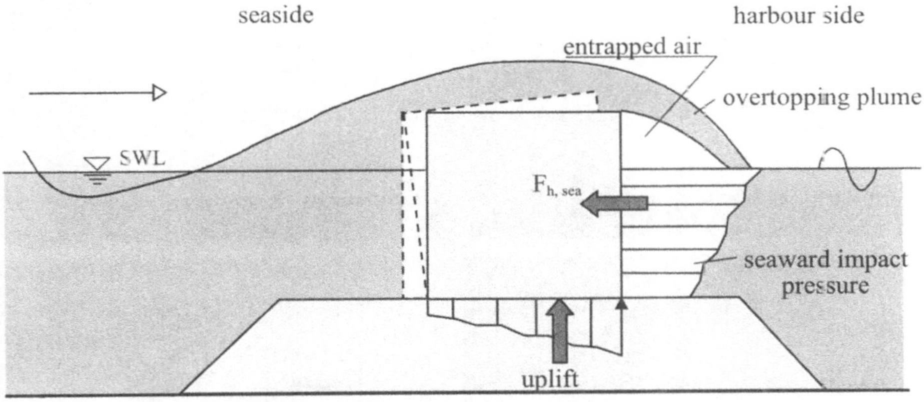

Figure 2-22. Seaward impact loading induced by wave overtopping.

To assess whether a reduction in caisson mass is safe against seaward failure, the designer may use the method given in section 2.4.2 to predict the pulsating seaward forces, but hitherto there has been no such guidance for the prediction of impulsive loads. Initial stages of development of such guidance has resulted from physical model tests at Edinburgh supported by numerical modelling at Bristol.

2.5.2.1 *Physical Model Tests*

The mechanism described above is illustrated for a sloping top caisson by the sequence in Figure 2-23. During the event shown, the maximum net seaward force recorded was ~40% greater than the largest quasi-static landward force recorded during wave runup before overtopping. Further, the largest seaward overturning moment about the seaward toe of the structure, which occurs at the time of entry of the plunging wave into the harbour, is larger than any landward overturning moment experienced during the event.

2.5.2.2 *Numerical Model Tests*

The effect of the impact of the overtopping wave into the harbour was modelled using pressure-impulse theory (see also section 2.5.5). The model first estimates the pressure impulse due to the plunging water, and then applies this to the surface of the harbour-side water, considered initially undisturbed. This results in a prediction of the distribution of pressure impulse over the rear face of the breakwater, and also along the bed on the harbour side behind the structure.

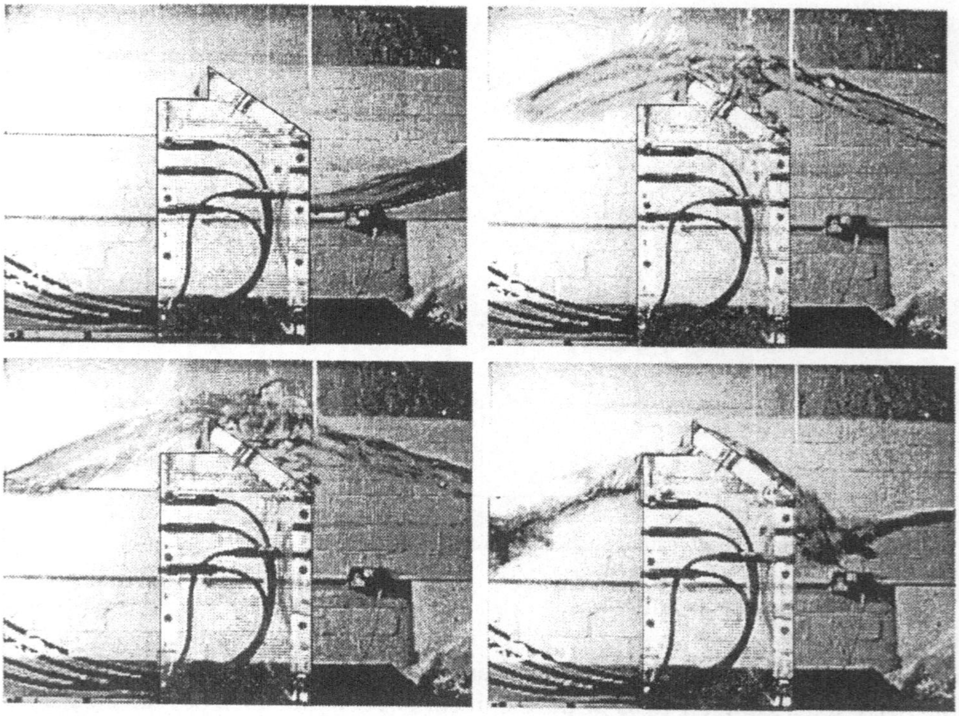

Figure 2-23. Frames from a video sequence showing the plunging overtopping wave.

The solution was found to be sensitive to the presence of a pocket of entrapped air under the plunging wave. However, once the effect of the air is accounted for, agreement with measurements from the physical model study is good, e.g. Figure 2-24, in which the crosses are the measured points and the solid line is the distribution predicted by pressure-impulse theory. The dotted line shows the prediction before correction for the effect of air. The success of the model depends upon the severity of the impact, which must be sufficiently violent for pressure impulse theory to be valid - a condition met by the extreme events studied.

2.5.2.3 *Initial guidance*

Conclusions drawn are limited by the scope of this first study into this failure mechanism. Further investigations will refine design guidance drawn from this work.. This study has however demonstrated that seaward failure as a result of impulsive loading on the rear face may be an important contribution to failure where large plunging overtopping events are likely (Walkden et al., 2000). Some information on overtopping of these structures is given by Franco (1994).

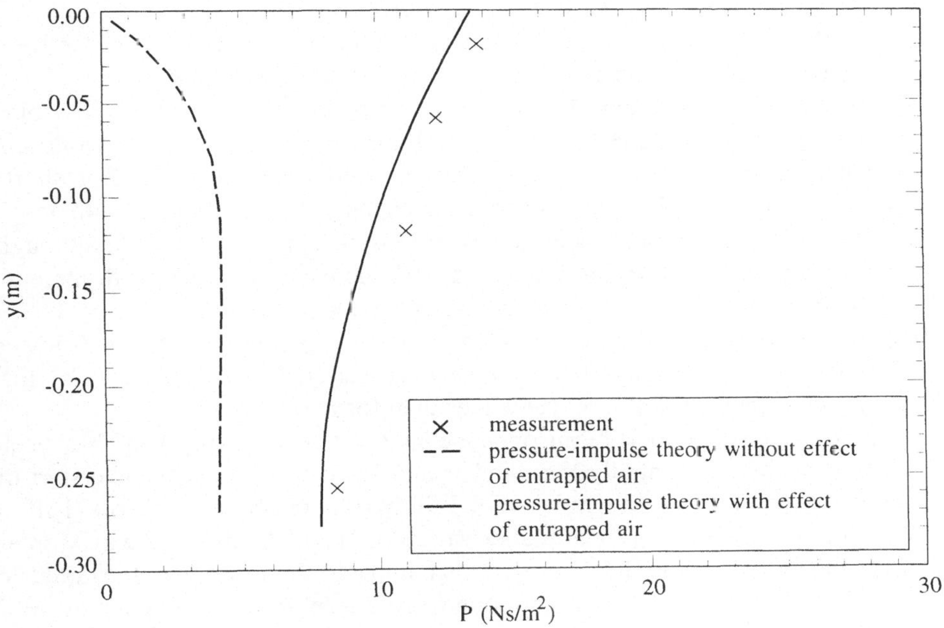

Figure 2-24. Predicted and measured pressure impulse on the rear face of the breakwater (Vol. IIa, section 5.2).

Structures of reduced design mass, e.g. Hanstholm–type structures have been identified as being particularly at risk as they have less mass to resist seaward forces generated due to the trough at the front face, and because they admit greater overtopping.

In addition attention should be drawn to the possible risk experienced by the shear keys of such structures, whose role may be relatively more important in these smaller structures. These keys may also be subjected to accelerated erosion in the cases of curved front structures - a phenomenon observed at Brighton marina breakwater, UK. Consequently their design should account for these coupled risks.

Pressure-impulse theory models the mechanism causing the impulsive force on the rear face due to the entry of a plunging overtopping wave, and good agreement between the model and experiment is found. However, this model was only compared with data from three violent impacts from this particular experimental set-up. Further comparisons are needed before extensions to prototype scale are possible. At present, physical modelling, perhaps supported by pressure-impulse methods, remain the main method to check the magnitude of these loads.

2.5.3 *Effects of 3-d wave attack on impact loadings*

Studies in the UK Coastal Research Facility at Wallingford (UKCRF) identified conditions which lead to impulsive loads and evaluated their occurrence.

Impacts under long-crested normal waves on plain vertical walls were compared with results from 2-d tests and showed little variation. The onset of impacts was reached at about H_{si}/h_s = 0.35 for long-crested waves at β=0°. For short-crested waves at β=0°, P_i again showed no changes in comparison with long-crested waves. Again, impacts begin at conditions close to H_{si}/h_s = 0.35, although there were some indications that impacts do not increase as rapidly with increasing H_{si} /h_s in short-crested waves as for long-crested waves.

For oblique long-crested waves, β = 15°, 30° and 45°, there were much fewer impacts than for normal long-crested waves. For larger waves, H_{si}/h_s > 0.35, impacts were less frequent with oblique waves than for normal attack, β=0°.

Addition of a small rock berm or slope in front of a simple wall has been shown to increase substantially the number and severity of impacts. The start of impact events on walls on low mounds, $0.3<h_b/h_s<0.6$, occurs at or above H_{si}/h_s = 0.20, with P_i increasing rapidly at higher values of H_{si}/h_s. Analysis of wave forces for long-crested waves at β=0° showed that impacts for low mounds started to occur at H_{si}/h_s = 0.30. Tests on high mound composite walls in the CRF at β=0° substantially confirmed that impacts increase further with high mounds, $0.6<h_b/h_s< 0.9$. Impacts started at smaller values of H_{si}/h_s, as low as 0.25.

2.5.3.1 *Horizontal forces*

Wave forces for long-crested normal waves agree well with results from 3-d tests by Franco et al (1996) where there are no impacts, only pulsating loads. Comparisons with Goda predictions show relatively good agreement over the pulsating zone.

An upper limit to wave impact forces under normal or long-crested oblique waves is given by the simple prediction by Allsop & Vicinanza (1996). For short-crested waves, forces show no significant change in local force for the range of conditions tested compared with loads generated by long-crested waves of the same height.

The influences of oblique long-crested waves on forces on any narrow strip of the caisson are more significant. Over the pulsating zone, $H_{si}/h_s \leq 0.35$, forces are very similar to those for normal approach, even though the component of force perpendicular to the caisson might have been expected to reduce. In the impact region however, wave loadings diminish considerably under oblique attack.

2.5.3.2 *Variability of impact forces along the breakwater*

These comparisons show consistent increases in $F_{h1/250}$, with reduced averaging. Allsop & Vicinanza's simple formula $F_{h(A\&V)}$ (Eq. 2-3) gives a reasonable representation of forces averaged over typical caisson widths of 10-20 m, but underestimates the "local" force over a single narrow strip, even for normal and long-crested wave attack.

Values of $F_{h(peak)} / F_{h(A\&V)}$ (measured over predicted forces) reached 1.2-1.3 for normal long-crested attack. Under long-crested oblique attack, most results were much lower, not exceeding $F_{h(peak)} / F_{h(A\&V)} = 1.15$, but with a single test giving 1.4. Under short-crested waves the ratio $F_{h(peak)} / F_{h(A\&V)}$ never exceeded 1.15, suggesting that peak forces are unlikely to exceed those analysed in this research by any substantial margin, except under conditions of normal attack.

2.5.3.3 *Effect of caisson length*

Battjes (1982) argued that oblique or short-crested wave attack on caisson of length l_c will give further reductions in effective force relative to normal and/or long-crested attack, and relative to loads on a narrow strip (modelled here as a single column of transducers.

Results from the UKCRF tests with vertical walls were combined with results from Franco et al (1996), which show little decay over caisson lengths l_c/L_p up to 0.4. Measurements from the UKCRF however show up to 10% decay, i.e. reduction factors of horizontal impact forces C_{Fh} down to 0.9 for non-impact conditions for relative caisson lengths up to l_c/L_p=0.15. Wave impact conditions (H_{si}/h_s>0.35), however, gave substantially greater reductions in the effective force, even over short caisson lengths, $0.005 < l_c/L_p < 0.2$. A simple regression line gives the reduction factor C_{Fh} in terms of relative caisson length with a coefficient $k_b = 1.35$ for long-crested waves and $\beta = 0°$:

$$C_{Fh} = 1 - k_b\,(l_c/L_p) \qquad (2\text{-}93)$$

Under slightly oblique attack, $\beta = 15°$, forces for non-impacting conditions show more significant reductions than for $\beta = 0°$, but there is only slightly greater change for impact conditions. The same simple form of regression line gives C_{Fh} in terms of l_c/L_p: for $\beta = 15°$, yielding $k_b = 1.69$ (Fig. 2-25).

At greater obliquities, the force reduction is more marked for pulsating conditions. Measurements at β= 30° show slightly greater reduction for impacts.

Short-crested waves show no significant effect of spreading between n=2 and n=6. The regression for $\beta = 0°$ gives $k_b = 1.56$, steeper than for long-crested waves at $\beta = 0°$, but less severe than for long-crested waves and $\beta = 15°$.

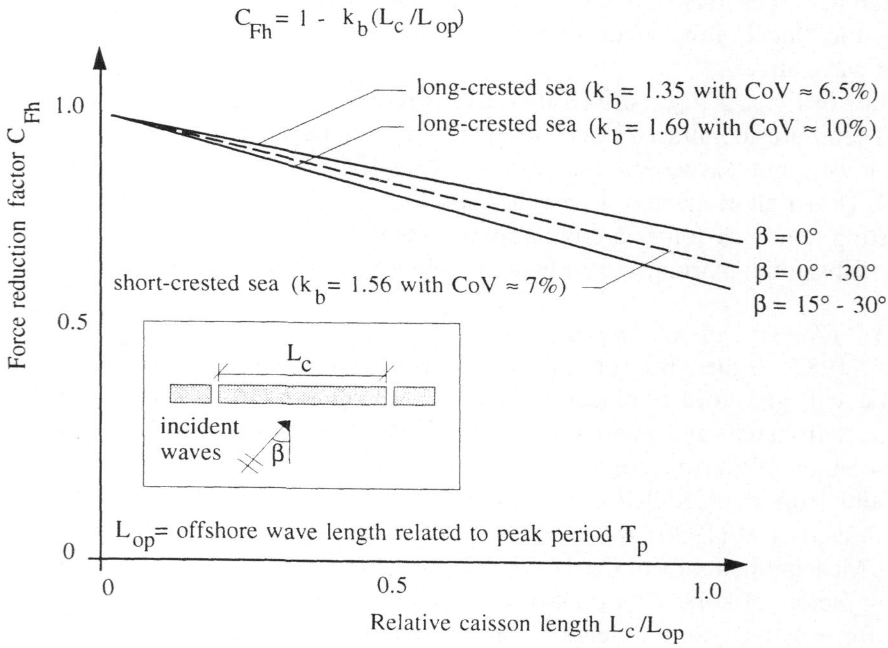

Figure 2-25. Effect of caisson length on impact forces for normal and oblique short-crested and long-crested waves (Schüttrumpf et al., 2000).

These results suggest that Battjes' model may be used to give conservative predictions in the pulsating zone, but that force reductions under impacts are much more significant than predicted by linear methods. Calculations of the mean decay function on F_h for impacting conditions can be summarised by the simple equation relating decay to relative caisson width, l_c/L_p given in equation (2-93) where coefficient k_b is defined for each test case below.

Table 2-8. Impact force reduction coefficients.

Wave Condition	**Coefficient k_b**	**Coeff. Var. [%]**	**Correlation r^2**
Long-crested, β = 0°	1.35	6.6	0.82
Long-crested, β = 15°	1.69	9.2	0.77
Long-crested, β = 30°	1.69	10.4	0.79
Short-crested, n= 2	1.55	10.6	0.76
Short-crested, n= 6	1.58	9.3	0.77
Short-crested, n= 2, 6	1.56	6.8	0.77

Under oblique or short-crested waves, the variation of peak forces relative to those averaged over a short length equivalent to a single caisson of about 20 m are relatively small, not exceeding a ratio of 1.2. The variation of peak force on a single narrow strip under normal wave attack is more substantial, with peak forces up to 1.3 times greater than the average.

For impact conditions, average forces reduce significantly with caisson length, giving reductions of 25% or so over relative caisson lengths of only 0.2. A simple reduction factor for F_h under impacting conditions as a function of l_c/L_{op} has been developed. Values of a coefficient k_b have been presented here in Table 2-8 for long-crested waves at different obliquities, and for short-crested waves.

These and other studies also suggest the following initial conclusions on spatial correlation of impact forces under oblique / short-crested waves:

- for heavy impacts ($F_{Impact}/F_{Goda} >> 2.5$), and small obliquity or spreading assume a typical coherence length $\leq L/16$;
- for light impacts ($F_{Impact}/F_{Goda} < 2$), normal wave attack ($\beta = 0°$) and little spreading assume a typical coherence length $\leq L/4$;

2.5.4 *Uncertainties and scale corrections*

2.5.4.1 *Uncertainties*

The concept of uncertainty analysis will be explained in chapter 5. In this chapter an uncertainty analysis will be presented on impact waves. All results summarised herein are based on the PROVERBS data obtained from hydraulic model tests from various scales. Horizontal impact and uplift forces were modelled by a Generalised Extreme Value (GEV) distribution (see section 2.5.1 for details). The statistical uncertainties of the GEV distribution were determined and upper bounds were calculated. Rise time and total impact durations were also modelled. For each data set the 95% upper bound values and the 10^{-2} quantiles were calculated (Volume IIa, section 5.4).

For rise times and durations the best fit was found for a new relation as given by Equation 2-76. The parameter c for this formula was found to be log-normally distributed. Upper bounds and 10^{-2} quantiles are also given in Volume IIa, section 5.4.

2.5.4.2 *Scale corrections*

First and simple scale corrections of impact wave have been suggested by Allsop et al. (1996) using results from field and model tests on hollow cube concrete armour units. The proposed correction factors are summarised in Table 2-9:

Table 2-9. Simple scale correction factors for impact loads after Allsop et al. (1996).

Non-exceedance level	Impact pressure correction factor	Rise time / duration correction factor
92%	0.44	9.0
95%	0.45	6.8
98%	0.43	5.0
99%	0.41	4.2

This approach is rather simple and possibly only valid for similar test conditions as for the HR tests. A more complete approach has therefore suggested by Oumeraci & Hewson (1997) based on detailed laboratory testing and a better understanding of the physics of impact loading (Fig. 2-26).

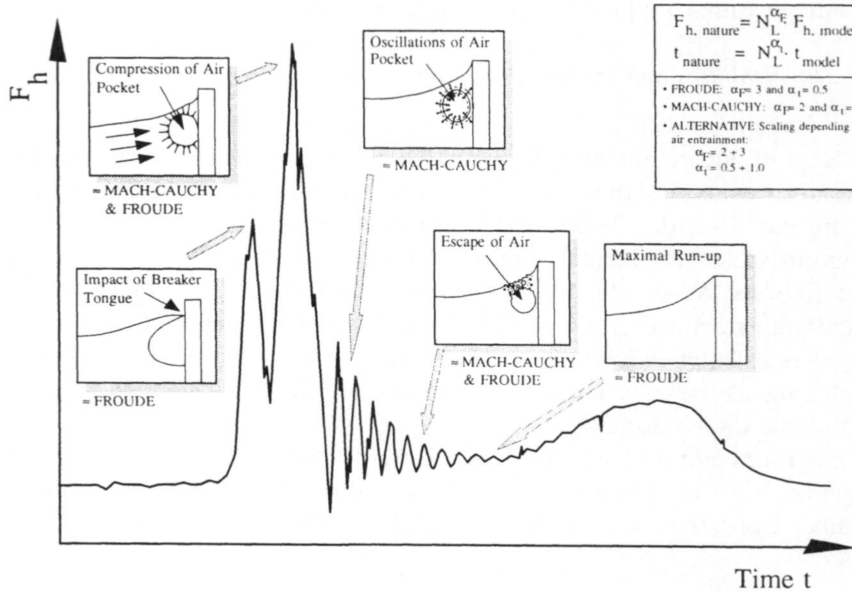

Figure 2-26. Physical processes involved in the wave load history and associated scaling problems (Oumeraci & Hewson, 1997).

In fact, the results have shown that, whether FROUDE or MACH-CAUCHY similarity laws or a combination of both has to be applied for the interpretation of the test results, will strongly depend on the level of aeration and the amount of entrapped air which both determine the compressibility of the impacting fluid mixture. Therefore, an improved scaling procedure based on the separation of the different components of the impact load history has been suggested (Fig. 2-27). For further details see Kortenhaus & Oumeraci (1999).

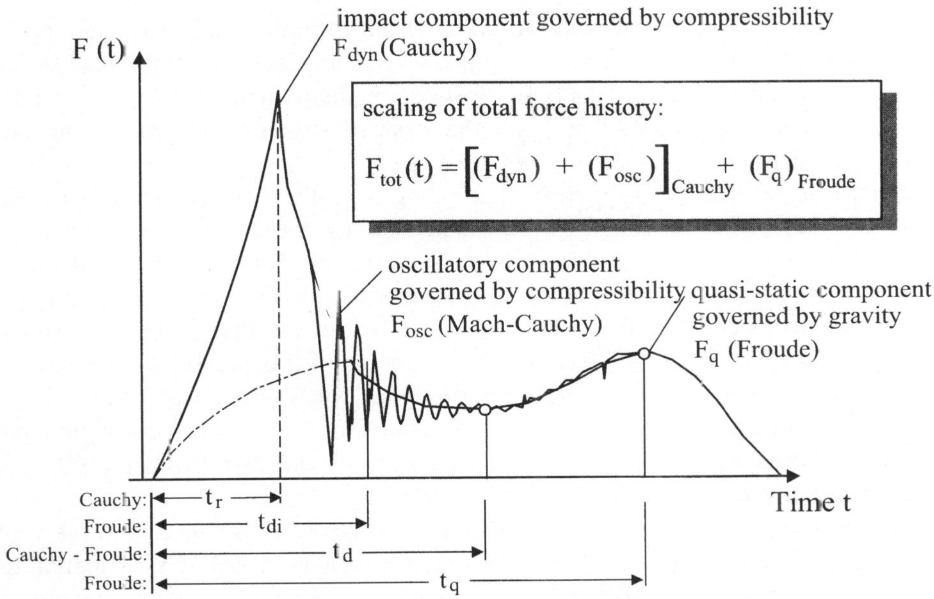

Figure 2-27. Suggested procedure for scaling the various components of the wave load history (Kortenhaus & Oumeraci, 1999).

A further more pragmatic approach based on correction factors for the impact load (reduction load factor) and the related rise time (inverse value of reduction load factor) has also been developed (see section 2.5.1.5) which are based on an empirical relationship between the level of aeration (in %) and the number of breaking waves within a given time. More details regarding this subject are given in Bullock et al. (1999).

2.5.5 *Use of numerical models*

As discussed in section 2.4.5, the complexity of breaking waves, together with complex shapes of many structures with flow in rubble mound berms and foundations means that within PROVERBS two approaches for numerical modelling are used. One is to idealise the motion and geometry so that a simple model may be used, pressure-impulse modelling. The other is the more detailed Volume Of Fluid model (VOF), Skylla, as discussed in section 2.4.5. For the pulsating wave loads the first track of VOF modelling is used and is discussed in section 2.4.5.

Pressure-impulse modelling, has been used to study the pressure impulse of a violent wave impacting on a vertical wall. Pressure impulse theory has the advantage of producing simple Fourier series expressions for complicated phenomena. Comparisons with experimental results have shown that this approach is successful, but the assumptions used to produce the formulae mean that the applicability is limited to the most violent impacts and to idealised structures. In Volume IIa, sections 4.6, 5.2 and 5.5 this form of mathematical modelling is used for pressure impulse prediction on both the vertical structure and along the berm taking into account a trapped air bubble.

Peregrine & Wood (1997) show that changes in the porosity of the berm have only a small effect on the pressure-impulse along the berm, hence the impact model with the air bubble, which has an impermeable berm, can be used for predictions even when a permeable berm is present. These models however require wave properties at the wall as input, so the effect of the berm on wave propagation to the wall is not accounted for. The extension of this pressure-impulse model to predict pressures along a berm is discussed in Volume IIa, chapter 5, section 5.5. Background information on pressure-impulse techniques which were extended here, is given in Cooker & Peregrine (1990) and Cooker (1992), and further applications are given in Wood (1997).

In addition to numerical models neural networks can also predict wave forces (PROVERBS paper by Van Gent & Van den Boogaard, 1998). Experimental data from the partners in PROVERBS were made available and used to develop the neural network. Figure 2-28 shows all data for horizontal forces made available, including situations with pulsating loads and wave impact loads. The method by Goda (1985) underpredicts measured forces (see the left part of Figure 2-28) while the neural network gives a better prediction although the scatter is still wide. This is not due to the performance of the neural network but caused by the scatter in the experimental data; however it demonstrates the necessity of a method to quantify the reliability of the predictions. This method has been developed and used in Task 4 for probabilistic designs of vertical breakwaters.

2.5.6 *Pressures on berms*

A visual analysis of three data sets has shown that the pressure distribution on a berm under impact wave loading is of a triangular or concave shape. For low ratios $F_{h,max} / F_{h,q}$ the pressure distribution at the moment of the maximum horizontal force is more or less triangular as it was for non impact conditions (see section 2.4.6). For large ratios, however, the shape is more concave. Unfortunately, it has not been possible to find a boundary value for the ratio $F_{h,max} / F_{h,q}$ to distinguish between triangular and concave pressure distributions.

Definitions of parameters related to the pressure distribution on the berm are given in Figure 2-29. The LWI prediction method developed for pressures along a berm under non-impact conditions (section 2.4.6) was applied to impact conditions, and was shown, as was expected, to be unsuitable. The measured values for impact waves showed an increase in magnitude and scatter when compared with non- impact waves.

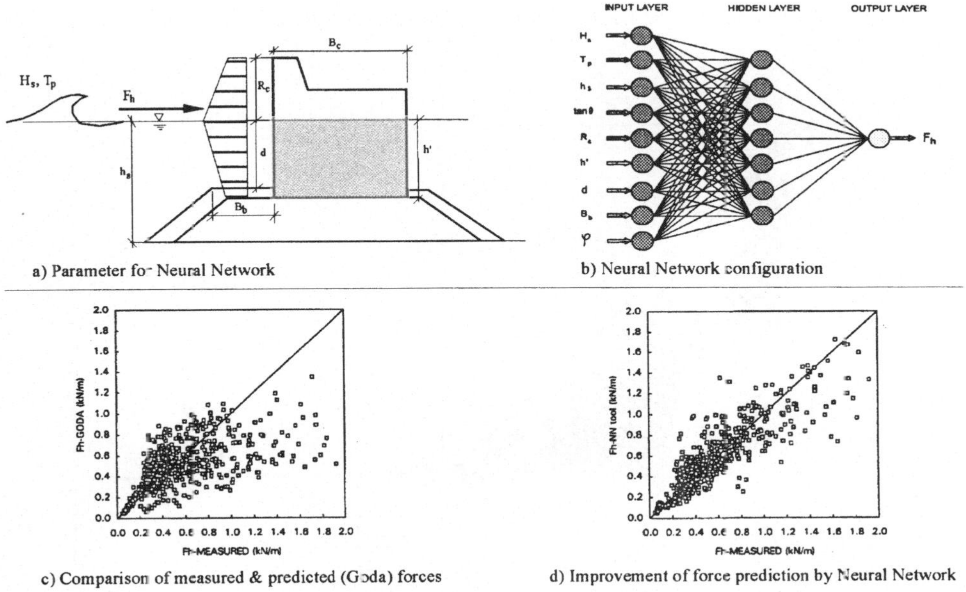

Figure 2-28. Comparison between neural network results and measurements (right) and a comparison between the Goda-method and measurements (left).

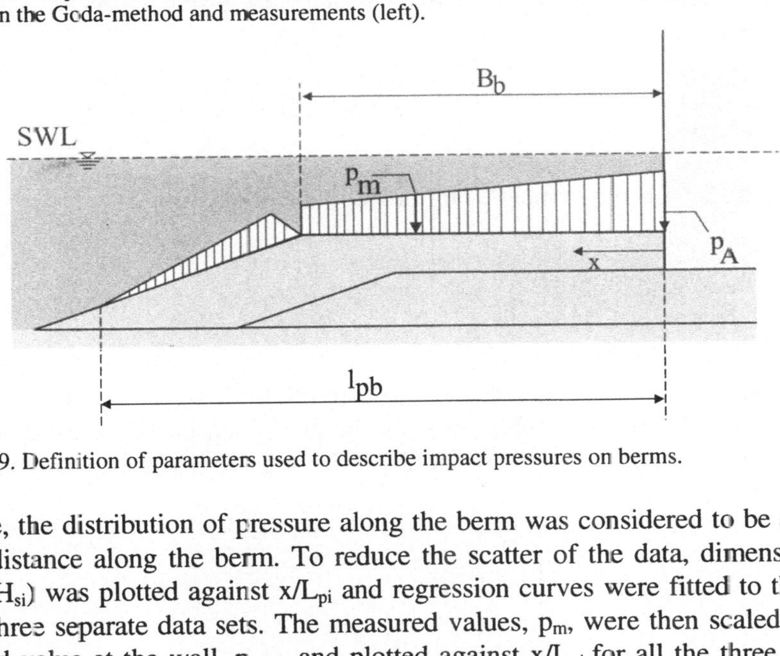

Figure 2-29. Definition of parameters used to describe impact pressures on berms.

Hence, the distribution of pressure along the berm was considered to be a function of distance along the berm. To reduce the scatter of the data, dimensionless $p_{1/50}/(\rho g H_{si})$ was plotted against x/L_{pi} and regression curves were fitted to the data for the three separate data sets. The measured values, p_m, were then scaled by the measured value at the wall, $p_{A,m}$, and plotted against x/L_{pi} for all the three sets of data together.

Due to the wide spread of the data the method of least squares was used to find a line approximating the trend in the data given in Equation (2-94) and shown in Figure 2-30.

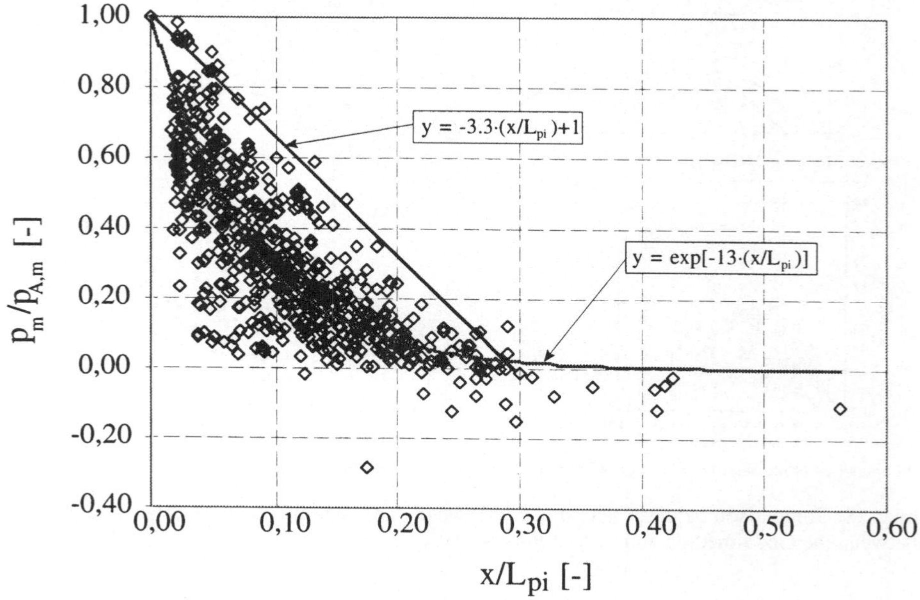

Figure 2-30. Relative berm pressure versus relative distance from wall under impact conditions (Vol. IIa, section 5.5).

Equation 2-94 is recommended for probabilistic modelling:

$$\frac{p_m}{p_{Am}} = \exp\left[-13 \cdot \left(\frac{x}{L_{pi}}\right)\right] \tag{2-94}$$

It should however be noted that the magnitude of p_m / $p_{A,m}$ is sometimes as much as a factor of three different from the prediction curve, so an upper bound has also been given (Eq. 2-95). The following equation is recommended for deterministic approaches:

$$\frac{p_m}{p_{Am}} = 1 - 3.3 \cdot \left(\frac{x}{L_{pi}}\right) \tag{2-95}$$

In both cases $p_{A,m}$ has to be estimated by available methods for impact loads (see section 2.5.1).

2.5.6.1 *Pressure-impulse modelling*

Pressure-impulse modelling was applied to small-scale data for suitably chosen impacts. The pressure-impulse technique (with the inclusion of an air pocket, bounce back model, see Wood, 1997) was extended by giving guidelines to the selection of the input parameters when velocity data and side-view video are not available. The comparisons between the pressure impulse bounce back model and the data were quite good.

It was found that sufficient conditions for using pressure-impulse theory were as follows:

a) Reflective pressures must be removed in the integration of pressures, where reflective pressures are the part of the pressure time graph which is due to reflection from the wall (often denominated as quasi-static pressures);
b) $F_{h,max} / F_{h,q} > 2.5$ and $F_{h,max} /(\rho g H_b^2 > 2.5)$, where F_{hmax} is the peak of the impact in the force history and F_{hq} is the quasi-static maximum for the same impact (see Annex 1);
c) For tests late in a run of waves, the model can only be used as an approximate upper bound;
d) $V\Delta t/\eta < 1$. (V = velocity scale, Δt=period of integration, η=height of water at the wall after initial impact)

In addition some ideas are explored with regards to converting pressure impulse to pressure at the time of maximum force, and one case was examined. This approach enabled a reasonably good prediction of the impact pressure, p_i, using the following relationship:

$$p_i = \frac{2P}{t_a - t_b} \tag{2-96}$$

where P is the pressure impulse, p_i is the maximum pressure on the berm above the background reflective pressure (the bounce back model assumes the reflective pressures are negligible), and t_a and t_b are the upper and lower boundary of the impulsive pressure area, respectively. It is thought that this method could be extended to predict pressure at the time of maximum force, p_{max}, by $p_{max} = p_i + p_{ref}$, where p_{ref} is predicted by methods described in Volume IIa, section 5.6 and p_i by pressure impulse theory. In Figure 2-31 a comparison of p_i for the bounce back model, the statistical equation (Eq. 94) and the measured values is shown. Both prediction methods over-estimate p_i, but the similarity between the two methods is remarkable.

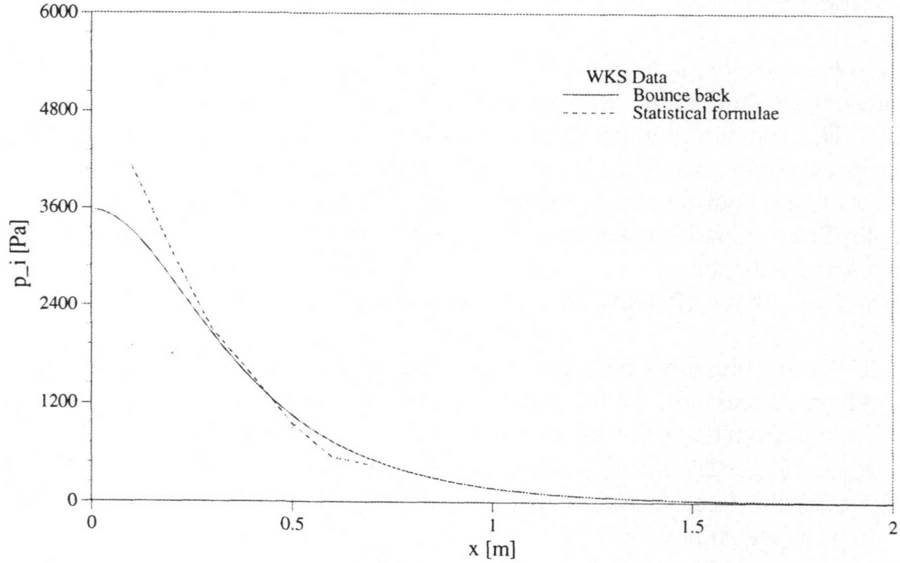

Figure 2-31. Comparison of p_i for the bounce back model, statistical equation and measured values (Vol. IIa, section 5.5).

2.6 BROKEN WAVE LOADS

2.6.1 *Strongly depth-limited waves*

The height of broken waves is depth limited which means that the relationship between H_{si}/d and F_h described in Equation (2-3) does not hold for the impacts of such events. Consequently a different method of calculating F_h is required. A field study of impacts loads caused by broken waves by Blackmore & Hewson (1984) has indicated that the following equation is appropriate:

$$p_{i\ max} = \lambda\ \rho\ T_p\ C_b^2 \tag{2-97}$$

Where λ is an aeration term that is discussed below, ρ is the water density, T_p is the wave spectral peak period, and C_b is the velocity of the breaker at the wall.

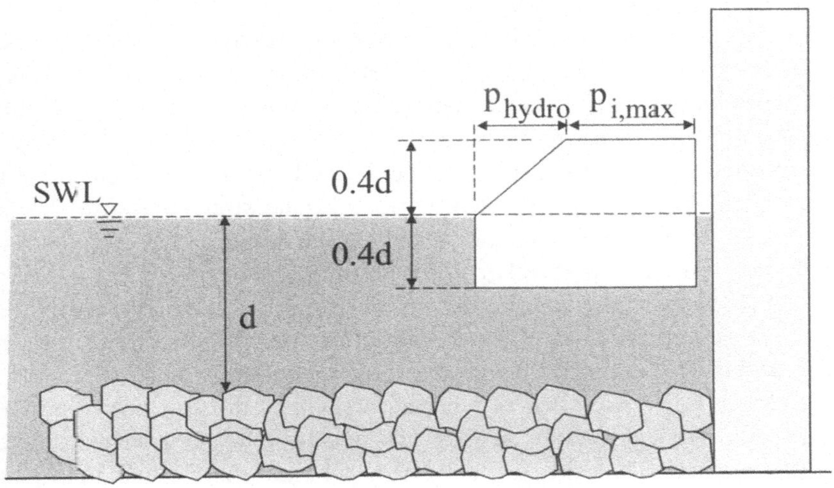

Figure 2-32. Pressure distribution induced by broken waves.

The simplest formula for the breaker celerity is obtained from shallow water wave theory ;

$$C_b = \sqrt{gd} \tag{2-98}$$

A more refined method to predict C_b can be found in SPM(1984). Two sources of values for λ in Equation 2-97 exist, the first is recommended.

For small scale waves, $\lambda = 10\ s^{-1}$
For prototype waves over an irregular foreshore, $\lambda = 0.3\ s^{-1}$
For prototype waves over a smoother foreshore, $\lambda = 0.5\ s^{-1}$

The second set of values are reproduced in Table 2-10.

Table 2-10. Values for the roughness term λ.

Foreshore slope	1/5 – 1/10	1/30 – 1/50	1/100
Foreshore conditions		$\lambda\ [s^{-1}]$	
Very smooth	1.5	0.9	0.7
Rough, rocky	0.5	0.3	0.24
Very rough, emergent rocks	0.13	0.18	0.14

2.6.2 *Wave loads on crown walls*

The variability of broken wave-induced forces on crown walls is rather less than of impact events on vertical breakwaters. Broken waves on crown walls show relative consistent responses, so a deterministic approach to the wave-to-force transfer function can be taken.

For design purposes, the sea state will be defined by the significant wave height at the toe of the breakwater, H_s, peak period, T_p and storm duration. For design methods using a single wave height, a calculation wave height, H_c, must be determined, recommended as $H_c = H_{99.8\%}$ for preliminary design (if $H_{99.8\%}$ can not be determined from the wave distribution in the design sea state, $H_c \cong 1.8\ H_s$ can be accepted.) For conditions where waves might be limited by depth, it must always be checked that H_c is compatible with the local water depth.

Next, it must be verified that waves can not plunge directly onto the wall generating impact pressures. Only surging, collapsing or broken waves are considered in this method. The criteria for deciding whether the method can be applied is the occurrence of shock impact events. Once selected the wave height (H_c) and wave period (T_p) are known, the impact event due to this wave does not occur, in the following cases:

1) If the design wave breaks before reaching the breakwater toe.
2) If $Ir > 3$, the wave breaks on the breakwater slope as a collapsing or surging breaker, where $Ir = \tan\alpha / s_c^{0.5}$, and $s_c = H_c / L_p$, and L_p is the peak wave length in deep water and α is the breakwater slope angle.
3) For other cases the method described in Volume IIa, section 6.2 (Fig. 2-33) identifies the regions of shock impact and non-impact events as a function of relative berm width (B_b/H_c) and relative berm crest height (A_c/H_c), where B_b is the berm length and A_c is the berm crest height above design water level.

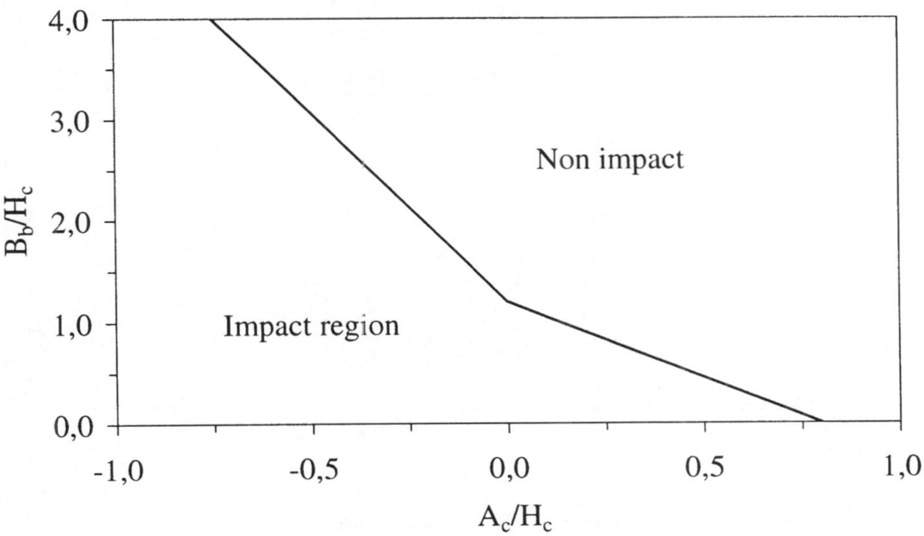

Figure 2-33. Definition of shock impact and non-impact regions (empirical).

It has been noted that a single wave may generate two peaks of force on the vertical structure. For wave impact events, the initial peak (impact force) is always larger than the second peak (pulsating force). For crown walls and other walls subjected to broken waves, whether the impact force is larger than the pulsating force depends on the wave and armour characteristics. Thus, the engineer must consider both load situations in the analysis, and select as design load the one which produces the lowest safety factor.

Simplified pressure distributions corresponding to the two load situations due to the calculation wave (defined by wave height H_c, peak wave period, T_p, and water depth h) are shown in Figure 2-34. A detailed description of the method and its basis can be found in Vol. IIa, section 6.2.

2.6.2.1 *Impact pressures*

The pressure distribution in this loading case is determined by P_{so} and C_{w2}. Over the unprotected region of the crown wall (above A_c) the pressure is:

$$P_i(z) = P_{so} = C_{w1}\, \rho_w\, g\, S_o \qquad \text{with} \quad A_c + S_o > z > A_c \qquad (2\text{-}99)$$

$$C_{w1} = 2.9\, [(R_u/H_c) \cos \alpha]^2 \qquad (2\text{-}100)$$

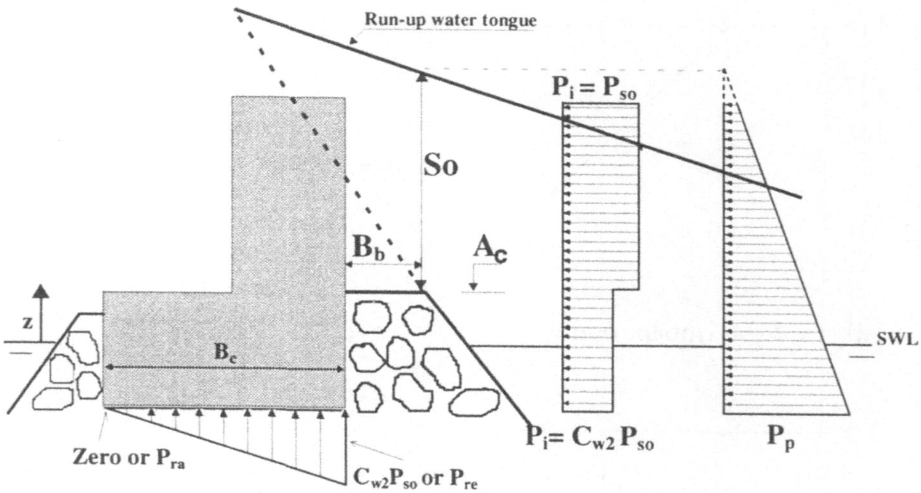

Figure 2-34. Pressure distributions, impact and pulsating (Vol. IIa, section 6.2).

$$S_o = H_c (1 - A_c/R_u) \tag{2-101}$$

Where z is the vertical co-ordinate, referred to a design SWL, positive upwards; R_u is the run-up height of the calculation wave (H_c, T_p) on a straight-infinite slope; A_c is the level of the armour berm above design SWL; ρ_w is the water density; and g is the gravitational acceleration.

To calculate R_u, Losada & Giménez-Curto (1981), based on experimental work under monochromatic waves and normal incidence, proposed the following expression for R_u on an infinite slope:

$$R_u/H_c = A_u \, [\, 1\text{-exp} \, (- B_u \, Ir \,) \,] \tag{2-102}$$

where, A_u and B_u are experimental coefficients, which depend on the type of armour unit and the Iribarren Number. A simple guide on selection of A_u and B_u is given in Figure 2-35.

Over the part of the crown wall protected by the armour berm, the pressure distribution is:

$$P_i(z) = C_{w2} \, P_{so} = C_{w1} \, C_{w2} \, \rho_w \, g \, S_o \quad \text{with } w_f < z < A_c \tag{2-103}$$

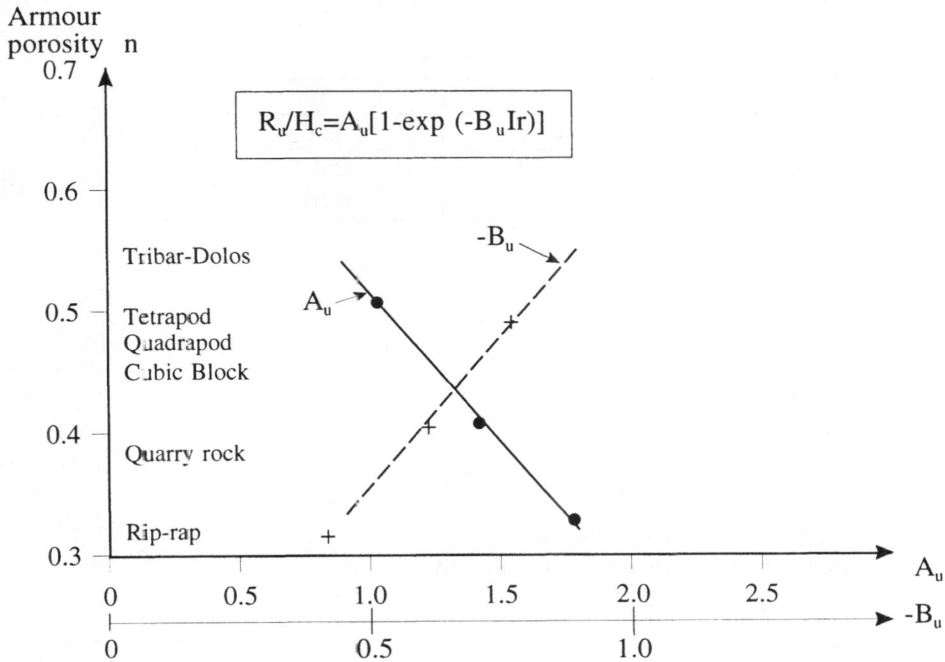

Figure 2-35. Run-up parameters A_u and B_u (after Losada, 1992).

where w_f is the foundation level of the crown wall (above SWL) and C_{w2} is an empirical non-dimensional parameter calculated for $0.03 < H_c/L_p < 0.075$, given by:

$$C_{w2} = 0.8 \exp(-10.9\, B_b/L_p) \tag{2-104}$$

where B_b is the armour berm width at A_c and L_p is the local peak wave length.

2.6.2.2 *Pulsating pressures*

The pressure distribution in this case is determined by:

$$P_p(z) = C_{w3}\, \rho_w\, g\, (S_o + A_c - z) \tag{2-105}$$

Where C_{w3} is a non-dimensional parameter evaluated empirically from monochromatic wave tests as:

$$C_{w3} = a \exp(Co) \tag{2-106}$$

$$Co = c\,(H_c/L_p - b)^2 \quad \text{with} \; (0.03 < H_c/L_p < 0.075) \qquad (2\text{-}107)$$

Table 2-11. Fitting coefficients for Eqs 2-106 and 2-107.

B_b/D_{n50}	a	b	c
1	0.45	0.068	259
2	0.36	0.069	357
3	0.30	0.073	383

where D_{n50} is the equivalent size of the armour units forming the berm.

2.6.2.3 *Uplift pressures*

The following values are adopted:

- Seaward edge:
  - impact pressure = $C_{w2}\,P_{so}$
  - pulsating pressure = $P_p(z = w_f) = P_{re}$
- Heel:
  - Negligible impact pressure, $P_{ra} = 0$
  - pulsating pressure = P_{ra}, from Figure 2-36.

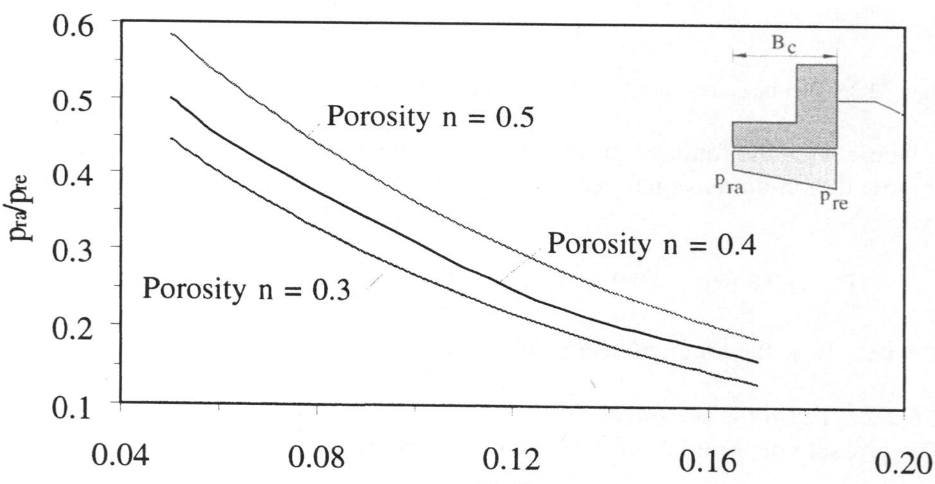

Figure 2-36. Relative pressures underneath a breakwater crown wall.

where B_c is the width of the crown wall foundation. For design purposes the porosity selected must represent the porosity of the material on which the crown wall is founded. A linear law between the seaward edge and the heel is proposed (see Vol. IIa, section 6.2).

2.6.3 *Wave loads on caisson on high mounds*

A new high mound composite breakwater (HMCB, Fig. 2-37) has been developed by the Port and Harbour Research Institute in Japan and studied by LWI (Muttray et al., 1999). This innovative breakwater consists of a high mound rubble foundation (h_b/h_s = 0.75 to 1.05) and an energy dissipating superstructure. The interaction between the high mound and breaking waves has been examined where the experiments focused on:

a) wave breaking and hydraulic performance in front of the structure;
b) dynamic and quasi-static wave loads on the superstructure due to breaking waves; and
c) the structural response of the superstructure.

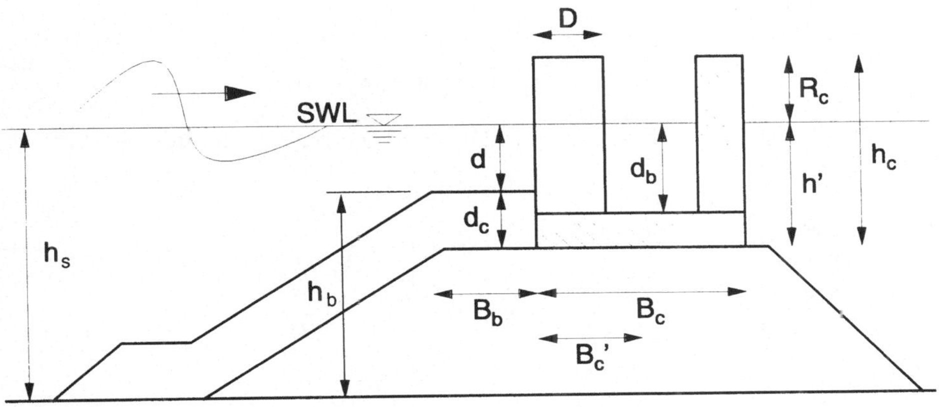

Figure 2-37. Definition sketch for characteristic lengths of the structure.

An important characteristic of the HMCB can be seen in Figure 2-38 where wave load (here given by pressures on the solid front wall) increases with increasing wave height up to a certain limit ($H < H_{b\text{-}b}$). Beyond this limit, further increasing wave heights cause almost constant wave loads ($H > H_{b\text{-}b}$). The largest wave loads have been observed at the transition from breaking to broken wave loads.

Due to this rather different response, the parameter map described in section 2.2.2 is of limited use for this type of breakwater. A better distinction of loading cases for a HCMB is achieved by defining critical wave heights $H_{n\text{-}b}$ (transition from non breaking to breaking wave loads) and $H_{b\text{-}b}$ (transition from breaking to broken wave loads) as given in Table 2-12. If the berm is above still water level the critical wave heights are zero.

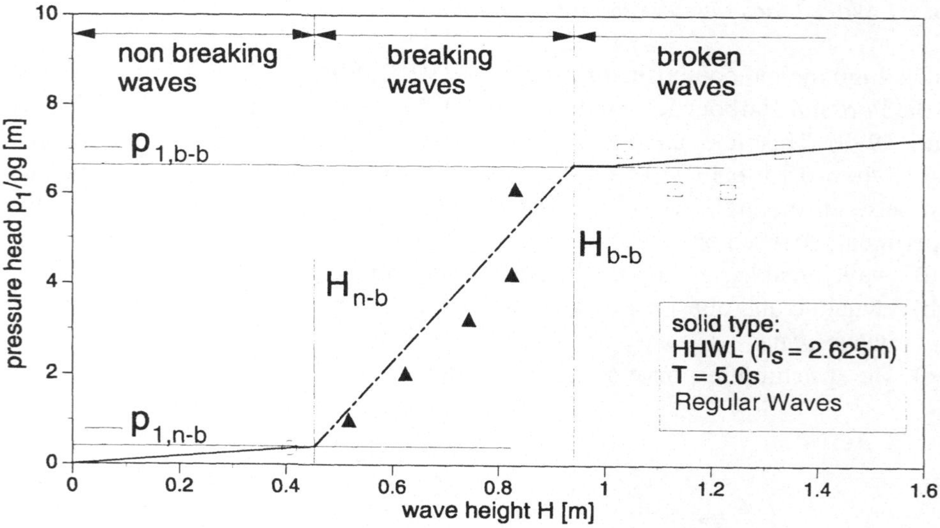

Figure 2-38. Wave pressure at the solid front wall for different load types (Muttray et al., 1999).

Table 2-12. Calculation of H_{n-b} and H_{b-b}.

Type of wave		**H_{n-b}/d [-]**	**H_{b-b}/d [-]**
Short waves:	$0.9 \leq 2\pi\, h_s/L < 1.2$	1.2	2.0
Intermed. Waves:	$0.6 \leq 2\pi\, h_s/L < 0.9$	1.0	1.8
Long waves:	$0.3 \leq 2\pi\, h_s/L < 0.6$	1.7	2.5

The load analysis is based on characteristic pressures which are used to predict the resulting wave forces on the superstructure and consider the influences of wave period, shallow water on berm, wave overtopping and permeability of the structure. The calculation procedure for the wave pressures and forces at the front wall for non breaking, breaking and broken waves is as follows for regular waves and a solid type wall. For random waves H can be replaced by $H_{1/3}$ or $H_{1/10}$, using the peak period T_p, then giving characteristic pressures $p_{1/3}$ or $p_{1/10}$ depending on which wave height was used.

2.6.3.1 *Critical wave heights*

H_{n-b} and H_{b-b} can be calculated from Table 2-12 where h_s is the water depth at the toe of the structure and L is the local wave length (L_{pi}) calculated from T_p and h_s as given in Equation 2-33.

2.6.3.2 *Critical wave pressures*

Calculate the corresponding critical pressures $p_{1,n-b}$ and $p_{1,b-b}$ by:

$$p_{1,\,n\text{-}b} = \alpha_{11} \cdot \rho\, g\, H_{n\text{-}b} \tag{2-108}$$

$$p_{1,\,b\text{-}b} = p_{1,\,n\text{-}b} + \alpha_{12}\, \alpha_{13}\, \rho\, g \left(H_{b\text{-}b} - H_{n\text{-}b} \right) \tag{2-109}$$

where p_1 is the wave induced pressure at the height of the still water level in front of the structure and the three pressure coefficients α_{11}, α_{12}, and α_{13} are given as follows:

$$\alpha_{11} = 0.6 + \left[\frac{4\pi\, h_s / L_{pi}}{\sinh\left(4\pi\, h_s / L_{pi} \right)} \right]^2 \tag{2-110}$$

$$\alpha_{12} = \begin{cases} 36 & \text{for } 0.9 \leq 2\pi\, h_s / L_{pi} < 1.2 \\ 16 & \text{for } 0.3 \leq 2\pi\, h_s / L_{pi} < 0.9 \end{cases} \tag{2-111}$$

$$\alpha_{13} = \min\left[1;\; R_c / H \right] \tag{2-112}$$

2.6.3.3 *Pressures and resultant force for non breaking waves*

The pressures and the resulting force for non breaking waves can be calculated as follows (ranges of validity are $0.25 < d/H \leq 1.25$):

$$p_1 = \alpha_{11}\, \rho\, g\, H \tag{2-113}$$

$$p_3 = p_1 \left[1 + \frac{z}{h'}\, \alpha_3 \right] \quad \text{for } -h' \leq z \leq 0 \tag{2-114}$$

$$p_4 = 9 \left(\alpha_2\, 2\pi / L_{pi} \right)^2 \rho\, g\, H \tag{2-115}$$

where $\alpha_2 = 1.5 \cdot H - R_c^*$ (but not smaller than zero); R_c^* equals the minimum of the freeboard and the water elevation η above SWL ($R_c^* = \min(R_c, \eta)$); α_3 is equal to 0.65 for $d \leq 0$ and equal to 0.8 for $d > 0$; h' is the distance between still water level and bottom of the superstructure; z is a variable in between 0 and h'; p_3 is the resulting pressure at the height of the base of the caisson; and p_4 is the pressure at the top front face of the structure. The resultant force F_h can then be calculated using the trapezoidal method:

$$F_h = \frac{1}{2}\, h'(p_1 + p_3) + \frac{1}{2}\, R_c^*(p_1 + p_4) \qquad \text{(2-116)}$$

2.6.3.4 *Pressures and resultant force for breaking waves*

For breaking waves the following formulae are valid within the range of $0.25 < d/H < 1.25$:

$$p_1 = p_{n\text{-}b} + \alpha_{12}\,\alpha_{13}\,\rho g(H - H_{n-b}) \qquad \text{(2-117)}$$

The pressures α_2 and α_3 can be calculated as given for non breaking waves. The resultant force takes into account the time lag between the maximum pressure at SWL and the top of the front wall:

$$F_h = \frac{1}{2}\, h'(p_1 + p_3) + \frac{d}{4}\, p_1 \qquad \text{(2-118)}$$

2.6.3.5 *Pressures and resultant force for broken waves*

For broken waves the pressure at SWL can be estimated as follows:

$$p_1 \leq p_{b\text{-}b} + \alpha_{14}\,\rho g(H - H_{b-b}) \qquad \text{(2-119)}$$

where α_{14} is 1.0 for the berm below SWL ($0 < d/H < 1.25$) and 3.0 for the berm being above SWL ($-0.25 \leq d/H \leq 0$). For slit-type structures (pillars in front of a vertical wall) the corresponding formulae are given in Volume IIa, chapter 6.3 and Muttray et al. (1999).

2.6.3.6 *Uplift forces*

As a first approach, uplift pressures and forces may be estimated by assuming a triangular pressure distribution underneath the structure where the pressure at the seaward side of the structure p_u equals p_3 (Eq. 2-114) and the pressure underneath the shoreward side may be assumed to equal zero.

However, when structure is overtopped or rocking occurs the pressure underneath the heel of the structure will increase up to the same pressure at the seaward side (rectangular distribution). This case has yet not been investigated in sufficient detail to give appropriate design formulae, but some further aspects are given in Volume IIa, section 6.3.

2.7 FIELD MEASUREMENTS AND DATABASE

During PROVERBS field measurements were carried out at five different field sites by five separate institutions. The breakwaters at each site were distinctly different in terms of construction, wave climate and tidal range, providing a wide cross section of field results. This section describes, briefly, each site and the data collected up until September 1998. Further details are given by Bullock et al. (1999).

2.7.1 *Dieppe*

The structure here is of Jarlan type caissons, which are perforated on both sides, standing on a rubble mound in 6m of water at MLW. The tide range is 12 m and the crest height is 2.4 m above MHW. The design wave height and period are $H_s = 12$ m and $T_s = 12$ s but so far the maximum measured data is $H_s = 2.1$ m and $T_s = 6.5$s. The loading expected is primarily pulsating with some transitional loading but the perforations will add complications.

2.7.2 *Porto Torres*

Here there is a direct comparison between plain and perforated caissons at prototype scale. The crest level is 8.0 m above MHW and the depth of water to the sea bed is 20 m. The prototype data was collected at a rate of 20Hz and the wave data obtained so far has maximum values of $H_s = 3.9$ m and $T_s = 9.2$ s. Because of the water depth and wave height expected only pulsating loads are likely. Model tests have been conducted at 1/20 scale. Both model and prototype measurements include uplift pressures.

2.7.3 *Las Palmas*

The structure is of caisson construction, 30 m high on top of a rubble mound, in 48 m of water. The tide range is 2.8 m. Accelerometer and pressure transducers are used to monitor the performance of the caissons. The data are logged at a rate of 25Hz. Model tests have been completed

2.7.4 *Gijon*

This is primarily an armoured breakwater, in 23 m of water, with a crest level 17 m above MHW and a tide range of 4 m. The concrete superstructure was instrumented with pressure transducers which were logged at a rate of 20 Hz and readings were completed in April 1998.

Two sets of model tests have been carried out one at 1/90 scale with regular and random waves. The second series used a 1/18 scale model (funded outside of PROVERBS) with regular, random and measured storm waves.

Wave data was obtained from a transducer array on the sea bed in front of the breakwater. The design wave height is $H_s = 10$ m and the maximum measured so far is 7 m.

Because of the nature of the structure the loading is unlikely to produce impacts but will generate pulsating loads or broken waves

2.7.5 *Alderney*

This is an old masonry structure built on a rubble mound in approximately 20m of water at MLW. The front face is stepped and slopes at about 30^0. The crest level is 6 m above MHW. The depth at the toe at MLW is 0.9 m and the tide range is 5.3 m. The loading, therefore, varies from broken through pulsating to impacts depending on the wave height and tide level at the time of recording. Prototype measurements of pressure and aeration were logged at 0.5kHz. Wave data was obtained from six pressure transducers located on the sea bed which gives wave height, period and direction (Fig. 2-39). Maximum wave height and period recorded at full scale are H = 2.9 m and T = 11.1 s, respectively.

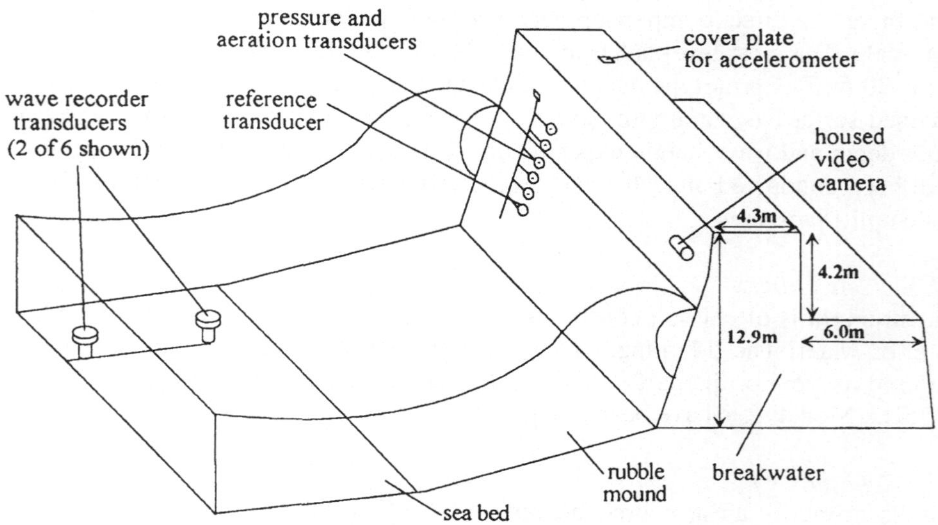

Figure 2-39. Field measurements of waves, impact pressure and aeration at Alderney breakwater.

Model test have been completed and were carried out at 1/25 scale with a logging rate of 2kHz with regular and random waves in both fresh and sea water.

2.7.6 *Field measurement database*

Because the great expenditure of effort and the time consumed in obtaining field data, unlike laboratory records, the data obtained has been carefully recorded and archived but it is unlikely that there will be enough impact events in a field record to enable a statistical analysis to be carried out.

The PROVERBS 'Field Work Database' has been progressively developed and adapted from the start of project to summarise the analysed data collected at each field site into the requisite parameters.

2.7.6.1 *Definition of database parameters*

Definitions of events and other parameters for field work were discussed in some detail during PROVERBS where it was decided that the same definitions and methods were to be used by all the field and laboratory workers for ease of comparing results. The wave and pressure trace parameters are defined in Annex 1 of this Volume.

Many of the field sites have varying record lengths and different methods of obtaining data. The present database format used here has been devised to take these factors into account and is in Excel format for ease of communication.

Analysis of the pressure data follows the methods, where possible, laid down by McConnell & Kortenhaus (1997).

Record Information: This is the basic information required to trace the record at the Partner Establishment. Only one set of these records are required per set of impacts or wave loading data. Where wave loading data sets are recorded at two locations at the same site e.g. Porto Torres, two sets of data entries are produced and distinguished per recording period as plain or perforated.

Wave Data: Essential details about the significant wave height, water depth at the toe (defined so that it is positive) when the toe is immersed at the time of the record and moment parameters needed to define the wave spectra for comparison with the model waves used in the laboratory.

Pressure Data: This is taken from the first pressure transducer (PT) above the still water level (SWL) at the time of recording. Only the first ten significant records are recorded and are listed in order of rank.The following columns to the right maintain this order

Entries have been left blank where data is not available. Where impact pressures do not occur this column is left blank and data is sorted according to maximum quasi-hydrostatic pressure (Pmwl,q) and corresponding columns left blank

- *Force data*: This is evaluated for each of the events picked out by the significant pressures. (It does not follow that the force data will be in order of rank). Units for Force are kN/m run and for Impulse are Ns/m.
- *Uplift Data*: Similar to the Force data. Not all field workers are recording uplift pressures.

- *Geometric Data*: Site plans, cross-sections, etc are included in the relevant section.

2.8 ALTERNATIVE LOW REFLECTION STRUCTURES

2.8.1 *Perforated vertical walls*

2.8.1.1 *Introduction*

Despite their increased complexity and cost of construction as compared to plain caissons, perforated caissons are becoming more and more popular not only for anti-reflective quaywalls inside sheltered harbours, but also for external caisson breakwaters, in order to partly overcome the typical drawbacks of vertical structures: large reflections, forces, overtopping and toe scour. Perforated vertical breakwaters are intended to absorb part of the wave energy within chambers or voids in the wall through various mechanisms, such as turbulence and resonance effects. The larger the water level difference at the two porous wall sides the larger the energy dissipation, which is therefore strongly dependent on the wave length L (variable efficiency with sea-states).

There are different structural solutions to provide this dissipation within a relatively narrow horizontal space and the most frequent are the caissons with absorbing chambers and the single/multiple perforated screens with variable porosity and screen spacing. Typically the walls have vertical or horizontal slits or a set of rows of circular or rectangular holes. The internal walls can also be perforated. The absorbing chambers are usually topped by a ceiling slab (often with venting holes for air escape) but they can also be open (as for multiple screens). The wall porosity, defined as the ratio of the area of the front wall openings to the total wall area, is typically within the range 15-40% and the chamber width B is typically within the range of 1/10 to 1/4 of the local wavelength L. The single or multiple screens are especially used in sheltered harbour waters to reduce wave transmission without excessive space occupancy and limitation of seawater exchange. Perforated caissons on rubble foundations are more commonly used in deepwater exposed sites.

The perforated wall for coastal engineering applications was introduced by Jarlan in the early 1960s with the first application at Comeau Bay (Canada). Various examples of perforated caisson breakwaters exist in Italy (Franco, 1994), France and Japan (Takahashi, 1996). Much theoretical and laboratory research has been carried out in the last 30 years in order to study the hydraulic performance of these structures and extensive bibliography is available. No general practical design criteria are however yet consolidated, due to the variability and complexity of the geometries and related hydraulic processes.

Further experimental model tests have been performed within PROVERBS by ENEL-Milan, Caen University and Leichtweiß Institute, Braunschweig, and field measurements at Porto Torres (Italy) and Dieppe (France). Several approaches are proposed to derive comprehensive formulae capable of allowing a computation of the pressures and forces acting on the structures under pulsating waves as well as to improve the physical understanding of the hydraulic performance of various structure configurations.

2.8.1.2 *Prototype measurements*

Field data from two perforated caissons have been considered: the Porto Torres (Sardinia, Italy) multi-chamber caisson and the one chamber Dieppe (Seine-Maritime, France) caisson.

The **Porto Torres** perforated caissons (three chambers with rectangular holes, with decreasing volume from seaside to harbour side) is shown in Figure 2-40. For nearly two years after Porto Torres harbour completion, in 1992, the performance of the new breakwater was monitored using 53 sensors: a perforated caisson and plain wall one, spaced 62 m, oriented in the same direction and with the same depth of 20 m, were monitored.

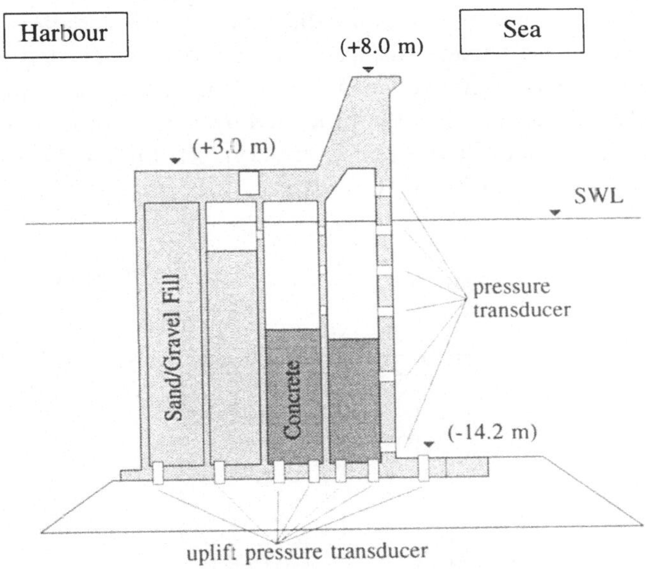

Figure 2-40. The perforated Porto Torres caisson with pressure transducers position.

In the two year operation, 10 sea states were recorded ($H_s \leq 3.5$ m; $6.0 < T_p < 9.2$ s). Runup levels, front face and uplift pressure measurements were analysed in both statistical and spectral ways in order to define the most important parameters. Interesting results were obtained comparing time histories of water levels of plain and perforated caissons, showing a 30% reduction of run-up height on the perforated caisson.

Horizontal and uplift forces were calculated for the plain caisson, at each instant of the data set (2 Hz sample frequency), integrating pressures with the trapezoidal rule (interpolating linearly data between two measurement points). It was found that Goda's models over-estimates both horizontal and uplift loads. Uplift pressures showed an almost flat trend compared with the triangular shape suggested by theory and by model tests. No interesting data could be extracted by the pressure recorded at the perforated caisson for the actual measured values showed very poor correlation with water level.

At **Dieppe** the 225 m long West breakwater made of double Jarlan caissons (Fig. 2-41) was inaugurated in 1992. The rectangular caissons are 25 m long, 17 m high and composed by 2 dissipation chambers separated by a wall. The perforated walls (28% porosity) have 0.90 m diameter circular holes, are 0.90 m thick and reinforced by internal transversal walls with 2 x 2.5 m rectangular holes. The sea side chamber width is B = 13.5 m with open ceiling. Each chamber is filled with concrete for a height of 2.8 m and so a typical 25 m caisson (but for the breakwater head) has a total mass around 8900 tons. The foundations were made of quarry stones while the friction surface was formed with selected pebbles (20-63 mm). A geotextile was placed between the existing soil (mud, sand and gravel) and the quarry layer. The caisson toe is protected by 1-3 ton blocks. Tide range is 10.80 m with depth in front of the breakwater varying from 6 to 17 m. The design wave parameters are $T_p = 12.0$ s and $H_s = 8.20$ m.

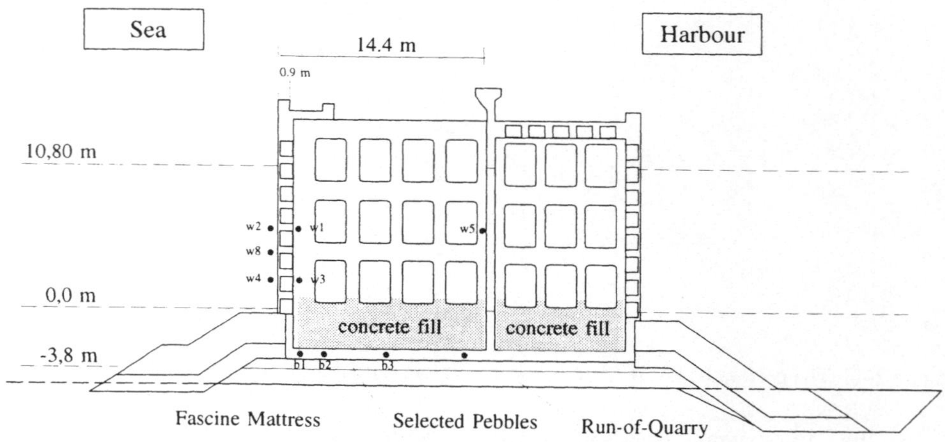

Figure 2-41. Cross section of Dieppe caisson and pressure cells location.

One caisson is instrumented with 7 pressure cells placed at the bottom and 8 similar cells on the vertical faces of the sea-side chamber (Fig. 2-41). Cells n°6 and 7 are located at the same elevation of n°2 at the outer wall. Sampling frequency was 3 Hz for 10 minutes duration.

The data analysed are essentially originating from the storm of 13th February 1997. The significant wave height calculated from pressure cells (no wave recorder available at the time) was equal to $H_s = 3.7$ m and the local wavelength was around 112 m (B/L = 0.12), the water level was 8.50 m above chart datum.

Figure 2-42 shows time series of the forces F_{ext}, F_{int}, F_r as well as the total force F_{tot}. F_r values are greater than F_{ext} values. This is due, in this case of storm waves, to the fact that pressure values are stronger on rear wall than on perforated one: $p_r = 21.66$ kPa while $p_{ext} = 20.92$ kPa.

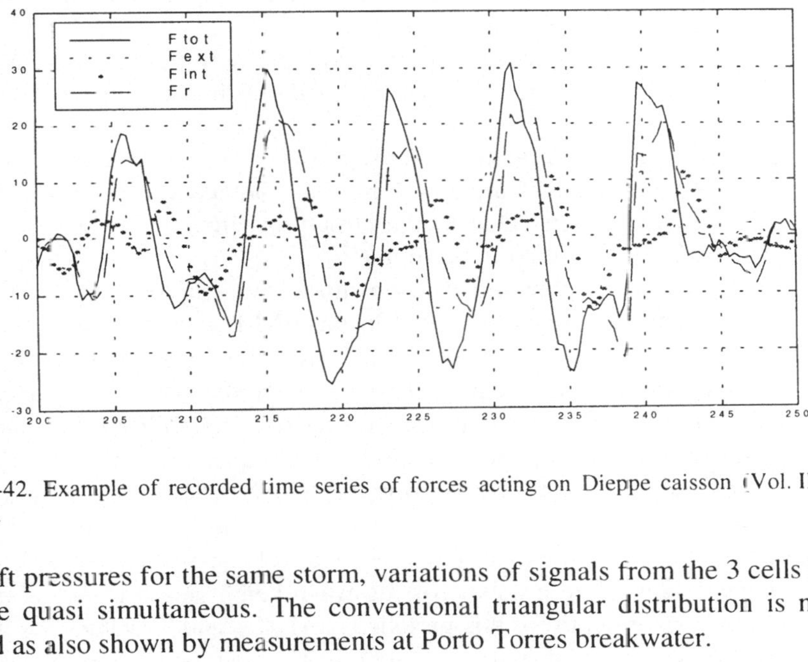

Figure 2-42. Example of recorded time series of forces acting on Dieppe caisson (Vol. IIa, section 8.1).

Uplift pressures for the same storm, variations of signals from the 3 cells b_1, b_2, b_3 were quasi simultaneous. The conventional triangular distribution is not respected as also shown by measurements at Porto Torres breakwater.

2.8.1.3 *Model tests*

Large and small scale model tests of perforated structures were conducted by LWI, ENEL and UC.

The **LWI tests** were conducted in the Large Wave Flume (GWK) of the Coastal Research Centre, a joint institution of the University of Hannover and the

Technical University of Braunschweig. The large-scale tests in the GWK were carried out firstly with single perforated screens (consisting of up to 29 horizontal elements made of 180x180mm square steel pipes) of 5 different porosities (porosities of 0% (impermeable), 11%, 20%, 26.5% and 40.5%).

The second part of the test programme was performed with chamber systems consisting of up to three perforated screens, combined with an impermeable rear wall ($C_t = 0$). The One Chamber Systems (OCS) was tested with porosity (n = 20%) of the seaward perforated wall but with two different chamber widths (B = 4.8 m and B = 7.8 m). Two Multi Chamber Systems were tested. The water depth was kept constant for all single permeable screens (d=4m) but was varied in three steps (d = 3.25, 4.00, 4.75 m) for the impermeable wall and the chamber systems. The incident waves, wave transmission and wave reflection were analysed by wave gauges which were grouped in three sets (one in the far field, one in front and one behind the structure) with 4 wave gauges each. Additional wave gauges were installed to measure the water surface elevation and the water level gradient directly at the permeable screen.

The resulting wave loads were measured with pressure transducers which were installed at 10 positions at the front and back side of the structure. Data were recorded at 200 Hz.

The tests were carried out with regular waves (H = 0.5-1.5 m and T = 4.5-12 s), random waves (H_s = 0.5-1.25 m, T_p = 4.5-12 s), solitary waves (H = 0.5-1.0 m) and transient wave packets. Due to the structure height of approximately 6 m (R_c/H about 1.33), overtopping is negligible.

The **Porto Torres perforated caisson model** was reproduced in 1:20 scale. The tests performed at the end of 1997 in the ENEL random wave flume were intended to measure simultaneously global horizontal forces and pressures. The model caisson was therefore equipped with dynamometer as well as pressure transducers. In order to give the necessary statistical reliability to this investigation, only tests performed with long random wave sequences were used.

For consistency with the Goda formula (Goda, 1985) the offshore $H_{1/250}$ ($H_{1/250,o}$) value for the required "highest wave in the design sea state" (H_{max}) has been used. $T_{1/250,o}$ is the wave period associated to $H_{1/250,o}$ (the average of periods of the highest 1/250 of zero-downcrossing wave heights) and $L_{1/250,o}$ is the wave length at the *h* water depth for this period. The $H_{m0,inc}$ and T_p values are the spectrum incident wave height and peak period. The test range conditions for the above parameters (in prototype terms) have been:

- $H_{1/250}$ =3.71÷8.08 m
- $T_{1/250}$ =6.22÷11.258.08 s
- $L_{1/250}$ =59÷143 m
- B/L=0.071÷0.172

The total width of the three perforated chambers of Porto Torres structure is B = 10.15 m, but the perforated chamber height is considerably reduced with respect to the simple caisson. All tests were performed with a water depth corresponding to 21.0 m.

The **Dieppe Caisson 1:25 Model Tests** were performed in the 1 m wide, 0.8 m deep, 25 m long wave flume at Caen University with the model caisson on a flat bottom or on a porous berm.

Conditions of tests were the following:

- Regular wave generation
- Chamber width: B = 0.54 m
- Water depth: h_s = 0.50 m for case a, and h_s = 0.40m for case b
- Perforation shapes: circular holes ϕ40mm giving total porosity n = 0.28
- Pressure devices: 11 pressure cells on the 3 faces for case a and 9 cells for case b.
- Signal sampling rate: 10 Hz per sensor.

The non-dimensional wave parameter ranges during the tests were the following:

- for the case of caisson without berm:
  - $0.066 < B/L < 0.30$
  - $0.004 < H/L < 0.057$
- for the case of caisson with a berm:
  - $0.096 < B/L < 0.30$
  - $0.008 < H/L < 0.054$

Experimental devices include two resistive sensors for measurement of run-up on the two walls of the caisson as well as 4 others put on the middle of the flume, each 1.50 m apart, for determination of reflection coefficient and incident wave height.

As water depth above the berm was d = 0.265 m, water depth in front of toe h_s = 0.40 m and berm width B_b=0.33 m (hence: $(h_s-d)/h_s = 0.33$ and $B_b/L < 0.2$) it can be assumed that no impact conditions apply and impact pressure coefficient can be set $\alpha_I = 0$ after Takahashi et al. (1994). This value is confirmed by the PROVERBS parameter map (Volume IIa, section 2.2) showing that when $0.04 < H_i/h_s < 0.25$ breaking waves occur without impacts and by actual observations of the 2D-model.

2.8.1.4 *Methods to predict forces for perforated caissons*

Overall forces on perforated caissons are reduced in comparison with plain wall caissons of the same dimensions for two main reasons:

- reduction of total wave height in front of the caisson due to a lower reflection coefficient due to dissipation of the energy ;

- phase lag between the maximum force timing of the incoming wave as the total force acting on the caisson is the resultant of the force on permeable wall(s) and impermeable rear wall.

These phenomena depend upon the wave characteristics and the geometry of the perforated caisson including chamber shape and dimensions, wall porosity, presence of vents (in case of closed ceiling which makes the determination of the wave forces upon the perforated breakwaters very complex).

Two approaches are most commonly used to calculate forces:

- Canel method deduced the hydrodynamic loading produced by an incoming wave of height *H* on a partially reflecting structure applying Goda model (which is specific to a plain wall caisson) with a suitably corrected design wave in which *Cr* is the expected reflection coefficient. Canel approach (Canel, 1995) which proposes a reduction of the design wave height $H_D = H_s \cdot (1 + C_r)/2$ to be introduced in the Goda formula based upon the reflection coefficient C_r; although very simple this approach is only useful for the designer when the structure reflection coefficient is known;
- Takahashi's modification of the Goda formula (Takahashi et al., 1994), which is based upon the calculation of the total force from the pressure distribution on the walls. To define the maximum forces, Takahashi's method defines the critical conditions for design out of three wave phases: the crest arrival at the perforated wall (crest I) or at the rear wall (crest IIa and crest IIb). Besides the fact that the method is relatively complex, the tests have shown that it either over or underestimate the total force according to the crest choice. This is due to the fact that the Takahashi formula does not take sufficiently into account the dephasing evolution between the actions on each wall.

To improve existing methods each partner proposed its own approach firstly intended to fit their data. All the proposed formula were then tested against all the data available to verify their prediction effectiveness.

Empirical formula developed from screen tests (GWK)

The sum of wave loads measured simultaneously on every successive perforated wall gives the total force on the structure. Measurements indicated that the overall load is reduced for wave crest and wave trough situations (pulsating loads) compared to the measured forces on the impermeable vertical wall (F_0). This force reduction is mainly due to the phase lag and in some cases due to opposite loading directions especially for large B/L ratios ("residence time"). The reduction factors (F_{tot}/F_0) are significant, particularly for the Multi Chamber Systems:

- One Chamber System: $F_{tot,OCS}/F_0 = 0.6\text{-}0.8$
- Multi Chamber System: $F_{tot,MCS}/F_0 = 0.5$

The overall load can be calculated as:

$$F_{tot} = 12 \cdot \tanh^{1.1} \cdot (0.009 \cdot FF_{tot}) \cdot \rho \cdot g \cdot H_i^2 \ \ [kN/m] \tag{2-120}$$

where FF_{tot} is a total load parameter defined as:

$$FF_{tot} = \frac{(d/B)^{2/3}}{H_i/L_d} \ \ [-] \tag{2-121}$$

which fits well the experimental data (Fig. 2-43) for non breaking waves.

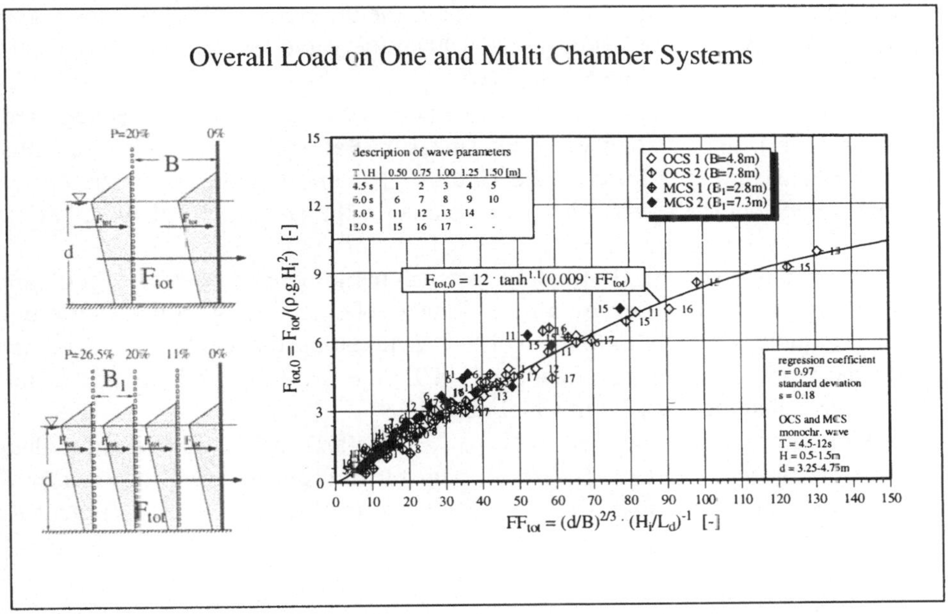

Figure 2-43. Overall loads on One- and Multi Chamber Systems (GWK, 1997) (Bélorgey et al., 1999).

The LWI approach was checked against other data sources where it proved to be very accurate for prediction of forces for the Dieppe caisson but rather poor (up to 50% under-prediction) for Porto Torres model data from 2D and 3D tests (ENEL and PM/DH respectively). This could be due both to the special configuration of Porto Torres breakwater (rubble foundation, stepped chambers) or because the former data refers to open ceiling and the latter to closed ceiling structures.

Empirical formula developed from Porto Torres model tests

ENEL approach has been addressed to find a modification of the design wave height involving only the dimensionless parameter B/L. The total horizontal force F_h by Goda model was calculated and compared with the corresponding measured values for each available perforated structure data set. As the data came from different experimental conditions, the application of Goda formula required the following assumptions about the wave parameters to be used:

- Porto Torres 2D random model tests (dynamometer data): the offshore values $H_{1/250,os}$ and the associated period $T_{1/250,os}$ were used; these wave parameters come from the statistical zero-crossing analysis of the signal recorded by a single probe placed offshore in the lateral of the three channels which the wave flume is divided in (de Gerloni et al., 1997); the wave length $L_{1/250,os}$ was calculated at 21 m water depth for $T = T_{1/250,os}$;
- Dieppe Caisson 2D model tests performed with regular waves (integrated pressure data): the incident wave H_{si} (measured at about 2 to 5 wave lengths from the caisson) was used as the Goda H_{max} value together with the corresponding period T; the wave length L related to the latter was calculated at 10 and 12.5 m water depth;
- PM/DH (Franco, 1996) 3D random tests (dynamometer data): the significant wave height H_s (measured by a directional probe according to maximum entropy standard DH method) times 1.8 was taken as the Goda H_{max} together with the corresponding significant period T_s; the wave length L related to the latter was calculated at 18.9 m water depth.

A reduction coefficient applied to $H_{1/250,os}$, to H_{si} and to $H_{s,os}$ for the three data sets respectively was calculated for each data set in order to make the measured and calculated force coincide. A relationship between this reduction coefficient K_r and B/L was found:

$$K_r = 1 - a\left(\frac{B}{L}\right) \qquad \text{for } B/L < 0.3 \qquad (2\text{-}122)$$

with $a = 1.43$, a standard deviation $\sigma_a = 0.08$ and a regression coefficient of $R^2 = 0.52$.

The proposed reduction coefficient and the Canel reduction coefficient were plotted in Figure 2-44 against B/L. The Canel reduction coefficient has been calculated from the reflection value measured in the tests.

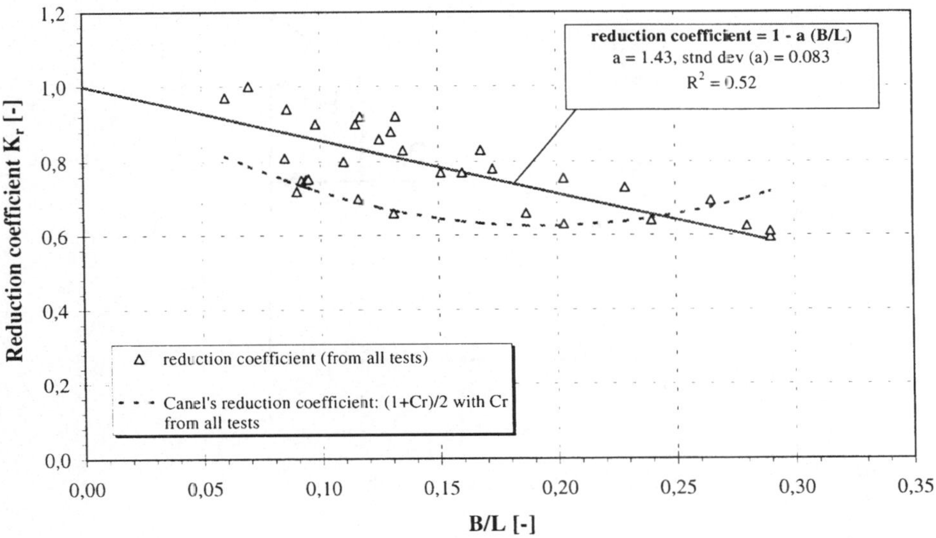

Figure 2-44. Comparison between proposed and Canel reduction coefficient against B/L (Vol. IIa, section 8.1).

<u>*Modification of Goda/Takahashi formula developed from Dieppe model tests*</u>
A modified calculation method was developed at UC/ULH to take into account the phase delay between the positive peak on the faces of the perforated wall as well as on the rear wall. This new method, based on Goda (1985) formula and applicable to pulsating wave conditions, uses modification factors in the calculation of forces on each wall and introduces a new χ factor tied to the phase delay. This improvement avoids the need to choose from the three wave phases presented by Takahashi.

This new calculation method (Tabet-Aoul & Belorgey, 1998), applicable for B/L < 0.35, can be processed as follows (see Fig. 2-45 for reference):

<u>*Calculation of wave run-up on the vertical walls*</u>
The parameter R_c*, including overtopping effect, is calculated using the Goda formula (Goda, 1985) as follows:

$$R_c^* = \min\{\eta^*, R_c\} \qquad (2\text{-}123)$$

Wave run-up can be modified taking into account details of the structure, but the effect of incidence wave angle β remains as proposed by Goda as 2D model test results do not allow for improvement of this.

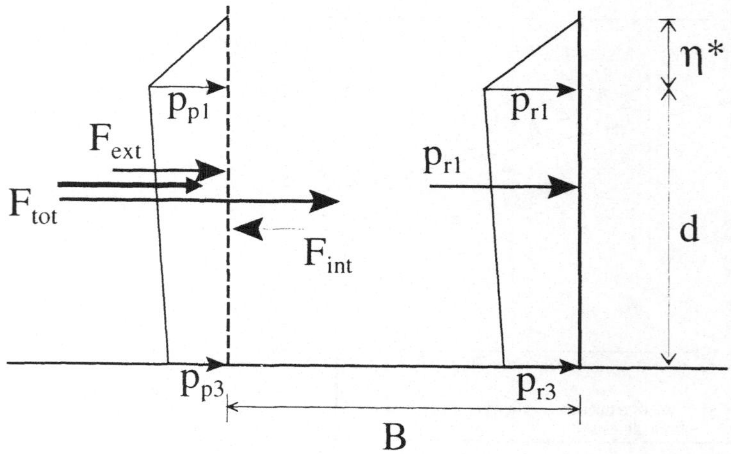

Figure 2-45. Peak pressure diagrams and resultant forces on the caisson.

$$\eta^* = 0.75\ (1+\cos\beta)\ \lambda_{i1}\ H_D \qquad (2\text{-}124)$$

where H_D is the design wave height, $\lambda_{i1} = \lambda_{p1} = 0.42$ for the perforated wall and $\lambda_{i1} = \lambda_{r1} = (0.7 - B/L)^2$ for the rear wall

Calculation of pressure parameters

$$\alpha_1 = 0.6 + 0.5[(2kh / \sinh(2kh)]^2 \qquad (2\text{-}125)$$
$$\alpha_2 = \min\ \{(1\text{-}d/h_b)(H_D/d)^2/3,\ 2d/H_D\} \qquad (2\text{-}126)$$
$$\alpha_3 = 1 - (h'/h)[1\text{-}1/\cosh(kh)] \qquad (2\text{-}127)$$
$$\alpha_4 = 1 - R_c^* / \eta^* \qquad (2\text{-}128)$$
$$R_c^* = \min\{\eta^*, R_c\} \qquad (2\text{-}129)$$
$$\alpha^* = \max\{\alpha_I, \alpha_2\} \qquad (2\text{-}130)$$

Calculation of pressure diagrams

- For perforated wall

$$p_{p1} = 0.5\ (1+\cos\beta)\ (0.42\alpha_1+(B/2L)\ (1+\alpha^*)\ \cos^2\beta)\ \rho g H_D \qquad (2\text{-}131)$$
$$p_{p3} = \alpha_3\ p_{p1} \qquad (2\text{-}132)$$
$$p_{p4} = \alpha_4\ p_p \qquad (2\text{-}133)$$

- For rear wall

$$p_{r1} = 0.5\ (1+\cos\beta)\ ((0.7\text{-}B/L)^2\alpha_1+(0.43\text{-}B/L)\ (1+\alpha^*)\ \cos^2\beta)\ \rho g H_D \qquad (2\text{-}134)$$
$$p_{r3} = \alpha_3\ p_{r1} \qquad (2\text{-}135)$$
$$p_{r4} = \alpha_4\ p_{r1} \qquad (2\text{-}136)$$

<u>*Calculation of forces*</u>

➢ Force on perforated wall

$$Fp = [(p_{p1} + p_{p3})d/2 + (p_{p1} + p_{p4})\, R_c^*/2]\, n' \qquad (2\text{-}137)$$

where n' = 1 - n (n : Porosity of the perforated wall)

➢ Force on rear wall

$$F_r = (p_{r1} + p_{r3})d/2 + (p_{r1} + p_{r4})\, R_c^*/2 \qquad (2\text{-}138)$$

➢ Total resultant force

$$F_{tot} = \chi\,(Fp + Fr) \qquad (2\text{-}139)$$

$$\chi = 1 - 0.36\,(B/L) + 7.4\,(B/L)^2 - 62.1\,(B/L)^3 + 116.3\,(B/L)^4 \qquad (2\text{-}140)$$

The new prediction method has been used with wave characteristics and geometrical parameters of the physical models in order to compare results with those measured in the model tests (60 cases without berm and 32 with berm). This comparison is shown in Figure 2-46, where:

➢ F_{meas} is the resultant horizontal peak force derived from pressure recordings; and

➢ F_{calc} is the horizontal peak force calculated by the proposed method

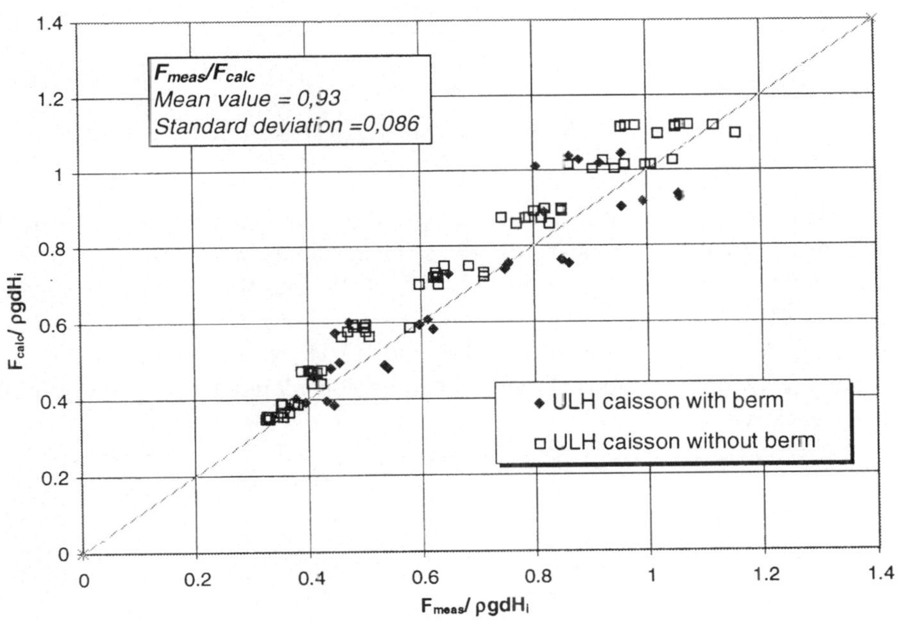

Figure 2-46. Comparison of horizontal forces calculated by the proposed method and those from Dieppe 2D model tests (Vol. IIa, section 8.1).

The ratio F_{meas}/F_{calc} for these data yields an average value of 0.93 with a standard deviation of 0.086.

In a second step, the proposed method has been applied to a larger data set from both model and prototype measurements as shown in Figure 2-47. The results of this comparison are shown in Table 2-13.

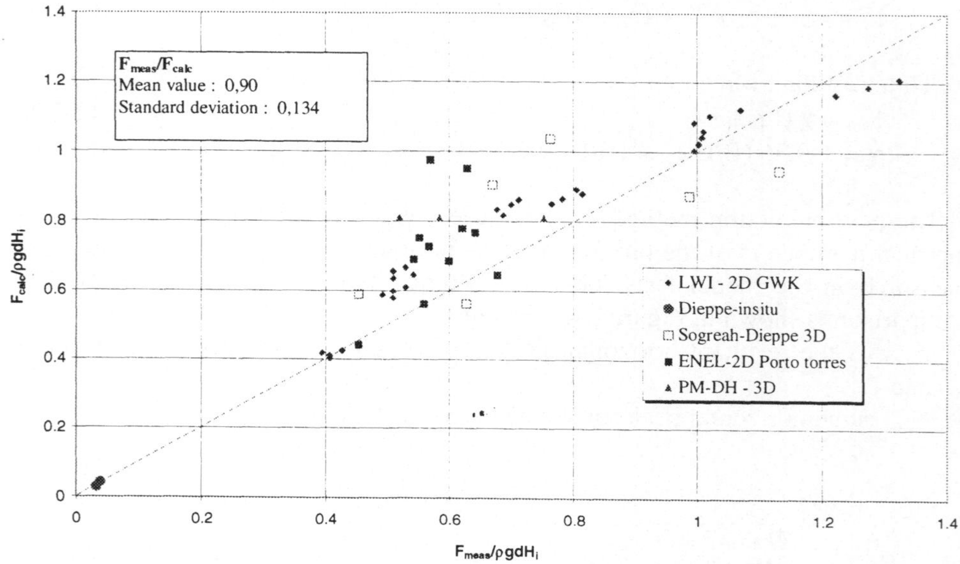

Figure 2-47. Comparison for validation of calculated forces and measured forces given by external sources.

The average values of the ratio F_{meas}/F_{calc} shows that the proposed method to determine the total horizontal force gives mean values which over-estimate the measurements from ULH model tests by about 7% and the data from other tests by about 10%.

Standard deviation are given as 0.086 and 0.134, respectively. Therefore, results derived from the proposed method can be considered as sufficiently reliable and conservative.

Table 2-13. Comparison of measured and calculated horizontal forces for perforated structures.

Data Source	**Data number**	$\mathbf{F_{meas}/F_{calc}}$	**Standard dev.**
Dieppe model data	92	0.93	0.086
Dieppe field data	2	1.0	n.a.
SOGREAH 3D Dieppe model	6	0.95	0.219
LWI large wave flume GWK tests	30	0.91	0.087
ENEL Porto Torres 2D model data	11	0.84	0.148
Other data of Porto Torres PM/DH	3	0.76	n.a.

2.8.2 *Other types of caissons*

Innovative structures are increasingly used to reduce loading on and hydraulic responses from vertical breakwaters leading to a reduction of the dimension and costs of the structure. The so-called 'Japanese-type' or armoured caisson breakwaters can be regarded as one of the better known types of structures to reduce the wave loading. This type of breakwaters consists of a protective layer in front of the vertical breakwater which is made of armour blocks. Especially in Japan, this type of breakwater is rather common so that they are very often named as 'Japanese type breakwater'. A typical cross section of this type of breakwaters is given in Figure 2-48. In the following, the results of large-scale model tests which are described in more detail by Kortenhaus & Oumeraci (2000) are summarised:

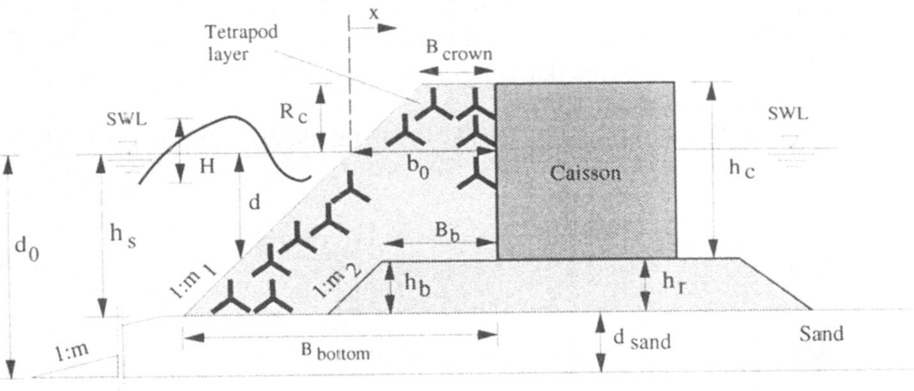

Figure 2-48. Typical section of horizontally composite breakwater (Japanese type).

2.8.2.1 *Physics of damping*

Based on the Bernoulli equation in combination with the continuity equation the decrease of pressure in the damping layer in the x-direction can be calculated by the following exponential expression:

$$p(x) = p_0 \exp\left(-C_d\, 2\pi \frac{x}{L'}\right) \tag{2-141}$$

where p_0 is the pressure at position $x = 0$ (seaward the damping layer), depending on the considered level and hydraulic conductivity of the protective layer in kPa, C_d is the damping coefficient, depending on the considered level, x is the horizontal distance from the surface of the damping layer to the considered point inside the layer in m, and L' is the wave length inside the damping layer calculated by

$$L' = \frac{1}{D_a}\frac{g T_p^2}{2\pi} \tanh\left(2\pi \frac{h_s}{L'}\right) \tag{2-142}$$

in which D_a denotes the added mass coefficient related to the damping layer.

2.8.2.2 *Analysis in time domain*

Large-scale model tests were performed in the Large Wave Flume (GWK) in Hannover, Germany. It was found from these tests that not only the dynamic horizontal forces but also the dynamic uplift forces are damped significantly. Furthermore, it was noted that the damping increased with increasing dynamic loading (Kortenhaus & Oumeraci, 2000).

Following the definition by Takahashi (1996) the damping ratio can be defined as:

$$\mu_{D,h} = \frac{F_h - F_{h,D}}{F_h} \tag{2-143}$$

for horizontal forces and in the same way for uplift forces:

$$\mu_{D,u} = \frac{F_u - F_{u,D}}{F_u} \tag{2-144}$$

where $F_{h,D}$ and $F_{u,D}$ are the damped horizontal and uplift forces, respectively; and μ_D is the damping ratio.

For **slightly breaking or quasi-standing waves** at the structure without damping layer the damping ratio is generally not higher than 50% for horizontal forces and 40% for uplift forces. It was found that the influence of the relative wave height is not very significant for the damping of the wave loads. Therefore, Equation 2-141 was adopted using the following simplifications:

- p_0 or $F_{h,0}$ is the pressure / force outside the damping layer; this will be set equal to the pressure / force at the wall in the undamped case;
- for calculation of the damping ratio in case of forces the effective length of the damping layer in height of the still water level b_0 will be used;
- the wave length in the damping layer L' can be substituted by the wave length at the toe of the structure L_{hs}.

Equation 2-141 can then be rewritten using the definition of the damping ratio for horizontal forces (Eq. 2-143):

$$\mu_{d,h} = 1 - \exp\left(-2\pi\, C_{d,h}\, \frac{b_0}{L_{hs}}\right) \tag{2-145}$$

and for uplift forces accordingly:

$$\mu_{d,u} = 1 - \exp\left(-2\pi\, C_{d,u}\, \frac{b_0}{L_{hs}}\right) \tag{2-146}$$

where b_0 is the effective length of the damping layer and can be calculated as follows:

$$b_0 = B_{crown} + h_s \frac{B_{bottom} - B_{crown}}{h_s + R_c} \tag{2-147}$$

B_{bottom} and B_{crown} are defined by Figure 2-48 and C_d is the damping ratio of the layer which is dependent on the type of material used to built the protection layer and the type of breaker at the structure. This ratio can now be used to adopt the curve to the cases observed in the GWK.

For deterministic approaches the following damping coefficients should be used to estimate the minimum damping ratio for horizontal and uplift forces under pulsating loads:

$$C_{d,h} = C_{d,u} = 0.4 \tag{2-148}$$

It should be noted that there is a difference between results of the GWK and the results described by Tanimoto due to the shorter effective length of the damping layer within the latter tests which results in smaller reductions of the forces. Generally, the measurements by Tanimoto seem to fit quite well with the new approach which accounts for variations in effective lengths of the damping layer.

For **impact waves** in case of no damping layer, the reduction is significantly higher and close to 80% to 90% for horizontal forces and 60% for uplift forces. For all waves investigated in this study no impact waves have been observed when a damping layer was installed in the tests.

It might be surprising to note that the force impulses ($\int F(t)\,dt$) will not decrease in the case with the damping layer, but will increase or remain the same in nearly all cases where impacts have been observed. This is due to the fact that rise times in the damped case (no impacts) are very much longer than in the undamped cases thus giving a high force impulse. Hence, impulses for damped wave forces must be calculated using the damped force and then multiplying the 'rise time' for non breaking waves.

The reduction of wave forces for both horizontal and uplift forces under impact loads was derived in the same way than for quasi-static waves leading to much higher damping coefficients as follows:

$$C_{d,h} = 1.5 \quad \text{for impact loads}$$
$$C_{d,u} = 1.0 \quad \text{for uplift loads} \qquad (2\text{-}149)$$

The number of waves for this analysis was relatively small so that only some estimates of the force reduction could be obtained. It should be stressed that under all conditions where impact waves were observed non impact waves do also occur. The method derived before can however be used to estimate the damping effect of a horizontally composite breakwater using Eqs. 2-145 and 2-146 and the damping coefficients for pulsating loads (Eq. 2-148) or impact loads (Eq. 2-149) depending on which loading case the structure has been designed for (in case of no damping layer).

2.8.2.3 *Statistical analysis*

Statistical distributions of forces were derived for the damped and undamped cases. A damping ratio can be derived directly comparing these distributions (Volume IIa, section 8.2). However, there should be a sufficient number of events for each distribution and impact and non impact waves should always be treated separately to deal with homogeneous data sets. Therefore, this approach will have to be applied to more extensive data sets.

Principally, the method based on the statistical distributions should be preferred as it can be applied to any distribution function for any test, i.e. it might be applied to other tests with different geometries as well. The latter is not possible when comparing the forces directly but needs to be investigated in detail for each test case. Further systematic investigations are needed to come to a final conclusion in this point.

ACKNOWLEDGEMENTS

Much of the work described here is based on studies under the PROVERBS project of the EU Mast programme under contract MAS3-CT95-0041, with some information produced by the MCS-project under contract MAS2-CT92-0047.

Substantial additional work has been completed by the PROVERBS partners under their own funding, or matching funds provided by other organisations. At HR Wallingford additional support to the research work presented here has been afforded by the UK Department of Environment, Transport & the Regions (DETR) under research contracts PECD 7/6/263, 7/6/312, and CI 39/5/96 and CI 39/5/99, from UK Ministry of Agriculture, Fisheries, and Food (MAFF) under Research Commissions FD02 and FD07, and from the UK Environment Agency (EA) under R & D project W5/006. Substantial additional support has been given by HR Wallingford and University of Sheffield.

Additional material or support was provided by:

- Leichtweiss Institute (LWI),
- University of Sheffield (UoS)
- University of Bristol (BrU)
- Delft Hydraulics (DH)
- University of Rome 3 / Politecnico di Milano (UR3 / PM)
- University of Plymouth (UoP)
- University of Le Havre (ULH)
- University of Caen (UC)
- University of Edinburgh (UE)
- Cantabria University (CU)
- Queen's University of Belfast (QuB)
- University of Naples (UoN)
- ENEL (ENEL)

Particular acknowledgement is given for assistance in assembling and editing this material by Andreas Kortenhaus (LWI), Daniela Columbo (ENEL), Kirsty McConnell (HR), Mario De Gerloni (ENEL), and Tom Bruce (UE).

REFERENCES TO VOLUME IIa

Chapter 1: ALLSOP, N.W.H. (1999): Introduction.

Chapter 2.1: ALLSOP, N.W.H.; DURAND, N. (1999): Influence of steep seabed slopes on breaking waves for structure design. 28 pp.

Chapter 2.2: MCCONNELL, K.J. (1999): Derivation, validation and use of parameter map. 5 pp.

Chapter 2.3: CALABRESE, M.; VICINANZA, D. (1999): Estimation of proportion of impacts. 15 pp.

Chapter 3.1: VOORTMAN, H.G.; HEIJN, K.M. (1999): Wave transmission over vertical breakwaters. 9 pp.

Chapter 3.2:ALLSOP, N.W.H.; BESLEY, P.; FRANCO, L. (1999): Wave overtopping discharges. 8 pp.

Chapter 3.3:ALLSOP, N.W.H. (1999): Wave reflections. 13 pp.

Chapter 4.1: VOORTMAN, H.G.; VAN GELDER, P.H.A.J.M.; VRIJLING, J.K. (1999): The Goda model for pulsating wave forces. 5 pp.

Chapter 4.2: FLOHR, H.; MCCONNELL, K.J.; ALLSOP, N.W.H. (1999): Negative or suction forces on caissons: development of improved prediction methods. 17 pp., 1 Annex.

Chapter 4.3: BURCHARTH, H.F.; LUI, Z. (1999): Force reduction of short-crested non-breaking waves on caissons. 17 pp., 3 Annexes.

Chapter 4.4: VRIJLING, J.K.; VAN GELDER, P.H.A.J.M. (1999): Uncertainty analysis of non breaking waves. 12 pp.

Chapter 4.5: VAN GENT, M.R.A.; TORENBEEK, R.V.; PETIT, H.A.H. (1999): VOF model for wave interaction with vertical breakwaters. 11 pp.

Chapter 4.6: LÖFFLER, A.; KORTENHAUS, A. (1999): Non breaking waves - pressures on berms. 23 pp.

Chapter 5.1: KORTENHAUS, A.; OUMERACI, H.; ALLSOP, N.W.H.; MCCONNELL, K.J.; VAN GELDER, P.H.A.J.M.; HEWSON, P.J. ET AL. (1999): Wave impact loads - pressures and forces. 39 pp.

Chapter 5.2: WALKDEN, M.; WOOD, D.J.; BRUCE, T.; PEREGRINE, D.H. (1999): Seaward impact forces. 25 pp.

Chapter 5.3: ALLSOP, N.W.H.; CALABRESE, M. (1999): Impact loadings on vertical walls in directional seas. 19 pp.

Chapter 5.4: VAN GELDER, P.H.A.J.M.; VRIJLING, J.K.; HEWSON, P.J. (1999): Uncertainty analysis of impact waves and scale corrections due to aeration. 12 pp.

Chapter 5.5: LÖFFLER, A.; KORTENHAUS, A.; WOOD, D.J. (1999): Wave impact loads - pressures on a berm. 22 pp.

Chapter 6.1: WALKDEN, M.; MÜLLER, G. (1999): Strongly depth limited waves. 4 pp.

Chapter 6.2: MARTíN, F.L.; LOSADA, M.A. (1999): Wave loads on crown walls. 36 pp.

Chapter 6.3: MUTTRAY, M.; OUMERACI, H. (1999): Wave loads on caisson on high mounds. 28 pp.

Chapter 7: CRAWFORD, A.R.; HEWSON, P.J. (1999): Field measurements and database. 4 pp.

Chapter 8.1: DE GERLONI, M.; COLOMBO, D.; BÉLORGEY, M.; BERGMANN, H.; FRANCO, L.; PASSONI, G.; ROUSSET, J.-M.; TABET-AOUL, E.H. (1999): Alternative low reflective structures - perforated vertical walls. 41 pp.

Chapter 8.2: KORTENHAUS, A.; OUMERACI, H. (1999): Alternative low reflective structures - other type of structures. 29 pp.

REFERENCES

Allsop, N.W.H.; McBride, M.W.; Colombo, D. 1994. The reflection performance of vertical walls and 'low reflection' alternatives: results of wave flume tests. *Proceedings 3rd Project Workshop, MAST II, MCS-Project: Monolithic (Vertical) Coastal Structures*, Emmeloord, The Netherlands, Paper 3.4, 22 pp.

Allsop, N.W.H.; McBride, M.W. 1996. Wave reflections from harbour breakwaters: development of new methods to assess vessel safety. *Proceedings International Harbour Congress*, vol. 11, Antwerp, Belgium, pp. 265-274.

Allsop, N.W.H.; Vicinanza, D. 1996. Wave impact loadings on vertical breakwaters: development of new prediction formulae. *Proceedings International Harbour Congress*, Antwerp, Belgium, vol. 11, pp. 275-284.

Allsop, N.W.H.; Calabrese, M. (1998): Impact loadings on vertical walls in directional seas. *Proceedings 26th International Conference Coastal Engineering (ICCE)*, ASCE, Copenhagen, Volume 2, pp. 2056-2068.

Allsop, N.W.H.; Calabrese, M. (1999): Forces on vertical breakwaters: effects of oblique or short-crested waves. *Report. Hydraulics Research Wallingford.*, Wallingford, U.K., SR 465.

Battjes, J.A. 1982. Effects of short-crestedness on wave loads on long structures. *Journal of Applied Ocean Research*, vol. 4, no. 3, pp. 165-172.

Bergmann H., Oumeraci H. (1999) Hydraulic Performance of Perforated Structures, *Proceedings 5th COPEDEC*, Cape Town, April 1999, pp. 1340-1349.

Besley P.B., Stewart T, & Allsop N.W.H. 1998. Overtopping of vertical structures: new methods to account for shallow water conditions. *Proceedings of Int. Conf. on Coastlines, Structures & Breakwaters '98*, March 1998, Institution of Civil Engineers / Thomas Telford, London, pp. 46-57.

Blackmore, P.A.; Hewson, P. (1984): Experiments on full scale impact pressures. *Coastal Engineering*, Amsterdam, The Netherlands: Elsevier Science Publishers B.V., vol. 8, pp. 331-346.

Bullock, G.N.; Hewson, P.J.; Crawford, A.R.; Bird, P.A.D. (1999): Field and laboratory measurement of wave loads on vertical breakwaters. *Proceedings Coastal Structures'99*, Santander, Spain: Balkema Publishers, Volume 1, pp. 613-622.

Burcharth, H.F. 1991. Introduction of partial coefficients in the design of rubble mound breakwaters. *Proceedings of the Conference on Coastal Structures and Breakwaters*, ICE, London, U.K.: Thomas Telford Ltd., pp. 543-565.

Bélorgey, M.; Rousset, J.M.; Tabet-Aoul, E.H.; Bergmann, H.; De Gerloni, M.; Colombo, D.; Franco, L.; Passoni, G. (1999): Perforated caisson breakwaters: wave loads and hydraulic performance. *Proceedings Coastal Structures'99*, Santander, Spain, Volume 2, pp. 603-612.

Canel, M. 1995. Dissipative monolithic breakwaters: some aspects of hydrodynamics. *Final Proceedings, MAST II, MCS-Project: Monolithic (Vertical) Coastal Strucures*, Alderney, U.K., Paper 4.11, 34 pp.

Cooker, M.J.; Peregrine, D.H. 1990. A model for breaking wave impact pressures *Proceedings International Conference Coastal Engineering (ICCE)*, ASCE, Delft, The Netherlands, no. 22, Volume 2, pp. 1473-1486.

Cooker, M.J. 1992. Theoretical study of wave impact. *Proceedings 2nd Project Workshop, MAST I, G6-S/Project 2: Wave impact loading of vertical structures*, Plymouth, U.K., vol. 2, Paper 9, 4 pp.

De Gerloni, M.; Colombo, D.; Pastori, S.; Boldrini, F. 1997. Wave forces on caisson breakwaters: optimization of hydraulic model test procedures. *Proceedings 1st Overall Project Workshop,*

MAST III, PROVERBS-Project: Probabilistic Design Tools for Vertical Breakwaters, Las Palmas, Gran Canaria, Annex 1F, 17 pp.

Durand, N.; Allsop, N.W.H. 1997. Effects of steep bed slopes on depth-limited wave breaking. *Waves '97, Proceedings of Conference*, ASCE, Virginia Beach, USA, Extended version of paper, 28 pp.

Franco, C.; Van der Meer, J.W.; Franco, L. 1996. Multidirectional wave loads on vertical breakwaters. *Proceedings International Conference Coastal Engineering (ICCE)*, ASCE, Orlando, Florida, USA, no. 25, Volume 2, Chapter 156, pp. 2008-2022.

Franco C.; Franco L. 1999. Overtopping formulae for caisson breakwaters with non-breaking 3-d waves. *Journal of Waterway, Port, Coastal & Ocean Engineering*, vol. 125, no. 2, March / April 1999, ASCE, New York, pp. 98-107.

Franco, L. 1994. Vertical breakwaters: the Italian experience. *Coastal Engineering, Special Issue on 'Vertical Breakwaters'*, Eds.: Oumeraci, H. et al., Amsterdam, The Netherlands: Elsevier Science Publishers B.V., vol. 22, nos. 1/2, pp. 31-55.

Fugazza, M.; Natale, L. 1992. Hydraulic design of perforated breakwaters. *Journal of Waterway, Port, Coastal and Ocean Engineering*, ASCE, vol. 118, no. 1, pp. 1-14.

Goda, Y. (1967): The fourth order approximation to the pressure of standing waves. *Coastal Engineering in Japan*, JSCE, vol. 10, pp. 1-11.

Goda. Y. 1969. Re-analysis of laboratory data on wave transmission over breakwaters. *Report of the Port and Harbour Research Institute Vol. 8.*

Goda, Y. 1975. Irregular wave deformation in the surf zone. *Coastal Engineering in Japan*, vol. 18, pp. 13-26.

Goda, Y. 1985. Random seas and design of maritime structures. Tokyo: University of Tokyo Press, 323 pp.

Hewson, P.J.; Crawford, A.R.; Walkden, M.J.A. 1998. Effect of aeration on wave impact forces. *Proceedings 2nd Overall Project Workshop, MAST III, PROVERBS-Project: Probabilistic Design Tools for Vertical Breakwaters*, Naples, Italy, Chapter 1.5a), 5 pp.

Hiroi, I. 1919. On a method of estimating the force of waves. *Journal of College of Engineering*, Tokyo Imperial University, vol. 10, no. 1, Tokyo, Japan, pp. 1-19.

Hull, P.; Müller, G.; Allsop, N.W.H. 1998. A vertical distribution of wave impact pressures for design purposes. *Research Report, MAST III, PROVERBS-Project: Probabilistic Design Tools for Vertical Breakwaters*, Belfast, Northern Ireland, 16 pp.

Ito, Y. 1971. Stability of mixed-type breakwater - a method of "probable sliding distance". *Coastal Engineering in Japan*, JSCE, vol. 14, pp. 53-61.

Kortenhaus, A.; Oumeraci, H. 1998. Classification of wave loading on monolithic coastal structures. Proceedings International Conference Coastal Engineering (ICCE), ASCE, no. 26, Volume 1, Copenhagen, Denmark, pp. 867-880.

Kortenhaus, A.; Oumeraci, H. 1999. Scale and model effects in breaking waves and impact loading of coastal structures. In: Evers, K.-U.; Grüne, J.; Van Os, A. (eds.): *Proceedings Hydralab Workshop on Experimental Research and Synergy Effects with Mathematical Models*, pp. 285-294

Kortenhaus, A.; Oumeraci, H. 2000. Damping of horizontal and uplift forces on horizontally composite breakwaters. *Proceedings International Conference Coastal Engineering (ICCE)*, ASCE, no. 27, Sydney, Australia, in print.

Losada, M.A.; Giménez-Curto, L.A. 1981. Flow characteristics on rough, permeable slopes under wave action. *Coastal Engineering*, Amsterdam, The Netherlands: Elsevier Science Publishers B.V., vol. 4, pp. 187-206.

McConnell, K.J.; Kortenhaus, A. 1997. Analysis of pressure measurements from hydraulic model tests and prototype measurements. *Proceedings 1st Overall Project Workshop, MAST III,*

PROVERBS-Project: Probabilistic Design Tools for Vertical Breakwaters, Annex C3, Las Palmas, Gran Canaria, 14pp., 1 Annex.

McConnell, K.J.M.; Allsop, N.W.H.; Flohr, H. (1999): Seaward wave loading on vertical coastal structures. *Proceedings Coastal Structures'99*, Santander, Spain: Balkema Publishers, Volume 1, pp. 447-454.

Miche, M.R. 1944. Mouvements ondulatoires de la mer en profondeur constante ou décroissante. *Annales des Ponts et Chaussées*, vol. 114. In French.

Mitsuyasu, H.; Tasai, F.; Suhara, T.; Mizuno, S.; Ohkusu, M.; Honda, T.; Rikiishi, K. 1975. Observation of the directional spectrum of ocean waves using a cloverleaf buoy. Journal of Physical Oceanography, vol. 5, pp. 750-760.

Muttray, M.; Oumeraci, H.; Shimosako, K.; Takahashi, S. 1999. Wave load on an innovative high mound composite breakwater. Results of large scale experiments and tentative design formulae. *Proceedings Coastal Structures'99*, Volume 1, Santander, Spain, Balkema Publishers, pp. 353-362.

Oumeraci, H.; Burcharth, H.F.; De Rouck, J.; Juhl, J.; Losada, M. 1995. Coastal structures – overview of MAST-2 projects. *XXVIth IAHR Congress-Hydra 2000*, Seminar 1: The EC Marine Science and Technology Programme (MAST), London, UK, 37 pp.

Oumeraci, H.; Hewson, P. 1997. Tentative recommendations for scaling wave impact loading. *Proceedings 2nd Task 1 Workshop, MAST III, PROVERBS-Project: Probabilistic Design Tools for Vertical Breakwaters*, Edinburgh, U.K., Discussion note, Annex 1.1.6, 8 pp.

Oumeraci, H.; Kortenhaus, A. 1997. Wave impact loading - tentative formulae and suggestions for the development of final formulae. *Proceedings 2nd Task 1 Workshop, MAST III, PROVERBS-Project: Probabilistic Design Tools for Vertical Breakwaters*, Edinburgh, U.K., Annex 1.0.2, 13 pp; 3 Annexes.

Owen, M.W. 1980. Design of seawalls allowing for wave overtopping. *Report. Hydraulics Research Station, Wallingford. EX*, Wallingford, U.K., EX 294.

Peregrine, D.H.; Wood, D.J. 1997. Impulsive pressures due to wave impact on a wall with a porous berm. *Proceedings 2nd Task 1 Workshop, MAST III, PROVERBS-Project: Probabilistic Design Tools for Vertical Breakwaters*, Edinburgh, U.K., Annex 1.3.1b, 7 pp.

Repko, A.; Van Gelder, P.H.A.J.M., Voortman, H.G., Vrijling, J.K. 2000. Bivariate statistical analysis of wave climates. Proceedings 27th International Conference on Coastal Engineering (ICCE), Sydney.

Sainflou, G. 1928.: Essai sur les diques maritimes verticales. *Annales des Ponts et Chaussees*, vol. 98, no. 4, pp. 5-48. In French.

Schüttrumpf, A ; Kortenhaus, A.; Oumeraci, H.; Martinelli, L. 2000. Pressure distribution at the front face and the bottom of a vertical breakwater in multidirectional seas. *Proceedings International Conference Coastal Engineering (ICCE)*, ASCE, no. 27, Sydney. Australia, in print.

SPM 1984. Shore protection manual. Chapters 6 through 8; appendices A through D, Vicksburg, Mississippi, USA, 4th edition, 800 pp.

Tabet-Aoul, E.H.; Belorgey, M. 1998. Progress report IV - new calculation method of horizontal forces on perforated caisson. *Proceedings 2nd Overall Project Workshop, MAST III, PROVERBS-Project: Probabilistic Design Tools for Vertical Breakwaters*, Naples, Italy, Class 2 Report, Chapter 1.8a), 8 pp.

Takahashi, S.; Tanimoto, K.; Shimosako, K. 1994. A proposal of impulsive pressure coefficient for the design of composite breakwaters. *Proceedings International Conference on Hydro-Technical Engineering for Port and Harbor Construction (Hydro-Port)*, Yokosuka, Japan, vol. 1, Part 1, pp. 489-504.

Takahashi, S. 1996. Design of vertical breakwaters. *Short Course International Conference Coastal Engineering (ICCE)*, Orlando, Florida, 85 pp.

Torrini, L.; McConnell, K.J.; Hugas, J. (1999): Simplified dynamic analysis of wave impacts on vertical/ composite breakwaters. *Proceedings Coastal Structures'99*, Santander, Spain: Balkema Publishers, Volume 1, pp. 433-438.

Van der Meer, J.W.; d'Angremond, K.; Juhl, J. 1994. Probabilistic calculations of wave forces on vertical structures. *Proceedings 23rd International Conference Coastal Engineering (ICCE)*, ASCE, Kobe, Japan.

Van Gent, M.R.A.; Van den Boogaard, H.F.P. 1998. Neural network modelling of forces on vertical structures. *Proceedings 26th International Conference Coastal Engineering (ICCE)*, ASCE, Copenhagen.

Vrijling, J.K.; Bruinsma, J. (1980): Hydraulic boundary conditions. Hydraulic aspects of coastal structures: developments in hydraulic engineering related to the design of the Oosterschelde storm surge barrier in the Netherlands, Part II: Design philosophy and strategy in the project related research, Delft, The Netherlands: Delft University Press, pp. 109-133

Walkden, M.J.; Wood, D.J.; Bruce, T.; Peregrine, D.H. (2000): Impulsive seaward loads induced by wave overtopping on caisson breakwaters, *Coastal Engineering*, paper accepted.

Weggel, J.R. 1972. Maximum breaker height. *Journal of Waterway, Port, Coastal and Ocean Engineering*, ASCE, vol. 98, no. WW4, pp. 529-547.

Wood, D.J. 1997. Pressure-impulse impact problems and plunging wave jet impact. *Ph.D. Thesis, University of Bristol, Faculty of Science*, Bristol, U.K., 136 pp.

CHAPTER 3

Geotechnical aspects

Maarten de Groot

3.1 INTRODUCTION

This chapter deals with the foundation of a vertical breakwater. All foundation design aspects will be discussed. Sufficient quantitative information will be presented to enable feasibility studies based on the content of this chapter. Quantitative information for preliminary and detailed design can be found in Volume IIb of this series of reports.

The input concerning hydraulic loads can be found in chapter 2 "Hydraulic aspects". Examples of application of the design methods discussed in this chapter are found in section 5.4 "Case studies".

A vertical breakwater usually consist of a wall and a foundation. The wall may be a caisson (with superstructure) or a block-structure. Which of both, is not relevant for the foundation: in all cases the wall may be considered for the foundation as a stiff monolithic structure.

The foundation usually consists of two parts: a rubble foundation and the subsoil. Sometimes the "rubble" is no more than gravel. Sometimes the "rubble foundation" is no more than a thin bedding layer between wall and subsoil. Still it will be referred to as "rubble foundation", as long as it is constructed of mainly granular material with a typical grain size of at least 10 mm. Armour layers on top of the foundation and filter layers underneath (also geotextiles), are all considered to be part of the rubble foundation and, consequently part of the foundation.

In many cases the subsoil is made up of the original seabed. In some cases part of the original soft seabed material is removed and replaced by (coarse) sand or fine gravel. In that case this sand or gravel is considered to be part of the "subsoil".

This chapter mainly deals with the geotechnical aspects. Consequently, just very limited attention will be paid to the following design aspects:

- Stability of individual armour blocks loaded by the waves;
- Quality of rubble, which is extensively discussed in CIRIA-83/CUR-154 (1991); and

➢ Filter stability.

Regarding the filter stability the following remarks should be noted: (i) in view of the strong currents in the rubble foundation it is advised to construct geometrically stable filters, i.e. filter interfaces should meet the ratio $D_{f15}/D_{b85} \leq 5$ (where "f" refers to the filter layer and "b" refers to the base layer which is protected against scour by the filter layer) and (ii) all materials should meet the requirement $D_{60}/D_{10} \leq 10$ in order to meet internal stability. More information can be found in CUR/RWS (1995).

The sections 3.2 to 3.8 of this chapter deal with the analysis of a given design. The last two sections discuss the most relevant failure modes as a function of boundary conditions and design and the possibilities for design improvement.

3.2 GUIDELINES FOR MODELLING

3.2.1 *Geotechnical failure modes*

The 4 main geotechnical failure modes are sketched in Figure 3-1:

I. Sliding along the base
II. Bearing capacity failure in rubble
III. Bearing capacity failure in subsoil
IV. Settlement by consolidation, creep or erosion of fine grained soil

Whatever phenomena occur in the foundation, resonance, fatigue, soil degradation or excess pore pressures, foundation failure can always be considered to take place according to one of these modes. Also several non-geotechnical failure phenomena, like wave induced instability of the berm, only bring about loss of the breakwater function after one of these 4 geotechnical failure modes.

Bearing capacity failure may take place in several forms (Fig. 3-2). With a shallow rupture surface underneath the harbour side edge of the wall, the wall will strongly rotate, which submode of mode II may be called "rotation failure". This submode looks similar to the often described mode of "overturning", e.g. Sekiguchi & Ohmaki (1992). "Overturning", however, will never occur without some bearing capacity failure. The rotation axis will never be at the very outer edge of the caisson bottom. It is therefore recommended to consider rotation failure instead of overturning.

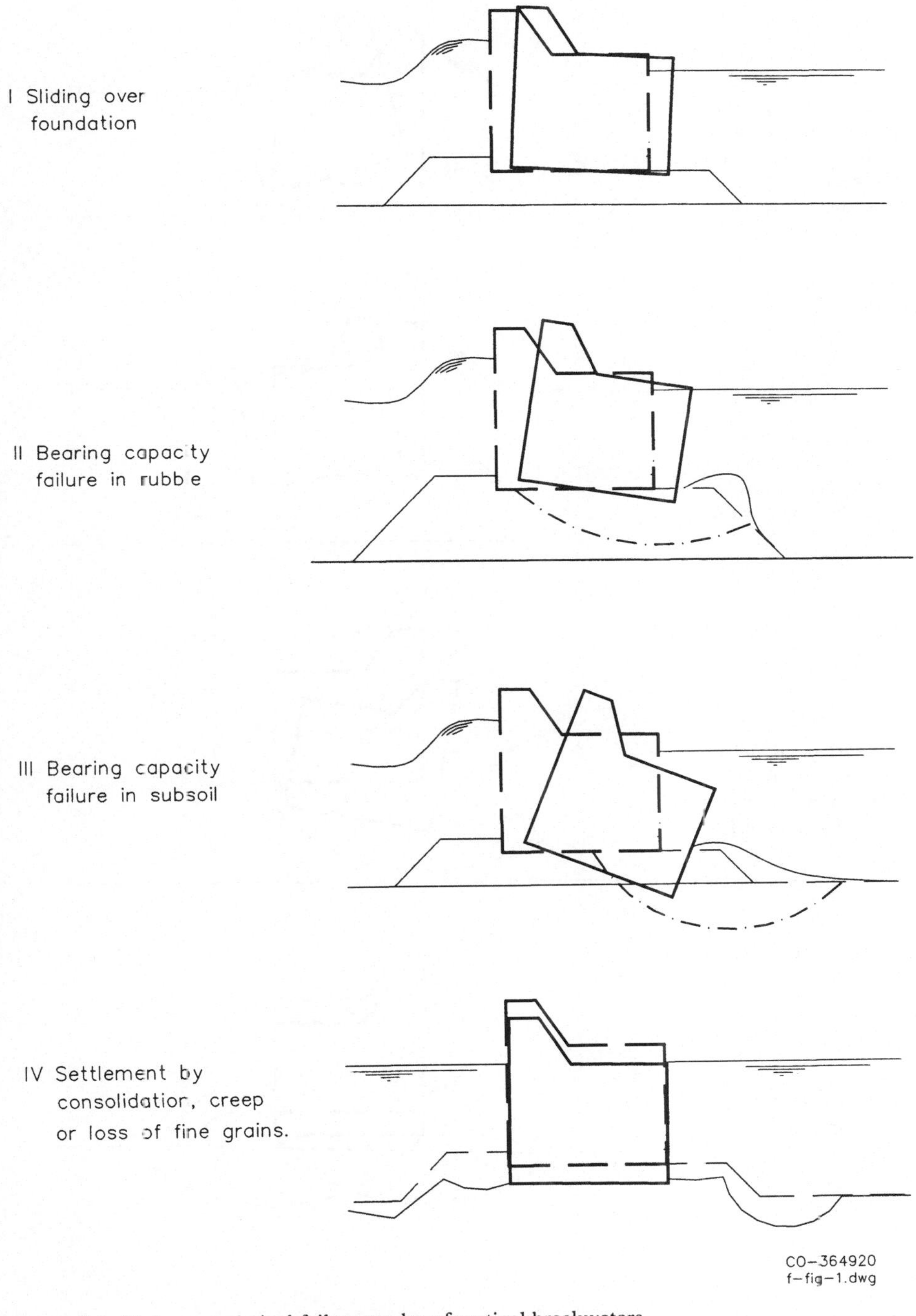

Figure 3-1. Main geotechnical failure modes of vertical breakwaters.

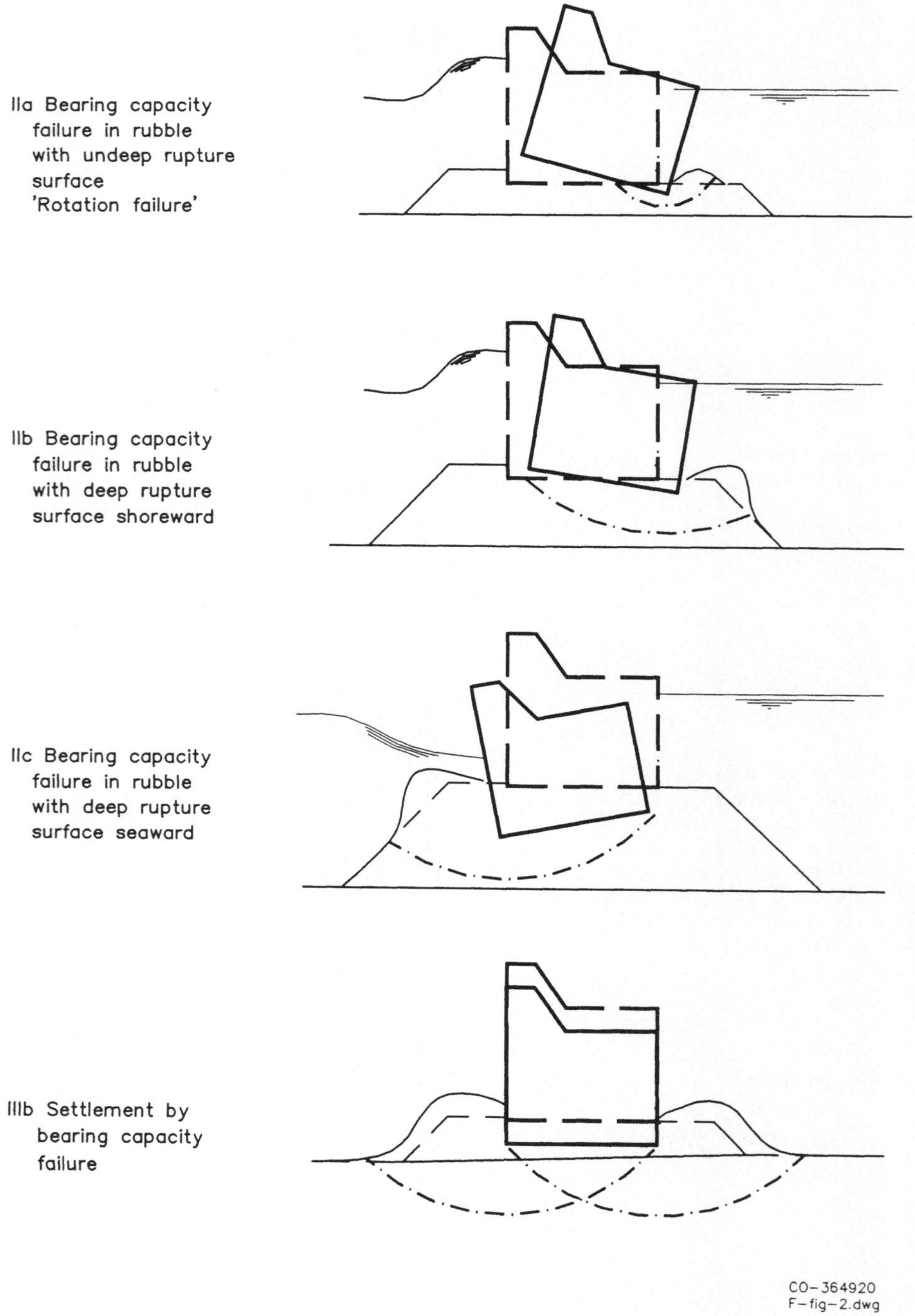

Figure 3-2. Several sub-modes of bearing capacity failure.

Settlement may be the result of bearing capacity failure with several rupture surfaces through the foundation (Fig. 3-2, submode IIIb). However, it can also be the result of weight induced compaction of soil layers or erosion of fine grained soil (failure mode IV). The last failure mode will not be discussed in this report.

The sliding distance along the base (mode I) or sliding along rupture surfaces through the foundation (mode II or III) can be so limited that the breakwater does not loose its function and failure is not yet the case. Repetition of such events due to many high wave loads and/or several storms, may yield an inadmissible sliding distance (Fig. 3-3). In several cases such "stepwise failure" might be the most dangerous. In other cases, however, just one exceptional load, not much larger than the highest previous load, may cause sliding over a large distance.

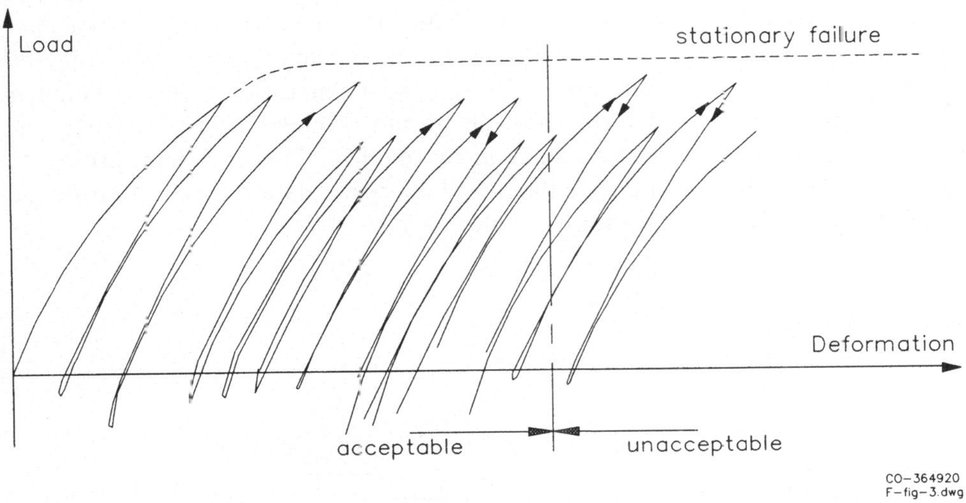

Figure 3-3. Stepwise failure due to repetitive loading.

3.2.2 *Relevant phenomena*

Many phenomena may be relevant for the foundation including its interaction with hydraulic loads and structure behaviour. Four groups of phenomena may distinguished:

- "DYNAMICS", i.e. the influence of the inertia of wall and added masses on the loads to the foundation.
- "INSTANTANEOUS PORE PRESSURES", i.e. pore pressures varying with the wave load and influenced by elastic compression of the pore fluid (air!), by elastic compression of the skeleton and by instantaneous dilation in combination with limited drainage.

- "DEGRADATION OR RESIDUAL PORE PRESSURES", i.e. the gradual change of the strength and stiffness of the soil due to repetitive loading or consolidation; the effect of repetitive loading on cohesive soil is usually expressed as loss of strength and stiffness ("degradation"); with non-cohesive soils this change is mainly expressed as "residual" pore pressure caused by gradual contraction and influenced by drainage.
- "INSTABILITY", i.e. the resulting permanent deformation, which, if large enough, yields failure according to failure mode I, II or III.

3.2.3 *Framework of analysis*

The distinction between the four groups of phenomena is partly artificial. There is also some interaction between the different phenomena. Nevertheless, the groups can be discussed more or less separately and the distinction yields a practical framework for analysis. Figure 3-4 illustrates how a quantitative analysis may be done. The circles at the left summarise the input parameters. The rectangular boxes represent the models for each of the groups of phenomena. The other circles indicate the (intermediate) output of the models. The lines with arrows in between boxes and circles indicate the most important relationships. Those should be taken into account with the most simple analysis.

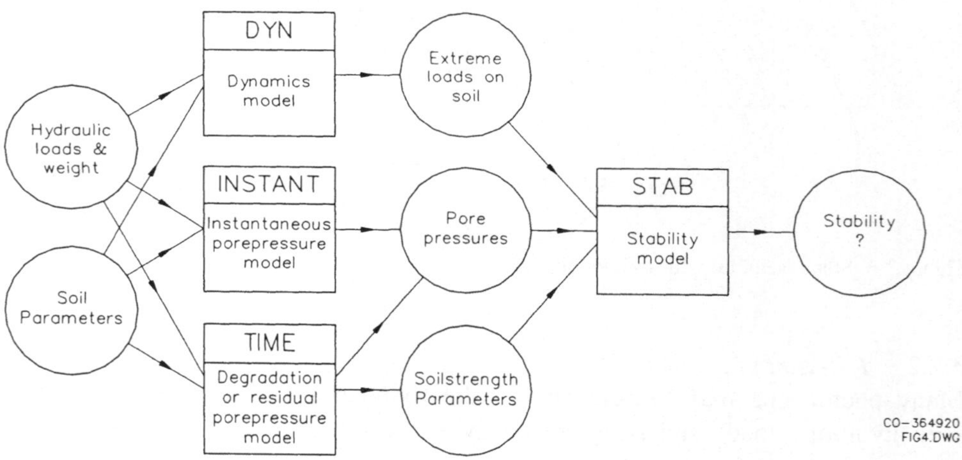

Figure 3-4. Framework of analysis.

In a more sophisticated analysis, also other relationships should be taken into account, such as the influence of the effect of dynamics on the instantaneous pore pressures or the effect of degradation on the elasticity and, consequently, the dynamics.

The input from hydraulic loads (upper left circle) can be found in chapter 2. The soil parameters (lower left circle) are discussed in section 3.3, whereas the subjects of the four rectangular boxes are subsequently discussed in sections 3.4 to 3.7. The uncertainties associated with the input parameters and with the models are discussed in section 3.8. The content of this section is an essential component of the failure mode analysis discussed in chapter 5.

3.3 SOIL INVESTIGATIONS AND SOIL PARAMETERS

3.3.1 *Strategy for soil investigations*

The collection of soil data for a new breakwater concerns information about the seabed soil where the breakwater may be built and information about quarries capable to deliver rockfill and other construction material. Last type of information is extensively discussed in CIRIA-83/CUR-54 (1991) and will not be discussed here.

The collection of sea bed soil data is usually done in phases, which correspond to the decision process. It would be a waste of money to make an extensive soil investigation, as long as no decision is made about whether a breakwater will be constructed and what its exact location will be. Nevertheless enough information should be collected in this phase to find out whether a vertical breakwater is feasible at the preferred location or the possible locations.

After such decisions are made, a much more extensive soil investigation for the preliminary design is justified. Also in this phase, however, it could be wise to start with a relatively global investigation and decide about the exact location of detailed site investigations and the number of samples at which specific laboratory tests will be done, after the preliminary design has made clear which soil properties at which locations are the most critical for the design. Even in the phase of detailed design additional investigations may appear to be justified in order to find the optimal design.

A feasibility study usually starts with a study of available documents, such as reports about previous soil investigations, experience with the construction of works in the neighbourhood and geological history. Subsequently, it is recommended to perform a seismic survey and cone penetration tests (CPT), preferably with porepressure measurements (CPTU-tests). Borings with soil sampling and classification tests may be an alternative (or a supplement) to the CPTU tests. In particular in unfamiliar subsoil conditions it is recommended to complement CPT's with borehole evidence. In strata with cohesionless materials SPT's should be carried out. The investigation should be extensive enough to establish a picture of the soil layering, the depth of the bedrock and special features, like infilled channels, presence of boulders etc. Some information on these elements of a soil investigation will be discussed below.

CPT(U)'s, sampling with classification tests and seismic (or other geophysical) surveys will also be done in the subsequent soil investigation phases, then at intermediate locations. But also other types of investigations may be done. This is discussed in more detail in Volume IIb.

3.3.2 *Seismic profiling*

The seismic profiling should cover the area that will be influenced by the structure. Continuous seismic reflection profiling with boomer equipment will most likely be the best. Depending on the soil conditions the use of a pinger (higher resolution, lower penetration) has to be considered as well.

3.3.3 *Interpretation of CPTU tests*

The results of the CPTU-tests can be used to estimate the soil layering and the soil parameters that are needed in the feasibility study.

Soil classification from the CPTU-tests can be done by means of the diagram in Figure 3-5. An example where the CPTU-tests are used to determine soil layering is given in Figure 3-6.

The relative density of sands may be estimated from Figure 3-7. The diagram is valid for pluviated, unaged, uncemented, clean (i.e. no silt content), quartz sand, and must be used with caution for other conditions.

The undrained shear strength, c_u, (also s_u) of clays can be estimated from the corrected total cone tip resistance, q_T. Experimental data are available for the expression:

$$c_u^{average} = (q_t - \sigma_{vo})/N_{KT} \tag{3-1}$$

where

$q_t = q_c + (1-a)u$	is the corrected total cone tip resistance
q_c	is the measured cone resistance
a	is the effective area ratio: a = area/total area which is used to compensate for the pore pressure acting downwards in the filter area above the cone tip. Usually $a \approx 0.6$ to 0.9
u	is the measured pore pressure
c_u^c	is the undrained triaxial compression shear strength
$c_u^{average}$	is the average of the undrained triaxial compression, DSS, and triaxial extension shear strengths, i.e. $c_u^{average} = (c_u^c + c_u^{DSS} + c_u^E)/3$

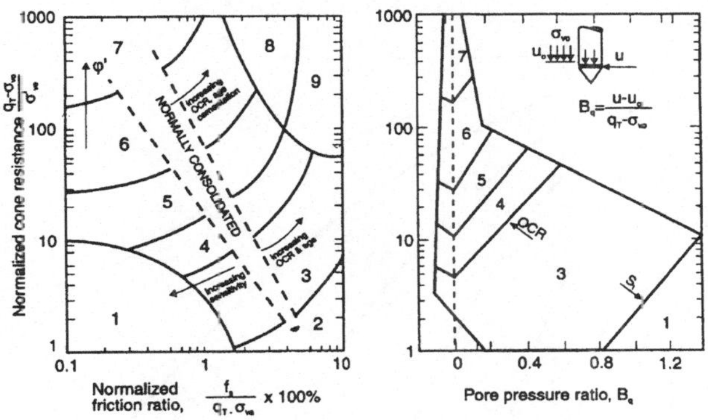

1. Sensitive fine grained
2. Organic soils - peats
3. Clays - clay to silty clay
4. Silt mixtures - clayey silt to silty clay
5. Sand mixtures - silty sand to sandy silt
6. Sands - clean sand to silty sand
7. Gravelly sand to sand
8. Very stiff sand to clayey * sand
9. Very stiff fine grained *

(*) Heavily overconsolidated or cemented

Figure 3-5. Soil type classification chart for CPTU (after Robertson, 1990).

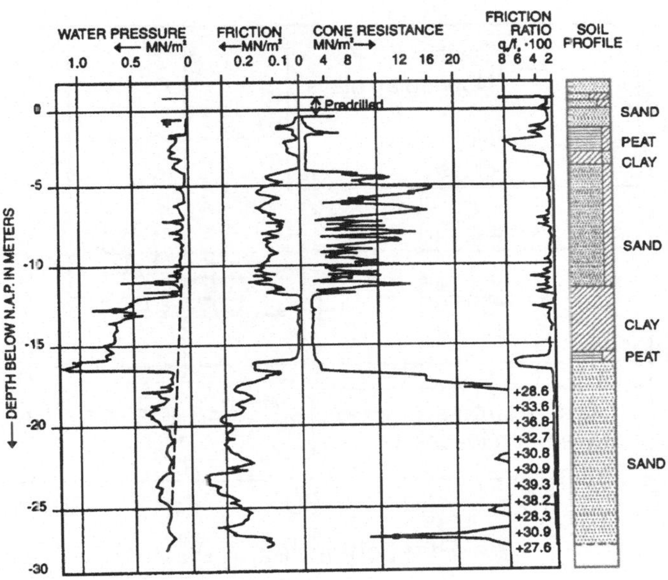

Figure 3-6. Example of use of piezecone (CPTU) to determine soil layering.

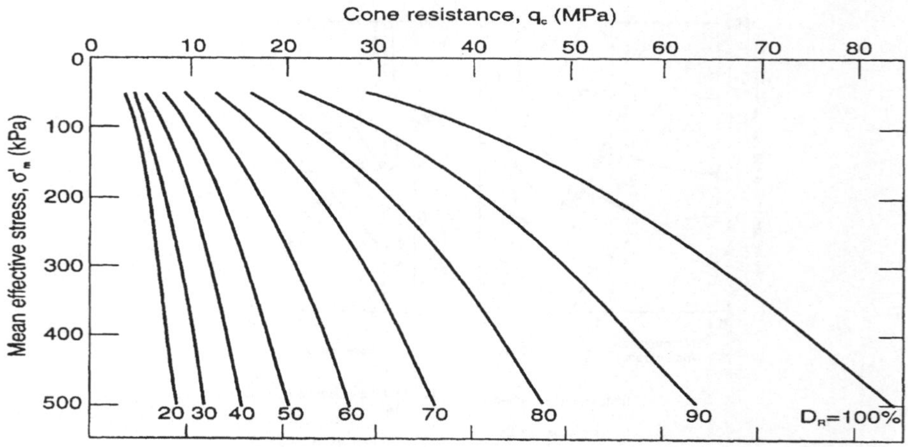

Figure 3-7. D_r - σ'_m – q_c relationship for CPT tests in sand (Baldi et al., 1986).

σ_{vo}	is the in situ vertical total stress
σ'_{vo}	is the in situ vertical effective stress
N_{KT}	is a cone factor which can be estimated from Figure 3-8.

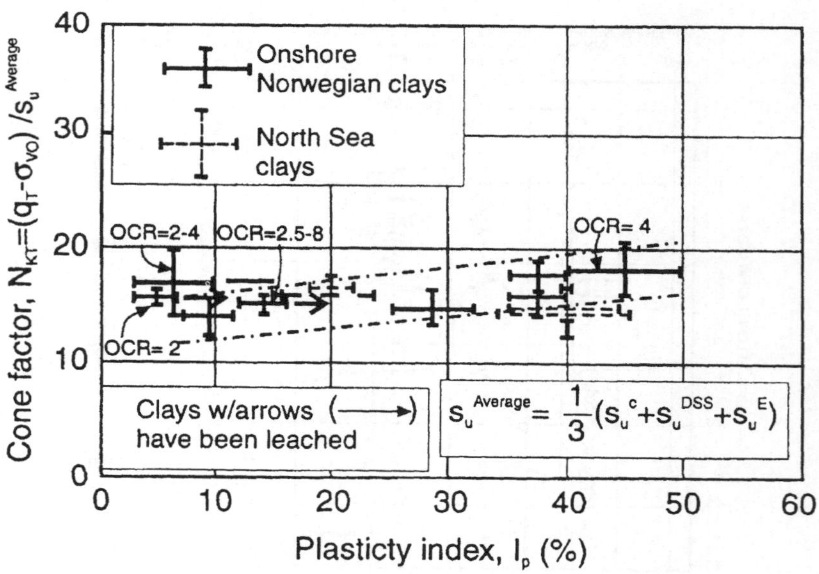

Figure 3-8. Cone factor, N_{KT}, as function of plasticity index (Aas et al., 1986).

The friction angle in unaged, non-cemented sand varies from 33° (for $q_t/\sigma'_{v0} \approx 50$) to 45° (for $q_t/\sigma'_{v0} \approx 500$).

3.3.4 *Borings, soil sampling and sample testing*

3.3.4.1 *Borings and soil sampling*

Several types of soil samplers exist, and one should generally use the type which gives the least sample disturbance. Piston samplers should be used in soft to stiff clays. When piston samplers cannot be used, thin walled push samplers should be tried. When this is not possible, as in dense sand, hammer samplers have to be used.

In some cases, like in boulder clays with stones, rock coring techniques may be necessary.

The way the sampling and the in situ testing can be carried out depends on factors like water depth, wave and wind conditions, and available equipment, as discussed by e.g. (Andresen & Lunne, 1986).

If the water depth is less than 30-40 m and the wave and wind conditions are not too severe, the operations can be carried out as on land by using an ordinary drilling rig from an anchored barge or platform.

In larger water depths or with severe water or wind conditions, 'offshore' type operations may be required. This may involve a special soil drilling vessel which is kept in place by anchoring or dynamic positioning and equipped with a heave compensating system. For offshore operations, in situ testing can be done either by penetration from the seabed or below the drill bit in a borehole using wireline technique. More details can be found in Lunne & Powell (1992).

3.3.4.2 *Soil classification from soil samples*

Classification tests should be performed to identify the soil.

The classification tests should include determination of grain size distribution and the soil density (for soils with cohesion: determination of the density of cohesionless borehole samples is highly unreliable). In addition the following tests should be done on cohesive soils: water content; Atterberger limits (plastic limit and liquid limit); fall cone, torvane or pocket penetrometer shear strengths; sensitivity. In addition the following tests should be done on non-cohesive soils: maximum, minimum and relative densities; angularity; mineralogy.

3.3.4.3 *Specific tests on soil samples*

More specific soil parameters are needed for models to be used in (preliminary) design phases than during a feasibility study only. Such parameters can be derived from further testing of the samples collected (and carefully preserved) during the feasibility study and samples collected in later soil investigations. Among those are triaxial tests, direct simple shear tests (DSS), oedometer tests and permeability tests. Another source of information is the Database, created in the framework of this project (Norwegian Geotechnical Institute, 1998).

3.3.5 *Character of soil parameters*

3.3.5.1 *Relationship between soil investigations and soil parameters*

The soil investigations discussed above are firstly used to find a general description of the soil layering: geometry of the layers and soil types. In the second place the investigations are used to determine the soil parameters needed in the analysis of any breakwater design. The soil parameters needed for a feasibility study and a preliminary design are discussed in the subsections 3.3.6 – 3.3.8.

3.3.5.2 *Soil types*

What parameters are needed and how they can be found, largely depends on the soil type. The most extensively discussed soil types are sand and clay. Sand is always supposed to be rather homogeneous: at least 80% of the mass should consist of sand particles and the typical sand behaviour is that of sand with a uniform grain size distribution (D_{60}/D_{10} < ca 2). Please note that the British Standards classifies a soil as a clay or silt when the percentage fines exceeds 35% rather than 20%. Most of the characteristics of sand parameters also apply for other non-cohesive soils: pure silt, gravel and good quality rubble.

As soon as a soil contains ca 20 % or more "lutum" (clay particles), it is considered to be a clay even if more than 50% consists of sand particles, and the soil parameters have the characteristics of a cohesive soil.

No information will be presented here about the characteristics of the soil parameters of other types of soil, like clayey sand, silty sand, cemented sand, peat, weathered or solid rock etc.

3.3.5.3 *Importance of density, stress level and stress history*

Many soil parameters are not only a function of the soil type, but also of the density of the soil, the stress level and the stress history. It has, for instance, little sense to mention the value of "the" shear modulus or "the" dilation angle of certain type of sand, if the relative density and the effective stress are not given. Another example is the undrained shear strength of a (normally consolidated) clay a few meters below the sea bed: its original value is just a fraction of the value the clay in the breakwater axis will have a year after construction.

The density of sands and other non-cohesive soils can best be expressed by:

D_r (relative density) = I_D (density index) = $(e_{max} - e)/(e_{max} - e_{min}) \approx (n_{max} - n)/(n_{max} - n_{min})$ with $n = e/(1+e)$, where e is the (actual) void ratio, n is the (actual) porosity and the indices "max" and "min" indicate the maximum and minimum values for that type of soil.

The density of clays can be expressed by the value of the soil density or by its water content in relation to its plastic limit and its liquid limit.

The stress level can be expressed by the vertical effective stress, which can often easily be found. The stress history of clays is expressed by the Over Consolidation Ratio, OCR. The properties of a clay with OCR > 1 approach those of a clay at a higher stress level than the present one.

3.3.6 *Permeability*

The permeability of rubble (and gravel) in which turbulent flow may occur, can be expressed by the values of "A" and the "B"in the Forchheimer equation:

$$i = A\, v + B\, v^2 \tag{3-2}$$

where i is the hydraulic gradient. With other soils the permeability can be expressed by the value of "k" in the Darcy equation:

$$v = k \cdot i \tag{3-3}$$

The previous equation can also be used for rubble, in which case a linearised k-value should be used. The "v" in both equations is the "seepage velocity", i.e. the discharge divided by the area of the soil.

The values of A and B can be estimated from a representative grainsize and the porosity with the following equations (Van Gent, 1993):

$$A = \alpha \cdot \frac{(1-n)^2}{n^3} \cdot \frac{\nu}{g\, D_{EQ}^{\;2}} \quad \text{and} \quad B = \beta \cdot \frac{(1-n)}{n^3} \cdot \frac{1}{g\, D_{EQ}} \tag{3-4}$$

where $\alpha \approx 1500$ and $\beta \approx 1{,}4$, at least for fairly uniformly distributed material, ν is the kinematic viscosity of the water ($\nu \approx 10^{-6}$ m^2/s), $D_{EQ} = \{6 \cdot m_{50}/(\pi \cdot \rho)\}^{1/3}$ and m_{50} is the median stone mass.

The k–value of sand or silt can also be found from a representative grain size and the porosity with k = 1/A and using van Gent's equation for A. For a continuously distributed sand or silt, D_{EQ} can best be replaced by D_{15} and $\alpha \approx 500$. A more accurate method to determine the k-value is a permeability test.

With clay such a test or an oedometer test are the most reliable methods. The permeability of clays usually varies between $k = 10^{-10}$ and 10^{-8} m/s.

3.3.7 *Stiffness*

Stiffness parameters describe the relationship between stress and deformation. Determination of stiffness parameters requires the distinction between the deformation found during virgin loading, which is largely plastic, and the deformation found during unloading or reloading which is mainly elastic.

3.3.7.1 *Virgin loading*

The stiffness of virgin loading can be expressed by the one of the following parameters for constrained deformation: "constrained modulus", M, the "oedometric modulus", E_{oed}, the "coefficient of volume compressibility", m_v, or the compression index, C_c, which are defined as follows:

$$M = E_{oed} = 1/m_v = \frac{(1+e_0)\sigma'_v}{0.435 C_c} = \frac{\partial \sigma'_v}{\partial \varepsilon_v} \tag{3-5}$$

where e_0 is the initial void ratio, σ_v' is the vertical effective stress and ε_v is the vertical strain.

Typical values are:

- Sand: $C_c = 0.003$ to 0.03
- Silt: $C_c = 0.04$ to 0.2
- Clay: $C_c = 0.04$ to 0.4

The lower values apply to soils with a high (relative) densities and consequent low compressibility; the higher to soils with low densities.

Apart from the plastic deformation described with one of these parameters also creep or "secondary" or "secular" deformation may be significant with clays.

3.3.7.2 *Unloading/reloading: elastic parameters*

In the same way the stiffness for unloading and reloading can be expressed by the "unloading/reloading modulus" M_r or the swell/reload index C_r. The value of C_r is often 15% to 30% of the value of M. The relationship with the 4 elastic parameters (shear modulus, G, Poisson ratio, Γ (or ν), compression modulus, K, and elasticity modulus, E), of which always two are needed for a complete description, is as follows:

$$G = \frac{E}{2 \cdot (1+\Gamma)} \tag{3-6}$$

$$K = \frac{E}{3 \cdot (1-2\Gamma)} \tag{3-7}$$

$$M_r = \frac{E \cdot (1-\Gamma)}{(1+\Gamma)\cdot(1-2\Gamma)} = K + \frac{4}{3}G \qquad (3\text{-}8)$$

With uniform rubble, sand and other non-cohesive materials, the values of G, K and M_r are roughly proportional to $(\sigma')^{0.5}$, where $\sigma' = (\sigma_1' + \sigma_2' + \sigma_3')/3$ is the mean effective stress (often $\sigma' \approx 0.7\ \sigma_v'$). The shear modulus of such soil at a certain relative density can be given as $G = G_{100}\cdot(\sigma')^{0.5}$ where G_{100} is the shear modulus at $\sigma' = 100$ kPa. For most of these soils G_{100} = 20 to 200 MPa, with the lower values valid for the lower relative densities and the higher for the higher densities. For the value of the Poisson ratio for non-cohesive soils, which behave drained, the following may be assumed: $\Gamma \approx 0.3$.

With normally consolidated clay G is roughly proportional to σ' and G_{100} = 5 to 50 Mpa. With overconsolidated clay G_{100} is usually higher. The shear modulus, however, does not increase as much with σ'.

It must be realised that the given shear moduli give the ratio between the shear stress and the shear strain at shear stress values well below the shear strength. As soon as the shear stress becomes higher than, say, half the shear strength, this ratio reduces significantly (e.g. to 50%). Then, however, also some plastic shear deformation occurs.

A second effect relevant for vertical breakwaters founded on clay is the reduction of the shear modulus of clay (especially overconsolidated clay) after cyclic loading with shear stresses higher than 30 to 50% of the undrained shear strength. Reduction down to 20% or 10% of the original value may occur, especially with overconsolidated clays and a large number of cycles.

3.3.8 *Strength*

3.3.8.1 *Non-cohesive soils*

The drained strength of non-cohesive materials can be expressed by the friction angle φ and the dilation angle ψ. Both depend on relative density (higher angles with higher relative density), the stress level (lower angles with higher stress level), the strength of the mineral (higher angles with higher strength), the angularity (higher angles with higher angularity) and the grainsize distribution. The friction angle is usually defined for triaxial compression. Lower angles are found with direct simple loading en triaxial extension loading.

The friction angle of rubble can be estimated by the procedure described in Barton & Kjaernsli (1981) and copied in CIRIA 83/CUR 154 (1991). The influence of stress level, mineral strength and angularity are quantified. No influence of relative density is reported, probably because it had not been varied. The friction angle usually varies between 35° and 50°. Limited information is available about the dilation angle (Vermeer 1990).

The friction angles and dilation angles of gravels, sands and silts can best be measured in (consolidated, drained) triaxial tests or DSS tests as a function of the density. The local density can be estimated from CPTU tests, which can also be used directly to estimate the friction angle (section 3.3.3).

In many parts of the world sand and often also gravel mainly consist of quartz, a mineral with large strength. The following is mainly valid for such soils. In some parts of the world, however, sand may consist almost entirely of carbonate particles, which are much more vulnerable to crushing at high stress levels.

The friction angles and dilation angles of quartz gravels, sands and silts usually vary between 30° and 48° (for uniform sands 33° and 45°). Most quartz sands have a moderate angularity. The variation in friction angle is mainly due to the relative density: with D_r = 100% the friction angle may be 10° higher than with D_r = 0%. The stress level is a 2nd factor of influence: with σ' = 100 kPa the friction angle may be 5° higher than with σ' = 500 kPa. A first estimate of the dilation angle of quartz sand can be found with: $\psi \approx \phi - 30°$ (see Vermeer 1990).

A first estimate of the undrained shear strength, $c_u^{average}$ (or $s_u^{average}$) of sands and silts can be found with the equation:

$$c_u^{average} \approx \sigma' \tan\phi \approx 0.7\,\sigma_v' \tan\phi \tag{3-9}$$

The value can, however, be higher due to dilation and consequent negative pore pressures (Volume IIb, section 6.5).

The undrained strength under cyclic loading may be expressed by a reduced c_u value (see section 3.6).

3.3.8.2 *Cohesive soils*

The drained strength (interesting for weight loads after complete consolidation) can be expressed by cohesion, c, and friction angle, ϕ. These parameters can best be measured in (time consuming consolidated, drained) triaxial tests or DSS tests on undisturbed samples. Many marine clays, if loaded purely drained, do not have a real cohesion and have friction angles of 15° to 30°. Usually no dilation is observed.

The undrained shear strength, c_u, can best be measured in (consolidated, undrained) triaxial tests or DSS tests on undisturbed samples. It is advisable to perform several tests with different consolidation stresses: the stress present just below the seabed and the stresses expected long after construction of the breakwater. The undrained shear stresses found in the last tests can be applied for stability calculations.

The present c_u value can also be estimated from CPTU test results (section 3.3.3). There are also correlations with the density (higher c_u values with higher densities). A first estimate of the values underneath the breakwater can be found by taking the highest value of the present value and of $c_u^{average} \approx 0.7\ \sigma_v'$ tan ϕ (e.g. $\approx 0.23\ \sigma_v'$ with $\phi = 18°$).

The undrained strength under cyclic loading needs the input of a reduced c_u value. The reduction is a function of the number of cycles, and the ratio between average and cyclic load. This will be discussed in section 3.6.

3.4 DYNAMICS

3.4.1 *Concept of equivalent stationary load*

The duration of wave impacts is rather short and may bring about a significant influence of the inertia of the wall, as well as the inertia of the adjacent water and soil that accelerate along with the wall: the load to the foundation differs from the load that would occur if a stationary hydraulic load with the same peak value would occur.

The effect can be expressed by the "dynamic load factor", ν_L, which is defined as follows (Oumeraci & Kortenhaus, 1994): the peak of the hydraulic load to the caisson wall, $F_{dyn,max}$, multiplied by the dynamic load factor, yields the equivalent stationary load, F_{stat}, i.e. the stationary hydraulic load that yields the same maximum load to the foundation. See equation (3-10).

$$F_{stat,equ} = \nu_L\ F_{dyn,max} \tag{3-10}$$

Instead a "dynamic response factor" or "dynamic resistance factor", ν_d, can be used, which is defined as the ratio between the maximum actual response (deformation) and the maximum static response, but has the same value or nearly the same value. This value can be estimated if the foundation is simplified to a system of linear elastic springs. Characteristic of the mass-spring system is the natural period, T_N. In its most simple form, the load/time curve is schematised to a "churchroof" pattern, of which only the triangular "tower", representing the wave impact, is taken into account (Fig. 3-9). The tower is characterised by 3 parameters: $F_{dyn,max} = F_{h,max} \cdot l_c$, maximum horizontal force per unit length times caisson length, the total impact duration, t_d, and the rise time, t_r. How this "churchroof" is derived from hydraulic model tests and/or wave data, is explained in subsection 2.5.1. The dynamic load factor of the mass-spring system subjected to such load can be expressed in a response curve which is only a function of ratios among T_N, t_d and t_r (Fig. 3-12). How the dynamic response factor can be found is briefly explained in sub-sections 3.4.2 to 3.4.4.

The mass-spring system can be approximated as a single degree of freedom system (1 DOF), with the spring only for the horizontal movement ("sway"). The rotational movement ("roll)", however, may have some importance as well. Therefore a two degree of freedom system (2DOF) for both, coupled movements is worked out here. The largest of the two natural periods of the coupled system, $T_{N,S}$, should be used for the response curve.

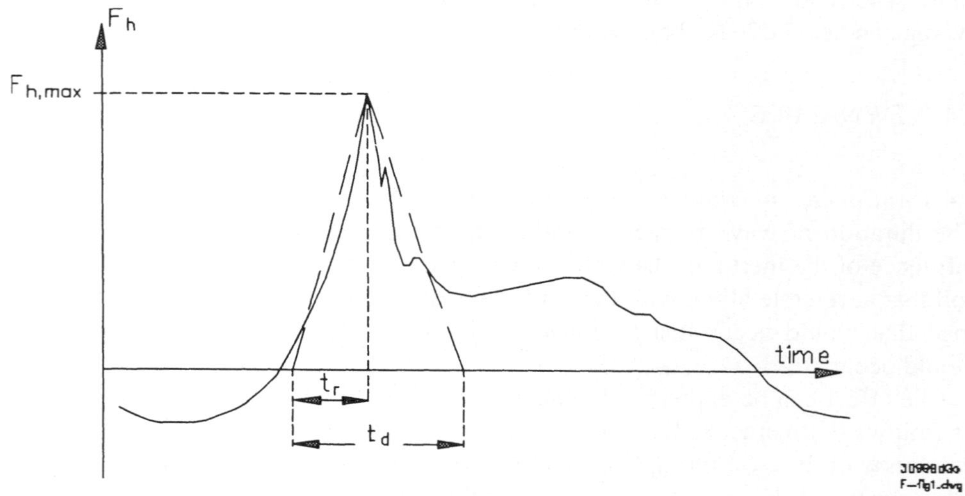

Figure 3-9. Real and simplified load to the wall. $F_{h,max} = F_{dyn,max} / L_c$

Damping can be introduced by extending the model with a dashpot (Oumeraci & Kortenhaus, 1994). The most common equations for damping are based on the theory of a structure on top of an elastic half-space and only take into account the "radiation damping", i.e the energy flux via pressure waves to half-space infinity. The influence of such damping on the dynamic load factor resulting from the single peak load is rather limited. There are indications that damping due to (pore) water flow and/or plastic deformation can be much larger (Meijers, 1994). Significant influence, however, probably only occurs if the foundation is near to failure. Then, plastic deformation dominates the process and dash-pots do not model the plastic deformation correctly, because the soil resistance does not depend on the velocity of the deformation. The situation of dominating plastic deformation will be discussed in sub-section 0. In the sub-sections 3.4.2 to 3.4.4 damping will be left out, which is a safe approach, or just a small amount of damping will be assumed.

The method summarised in this section is largely based on studies summarised in Oumeraci & Kortenhaus, 1994, and worked out in de Groot et al., 1996. The equations for determining the stiffness parameters, presented below, have been verified with hindcasts of several large-scale and full-scale tests performed in the Netherlands on behalf of the Oosterschelde works (Meijers, 1994). The complete method for prediction of the dynamic response, is supported by the following studies performed in the framework of the MAST/PROVERBS project:

- ➢ a hindcast of large scale tests performed on a vertical breakwater in the Large Wave Flume in Hannover (Hölscher et al., 1998 or de Groot et al., 1999).
- ➢ full scale experiments performed on the vertical breakwaters in Genoa Voltri and Brindisi (Volume IIb, chapter 3).

A more refined analysis requires the application of a numerical mass-spring model with the actual load time-curve, instead of the schematised one (Fig. 3-9), as input. How such an analysis can be made is explained in de Groot et al. (1996) and Oumeraci & Kortenhaus (1994).

All these analyses are valid for one isolated caisson or a few isolated caissons, because wave impacts hit the wall over a limited length, often no more than one or a few caissons, and the rest of the breakwater is not considered. A second refinement would be the introduction of the influence of the adjacent caissons in the array of caissons making up the breakwater. The full scale experiments performed in Genoa Voltri and Brindisi have shown that a 2DOF system with 2 sway modes, one for the hit caisson and one for the adjacent ones, might be more justified than a 2DOF system with one sway mode and one roll mode of the isolated hit caisson. This system is described in chapter 3 of Volume IIb. Graphs for the dynamic response factor, similar to the one of Figure 3-12, but based on this 2DOF system, are also presented in that chapter.

3.4.2 *Basic assumptions of mass-spring(-dashpot) model*

Caissons represent an essentially monolithic and rigid element, in the sense that its deformations are much smaller than that of the surrounding media (water and foundation) and its strength is normally sufficient to transfer even the strongest impact loads applied by waves to the foundation and distribute it over the length of at least one caisson. Depending on the connections between the caissons part of the wave load will be distributed over greater length. Here, however, just one caisson will be considered, not connected to its neighbours and hit as the only one by a wave impact.

Block walls will be modelled in the same way. The capacity of such walls to distribute the wave load in the longitudinal direction may be different, but the part of the wall with a length of at least the part hit by a wave impact, may be treated in the same way as one caisson.

Isolated caissons dynamics can be represented by a Mass, Spring and Dash-pot model where contributions to mass, stiffness and damping are partially due to rubble foundation, subsoil and seawater.

The caisson makes up the main part of the mass to be accelerated. Added should be at least the mass of the water that accelerates along with the caisson and which point of gravity can be considered equal to that of the caisson: "hydrodynamic mass". Often also the mass of part of the soil is added: "geodynamic mass". This is not correct if "radial damping" is applied, because this is an alternative way to model the acceleration of the soil mass. If some soil mass is included in the mass of the system, there is a (small) difference between the dynamic load factor, ν_L, and the dynamic response factor, ν_d .

As a rigid body every caisson has 6 degrees of freedom (3 translations and 3 rotations); caissons actually form long arrays and since the action of waves (horizontal and vertical forces) is intrinsically normal to the array axis, forces do not excite the longitudinal displacement and can, at least as a first approximation, be thought as applied at the mid length of each caisson, so that rotation along transversal axes (vertical and harbour directed) can be neglected. In conclusions the degrees of freedom of the caisson of interest are three, relative to the movements in the plane perpendicular to the longitudinal direction.

The idealised mass-spring-dashpot-system for calculation of oscillatory motions is sketched in Figure 3-10. P_x, P_z, and P_θ represent the resistance of the foundation. P_x is the horizontal force response to a horizontal displacement of the caisson bottom in the direction of the harbour, modelled as linear elastic spring with spring coefficient K_{11} and a dashpot with damping coefficient D_{11}. P_z is the vertical force response to a purely vertical downward displacement of the caisson. P_θ is the moment response to a rotation of the caisson bottom. Remarks: a rotation around the origin ($x = 0$, $z = 0$) calls upon a P_θ, but also a P_z; a rotation around the centre of gravity, COG, also calls upon a P_x; if $a_x \neq 0$, a vertical force through COG will yield both a vertical displacement and a rotation of the caisson bottom, thus both a P_z and a P_θ.

The equations of motion can be written as:

$$M\ddot{q} + D\dot{q} + Kq = F(t) \qquad (3\text{- }11)$$

Where q is the displacement vector (Lagragian co-ordinates; $q_1 = \Delta x$; $q_2 = \Delta\theta$; $q_3 = \Delta z$), M is the mass matrix, D the damping matrix, K the stiffness matrix and F(t) is the hydraulic load vector as a function of time.

For the dynamic modelling the origin (pole) is assumed to be at base centre, with $x = ½\ B_c$ and $z = 0$, in order to have the simplest description of the mass matrix (hydrodynamic and geodynamic terms are naturally given for such origin) and of the stiffness matrix (which results diagonal). See the left part of Figure 3-11.

It should be noted that often in literature the geodynamic added mass and stiffness matrices relative to rototranslation around a pole placed at the caisson base centre are erroneously applied to different co-ordinate systems. The same problem is enhanced for the hydrodynamic added mass terms, since the mixed term is not considered although it has the same order of magnitude of the others.

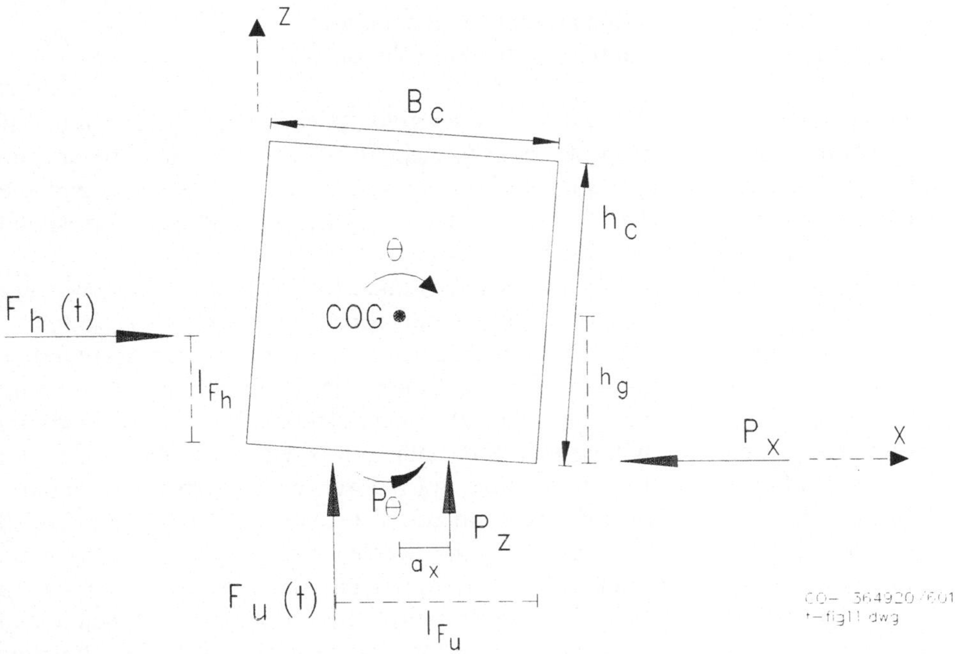

Figure 3-10. Idealised mass-spring-dashpot-system for calculation of oscillatory motions.

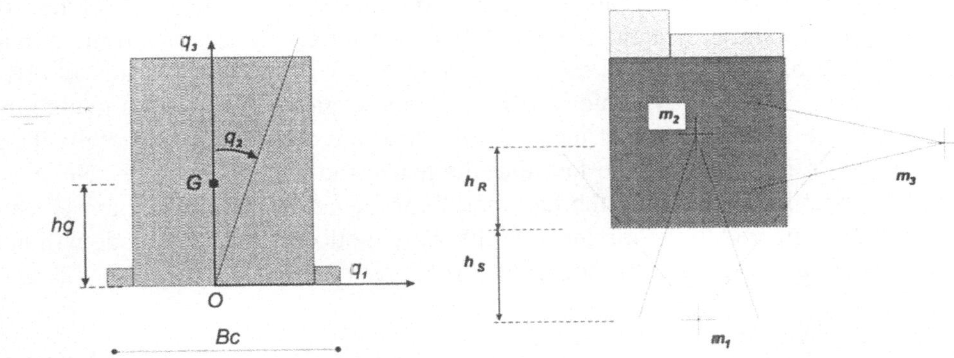

Figure 3-11. Lagrangian co-ordinates describing the isolated caisson movements (left) and natural oscillation modes (right).

The displacement can alternatively be expressed as the combination of 3 rotations round the centres m_1, m_2, and m_3 as presented in the right part of Figure 3-11. It can be shown that there is a special location for each centre, such that the 3 rotations correspond to the 3 independent oscillations:

1. the *sway mode* (m_1): an almost horizontal translation (or rotation around a low centre);
2. the *roll mode* (m_2): a rotation around a higher centre;
3. the *heave mode* (m_3): an almost vertical translation.

The equations can be solved, if values are given to each matrix element and if the hydraulic load F(t) with peak value $F_{dyn,max}$, is known. The outcome are the oscillating horizontal motion $q_1(t)$, the oscillating vertical motion $q_3(t)$ and the oscillating rotation $q_2(t)$. The maximum values of $q_1(t)$, $q_2(t)$ and $q_3(t)$ correspond to the maximum elastic load to the foundation.

The dynamic load factor, ν_L, can now be found by calculating the stationary load, F_{stat}, that yields the same maximum values for q_1. Then $F_{stat} = F_{stat,equ}$ and ν_L follows from equation (3-10). The dynamic response factor, ν_d, can be found as the ratio between the maximum of $q_1(t)$ and the static deformation q_1 that would result if $F_{stat} = F_{dyn,max}$. Given the assumed linear relationship between load and response, the same values will be found for both factors. However, the actual total maximum load transmitted to the foundation (in absence of sliding) also includes the force necessary to accelerate the foundation, i.e., in our scheme, the added geodynamic mass matrix multiplied by the acceleration vector. The maximum load transmitted to the foundation is the equivalent static force. If the inertia forces of the foundation are negligible, the ratio between maximum dynamic and equivalent static forces, ν_L, is equal to the ratio between the maximum dynamic and equivalent static displacement, ν_d.

Although the maximum values of the 3 components of the hydraulic load essentially occur almost simultaneously, this is not always the case with the maximum values of $q_1(t)$, $q_2(t)$ and $q_3(t)$. This yields a principal complication, as does the fact that 3 different dynamic response factors would be found if the maximum values of the 3 components of the hydraulic load are considered separately. This complication is limited by the fact that the heave mode has very little relevance under impact loading and by the fact that the influence of the roll motion is usually small compared to the influence of the sway motion. The heave mode will not be considered any more in the subsequent sections.

3.4.3 *Prediction of natural periods*

The two-degree of freedom system formed by the Lagrangian co-ordinates $q_1(t)$ and $q_2(t)$ is characterised by 2 natural periods for the sway and the roll modes, $T_{N,S}$ and $T_{N,R}$. They are functions of the 2x2 symmetric mass matrix M_{tot}, and the 2x2 diagonal stiffness matrix, K. The five independent components are formulated below for a schematic parallelepiped caisson of length L_c, width B_c, height h_c and gravity centre at $h_c/2$. This means (Fig. 3-10) that $a_x = 0$ and $hg = h_c/2$.

The total mass matrix M_{tot} is obtained by considering the mass of the caisson, M_{cai}, the hydrodynamic mass, M_{hyd}, and the geodynamic mass, M_{geo} (equation 3-12):

$$M_{tot} = M_{cai} + M_{hyd} + M_{geo} \qquad (3\text{-}12)$$

The matrix components along q_1 and q_2 are:

$$M_{11,cai} = \rho_c\, h_c\, B_c\, L_c \qquad \text{unit: kg} \qquad (3\text{-}13)$$

$$M_{12,cai} = \rho_c\, B_c\, L_c\, h_c^2/2 \qquad \text{unit: kgm} \qquad (3\text{-}14)$$

$$M_{22,cai} = \rho_c\, h_c\, B_c\, L_c\, (h_c^2 + B_c^2/4)/3 \qquad \text{unit: kgm}^2 \qquad (3\text{-}15)$$

According to Martinelli, 1998, $M_{11,tot}$ should be increased with ca 40% of $M_{11,cai}$ in view of the hydrodynamic added mass and with 15% in view of the geodynamic added mass in the case of the main Genoa Voltri breakwater. The percentage for the hydrodynamic mass is roughly proportional to $d^2/(h_c B_c)$, which had the value of 0.56 in Genoa. Details for their evaluation are provided in chapter 3 of Volume IIb. The values of M_{12} and M_{22} should be increased with no more than 5% of the values for the caisson alone. More accurate values can predicted with the equations presented in Oumeraci & Kortenhaus (1994) or in section 6.3.3 of de Groot et al (1996).

From last reference the following stiffness diagonal mass matrix terms can be derived:

$$K_{11} = 3\,G\,(B_c\, l_c)^{0.5} \qquad \text{unit: N/m} \qquad (3\text{-}16)$$

$$K_{22} = 0.8\,G\,B_c^2\, l_c \qquad \text{for } 0.1 < B_c/l_c < 1 \qquad \text{unit: Nm} \qquad (3\text{-}17)$$

$$K_{22} = 0.8\,G\,B_c^{2.5}\, l_c^{0.5} \qquad \text{for } 1 < B_c/l_c < 10 \qquad \text{unit: Nm} \qquad (3\text{-}18)$$

where G is the shear modulus. The equations can be used here, although they are developed for the situation of a caisson placed directly on a horizontal bed of a homogenous soil with everywhere the same linear-elastic response. With a thick rubble foundation ($h_r > 0.5\, B_c$), the average values of the (drained) rubble can be used. With a thin bedding layer, the values of the (drained or undrained) subsoil underneath the heart of the breakwater averaged over a depth of ca $0.5\, B_c$. In all cases the unloading/reloading values should be used, not the values for primary loading.

The natural frequencies are:

$$\begin{Bmatrix} \omega_S^2 \\ \omega_R^2 \end{Bmatrix} = \frac{M_{11}\,K_{22}+M_{22}\,K_{11} \mp \sqrt{(M_{11}\,K_{22}-M_{22}\,K_{11})^2+4\,M_{12}^2 K_{11} K_{22}}}{2\,(M_{11}M_{22}-M_{12}^2)} \tag{3-19}$$

where and ω_S (corresponding to the - sign; lower eigenvalue) is the frequency of the sway mode and ω_R (corresponding to the + sign) that of the roll mode.

The sway natural period (and, similarly, the roll natural period) can be found with the following equation:

$$T_{N,S} = \frac{2\,\pi}{\omega_S} \tag{3-20}$$

In a first approximation $\omega_S << \omega_R$ may be assumed and:

$$T_{N,S} \approx 2\pi\sqrt{\frac{M_{tot}}{K_{11}}} \tag{3-21}$$

The system eigen-vectors define the heights of the rotation centre h_S of the sway mode (placed below the base) and h_R of the roll mode:

$$h_S = \frac{-\,M_{12}\,K_{22}}{\omega_S^2\,(M_{11}\,M_{22}-\,M_{12}^2)\,-\,M_{22}\,K_{11}} \tag{3-22}$$

$$h_R = \frac{M_{12}\,K_{22}}{\omega_R^2\,(M_{11}\,M_{22}-\,M_{12}^2)\,-\,M_{22}\,K_{11}} \tag{3-23}$$

If the system movements are described using the natural modes as co-ordinate system, the dynamic equations are uncoupled; they are generally called "normal equations" (Flügge, 1962); assuming that the impact wave force is applied at l_{Fh} over the base, the equation for the undamped rotation (θ_S) around the low centre (h_S) or sway is:

$$\ddot{\theta}_S + \omega_S^2\theta_S = \frac{-M_{22}+M_{12}(l_{Fh}+h_R)-l_{Fh}h_R M_{11}}{(M_{11}{*}M_{22}-M_{12}^2)\,(h_R+h_S)}\,F(t) \tag{3-24}$$

If a damping factor $\exp(-\omega_S\,\zeta_S\,t)$ is applied to the impulse response of equation (3-24), the simplified dynamic of an isolated caisson with only 1 DOF (single Degree Of Freedom) is represented; for instance if the sway rotation is the only present mode, the rotation and the displacement at the base are given by $q_2(t) = \theta_S(t)$ (the sway rotation) and $q_1(t) = h_S\,\theta_S(t)$ (sway rotation multiplied by distance of the base from the sway rotation centre), respectively. A normal equation similar to equation (3-24) holds for the rolling mode but, if the point of application of the force is close to the height of the roll centre h_R, the roll mode oscillation is irrelevant and the result derived from equation (3-24) is similar to the 2 DOF mass spring dash-pot model by Oumeraci and Kortenhaus (1994).

When also the effect of the elastic force exchanged by two adjacent caissons, directly or through the foundation, is considered, the solution becomes slightly more complicated (see chapter 3 of Volume IIb).

3.4.4 *Prediction of dynamic response factor*

The dynamic response to a single degree of freedom mass-spring-dashpot system to forces of different duration is presented. Equation (3-24) can be rearranged considering an appropriate damping ζ, the homogeneous equation being:

$$\ddot{\theta}_s + 2\xi\omega\dot{\theta}_s + \omega 2\theta_s = 0 \tag{3-25}$$

A reduction of amplitude per oscillation cycle slightly less than 50% was observed to be typical of breakwater caissons; the corresponding relative damping ζ=0.1 is therefore adopted as a default.

Two extreme load cases are actually applied, an asymmetrical (t_r = 0) and a symmetrical impulse shape ($t_r = 0.5t_d$).

Figure 3-12 shows that the maximum displacement is lower than the static response for t_d/T_n< 0.3 to 0.5. The response to every impulsive force shape is similar and only moderately reduced by damping. For t_d/T_n> 2 the response to a symmetric loading is well represented by the static model: $v_d \approx 1$. For intermediate values of t_d/T_n the maximum displacement is greater than the static response (depending on system damping): "overshoot". The same is found for the combination of t_d/T_n> 0.3 to 0.5 and t_r = 0.

Two asymptotic cases are considered in more detail in the following figures, aiming to show the effect of damping and of rise time:

- a "step loading" with finite rise time
- an "impulse loading" with zero rise time, but finite impulse.

Figure 3-13 shows the maximum damped response to step forces of variable rise time, t_r. If the rise time is shorter than the natural oscillation period, a relevant overshoot takes place (it would be 2 in absence of damping), whereas for longer rise times, the response is almost static.

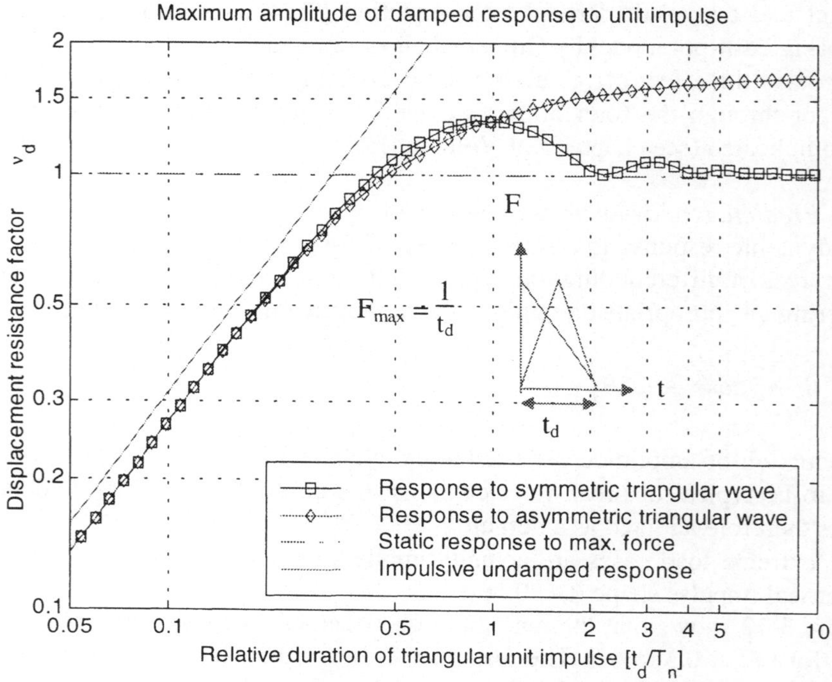

Figure 3-12. Dynamic response factor, v_d, as a function of the ratio between impact duration, t_d, and natural period, T_N, for two values of impact rise time t_r with and without some damping.

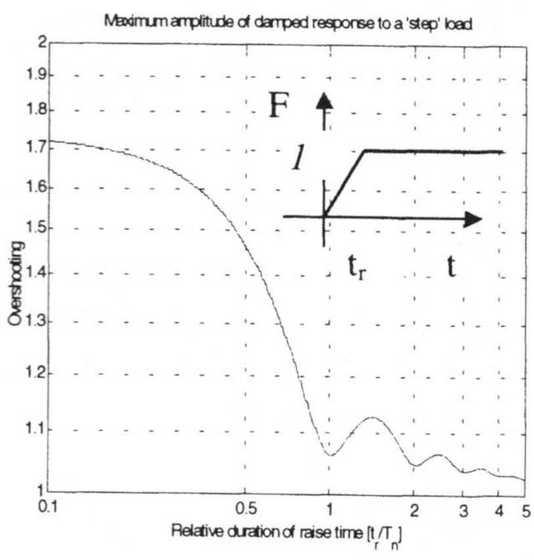

Figure 3-13. Dynamic response factor ("overshoot") as function of ratio t_r/T_N for 'step' loading with $\zeta = 0.1$

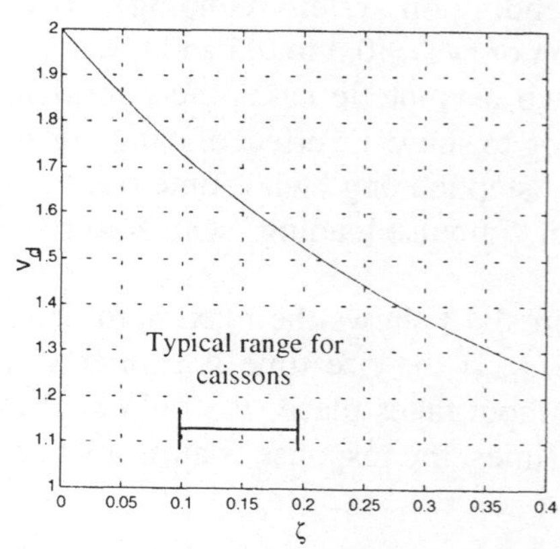

Figure 3-14. Dynamic response factor as a function of damping for 'step' loading with $t_r = 0$

Figure 3-14 shows the effect of damping for a step loading with $t_r = 0$. For the assumed standard damping ratio ($\zeta = 0.1$), the maximum displacement does not differ much from the undamped case: the dynamic response factor is 1.72 rather than 2.0 (undamped case).

Figure 3-15 shows the effect of damping for an impulse loading with $t_r = 0$. The relative reduction (ratio between damped and undamped case) of the maximum response is similar ≅14%, and remains moderate for any damping ratio reasonable for caissons.

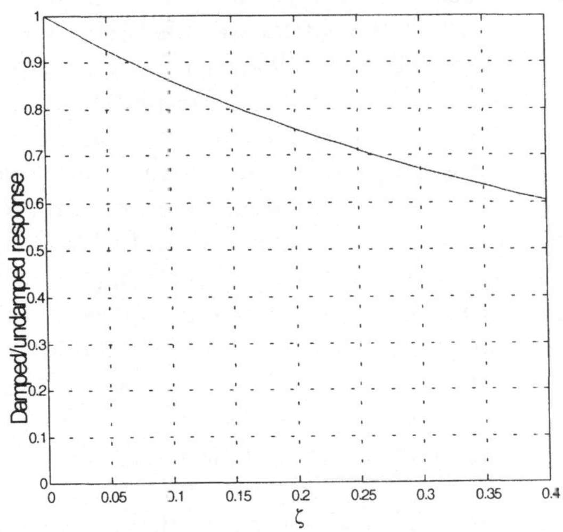

Figure 3-15. Dynamic response factor as a function damping for impulse loading with $t_r = 0$

3.4.5 *Inertia with plastic deformation*

The concept of equivalent stationary load, worked out in the subsections 3.4.1 to 3.4.4, is based on the assumption of (linear) elastic response of the foundation and no plastic deformation. If the resulting foundation load approaches the foundation strength (section 3.7), these assumptions are not realistic: the soil stiffness reduces considerably (non-linear) and simultaneously considerable plastic deformation takes place. Response factors larger than 1 may not be realistic in those cases, because the soil stiffness reduction causes an apparent increase in natural period and the plastic deformation has the same effect as considerable damping.

Plastic deformations, as long they remain small and occur only with extreme and very rare wave loads may be acceptable. Inertia may help to keep the plastic deformations small, if the extreme load exceeds the foundation strength during a very short time. This should be modelled in a different way.

A simple model for permanent displacement of caisson breakwaters under impact loads for the failure mode of "sliding over the base" is described in Appendix VIII of (De Groot et al., 1996). A similar model for bearing capacity failure through an undrained subsoil is described in Volume IIb, chapter 6, Annex B.

3.5 INSTANTANEOUS PORE PRESSURES AND UPLIFT FORCES

3.5.1 *Relevant phenomena*

The content of this section is based on extensive studies among which measurements underneath a real breakwater, large scale model tests, centrifuge tests and hindcasts with sophisticated mathematical models. See Volume IIb, chapter 4, Kortenhaus et al (1994), Hölscher et al (1998) and Zwanenburg et al (2000).

The instantaneous pore pressures, i.e. the pore pressures in rubble foundation and in the subsoil that fluctuate during each wave cycle, are caused by the wave induced fluctuations of the water pressures along seabed and seaward boundary of the rubble foundation and by the rocking motions of the wall. Several phenomena may be relevant, depending on the wave and soil characteristics, one quasi-stationary phenomenon and 4 non-stationary phenomena:

1. Spatial variation of the pore pressure head induced by spatial gradients of water pressures along seabed and boundary of the rubble foundation and by spatial variation of flow resistance. This is a quasi-stationary phenomenon.
2. Inertia of pore fluid and skeleton.
3. Elastic storage in the pores through compression and decompression of the pore water
4. Elastic storage in the pores through fluctuations of the pore volume due to isotropic compression and decompression of the skeleton
5. Elastic storage in the pores through fluctuations of the pore volume due to the elastic component of dilation or contraction by shear stress fluctuations (the plastic component of contraction is discussed in section 3.6).

The first phenomenon is relevant in most cases. The second only if the characteristic period for the propagation of a sound wave, T_{sound}, is larger than the load duration or has the same order of magnitude. One or more of the last 3 phenomena are relevant, if the dissipation of the instantaneous excess pore pressure is limited during the wave load. And that is the case if the characteristic drainage period for the phenomenon is larger than the load duration or has the same order of magnitude. The characteristic periods can be defined as follows:

- phenomenon 2: $T_{sound} = A/c_p$
- phenomenon 3: $T_{ESP} = A^2/c_{vp}$
- phenomenon 4: $T_{ESS} = A^2/c_{vs}$
- phenomenon 5: $T_{ESD} = A^2/c_{vd}$

where A [m] is the characteristic distance (e.g. drainage distance, layer thickness or depth of rupture surface or 0.5 B_c), c_p [m/s] is the sound propagation velocity and c_{vp}, c_{vs}, c_{vd} [m^2/s] are the consolidation coefficients for elastic deformation. The consolidation coefficients are functions of the stiffness parameters, K_w (compression modulus of the pore water), K (compression modulus of the skeleton) and G (shear modulus of the skeleton) and the permeability, k, and the dilation angle, ψ:

- $c_p =$ the smallest value of $c_{p1} = \sqrt{(K_w/n + K + 4/3G)/(n\rho_w + [1-n]\rho_s)}$ and $c_{p2} = \sqrt{(K_w/\rho_w)}$
- $c_{vp} = k \cdot K_w/(n \cdot \gamma_w)$
- $c_{vs} = k \cdot (K + 4/3G)/\gamma_w$
- $c_{vd} = k \cdot G/(\tan\psi \cdot \gamma_w)$.

The load duration for pulsating wave loads $\approx T/\pi$; for wave impacts = t_d. According to an alternative, but similar approach, a “characteristic length” is defined for each phenomenon, instead of a “characteristic period”. This length is found from the above definitions of characteristic period, by replacing the characteristic period by the load duration and “A” by the characteristic length. The characteristic length can be compared to the characteristic distance, A, to find the relevance of the phenomenon. The phenomenon is only relevant if its characteristic length is smaller than A or roughly equal to it. The characteristic length of the first phenomenon can be set equal to the characteristic length of the spatial variation.

3.5.2 *Quasi-stationary flow in the rubble foundation*

The characteristic period for the propagation of a sound wave and the characteristic drainage periods are usually very short in the rubble foundation (see section 3.5.4). Thus, quasi-stationary flow dominates in the rubble foundation.

The water pressures along the seaward slope of the rubble foundation fluctuate during any wave cycle, whereas the water pressures at the harbour side remain nearly constant. The corresponding fluctuating pressure gradients cause a fluctuating pore flow through the rubble foundation and simultaneously fluctuating pore pressures in the rubble foundation. According to the quasi-stationary approach, pressure gradients are assumed to be completely balanced by flow resistance in each phase of the wave cycle and no storage of water occurs in any part of the rubble foundation.

Hindcasts are made of tests on a large scale vertical breakwater model in Hannover and on measurements performed underneath the breakwater in Porto Torres. Based on these hindcasts and associated analysis, it is concluded that the (quasi-) stationary approach is justified for pulsating wave loads during wave crest and wave trough, unless the rubble foundation material is very fine. The approach can also be used in other conditions, although some corrections may be needed as discussed in next section.

In the following only the most relevant phases of the wave cycle are considered: when $F_h = F_{h,max}$ or $F_h = F_{h,q}$ or $F_h = F_{h,min}$ (see Figure 4 in Annex 1).

The prediction of the pore pressures starts with the water pressures at the seaward boundary of the rubble foundation, in particular the pressure at the lower seaward corner of the wall. Goda presents two values: p_3 and p_u. Here it is suggested to use p_u (see Figure 3-16).

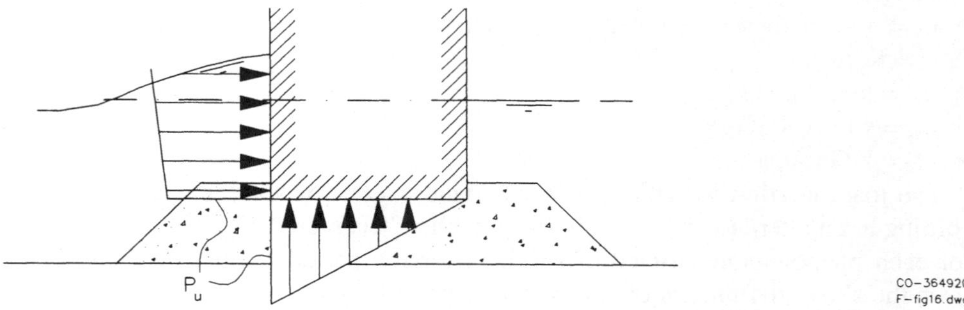

Figure 3-16. First approximation: linear distribution of excess pore pressure.

As a first approximation a triangular pressure distribution along the bottom slab (Fig. 3-16) may be assumed and hydrostatic distribution in vertical direction. Such a distribution is often found in flume tests. The following phenomena may cause deviations from this distribution:

- effects of locally varying grainsizes; if the inner top corner of the rubble foundation consists of relatively fine material and the rest of relatively coarse material, pressure distributions as indicated in Figure 3-17 may occur; such distribution was also found underneath the Porto Torres breakwater;

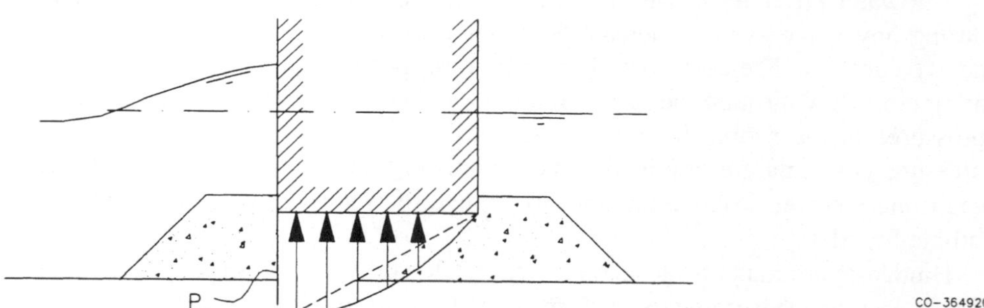

Figure 3-17. Excess pore pressures with unfavourable flow resistance distribution.

- non-flat top of the rubble foundation leaving space locally underneath the caisson bottom, which may cause similar effects;

- 2-dimensional character of the flow, as illustrated by the lines of equal pressure head and the flow lines in Figure 3-18, resulting in flow concentrations around the corners and lower horizontal gradients at lower levels;

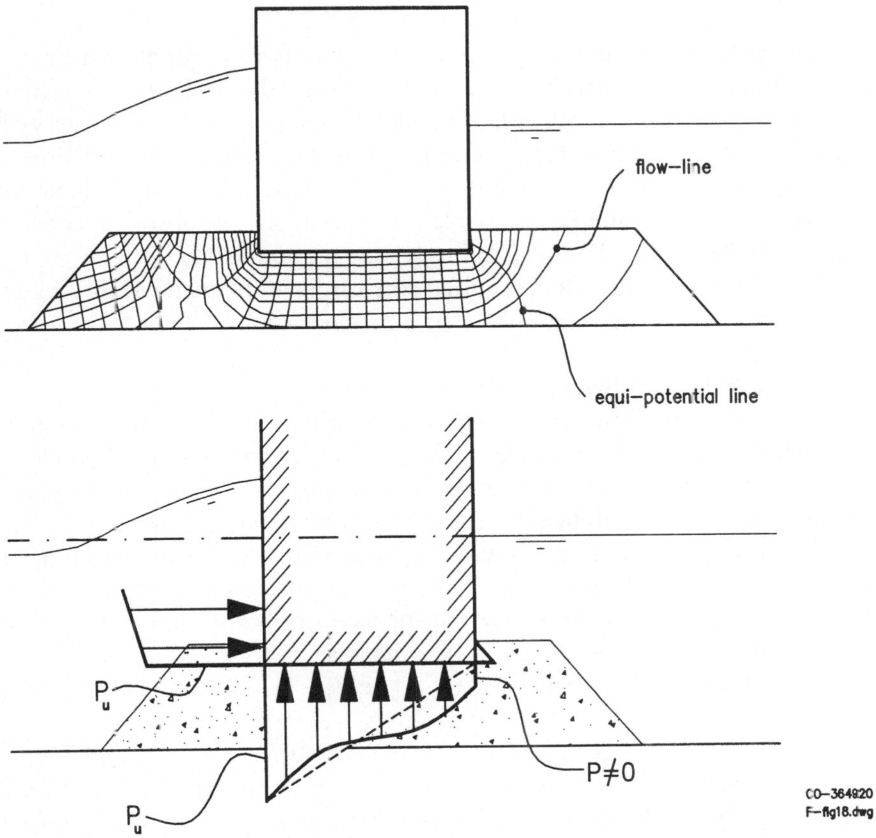

Figure 3-18. Flow concentration around corners, yielding locally additional flow resistance.

- turbulence, which may intensify the pressure head loss around the corners;
- additional pressure head loss around any apron slabs, if these are placed directly adjacent to the wall and do not have large holes in it.

An example of these effects is presented in subsection 4.1.3 of Volume IIb for the rubble foundation of the Porto Torres breakwater, where measurements have been performed.

3.5.3 *Uplift force, downward force and seepage force in rubble foundation*

The following approximation for the uplift force by the pore water in the rubble foundation on the bottom of the structure can be used for feasibility studies for rather common layout of the rubble foundation if no special measures are considered:

$$F_u = 0.5\ B_c\ l_c\ p_u \tag{3-26}$$

$$l_{Fu} = 0.67\ B_c \tag{3-27}$$

The seepage force by the pore water flowing through the rubble acting on the rubble corresponds approximately to the horizontal gradient in the rubble foundation, which can be assumed according to the triangular distribution: everywhere underneath the wall equal to p_u/B_c. The resulting horizontal seepage force, F_{hu}, can be found with equation (3-32) and (3-33) in section 3.7.1. The vertical downward force by the pore water in the rubble foundation on the subsoil, if considered undrained, may be assumed correspondingly (see section 3.7.1).

A more sophisticated approach requires corrections to be made according to the approaches presented in the sections 3.5.1, 3.5.2 and 3.5.4.

3.5.4 *Non-stationary flow in rubble foundation*

An extensive hindcast of the tests performed on a large scale vertical breakwater model in Hannover has been made with the help of the numerical model "TITAN", capable to model non-stationary, two-phase flow. With the help of this hindcast and additional analytical models, the following was found.

The effects of non-stationary flow during wave crest or wave trough are only relevant if one or more of the characteristic periods defined in subsection 3.5.1 are larger than the duration of the wave load or have the same order of magnitude. And this can only be the case during wave-impacts.

The understanding of these effects is helped by distinguishing between 2 components (Fig. 3-19):

DIRECT: pore pressures generated through pore pressure propagation from water pressure variation at the sea side, *for as far not influenced by the movement of the wall*

INDIRECT: pore pressure generation by the movements of the wall.

The first approximation of the direct component is the quasi-stationary pore pressure distribution, as discussed above. However, inertia (phenomenon 2) and elastic compressibility (phenomena 3 and 4) influence the direct component during wave impacts with short duration: the pressure wave induced at the sea side needs some time to pass through the rubble foundation.

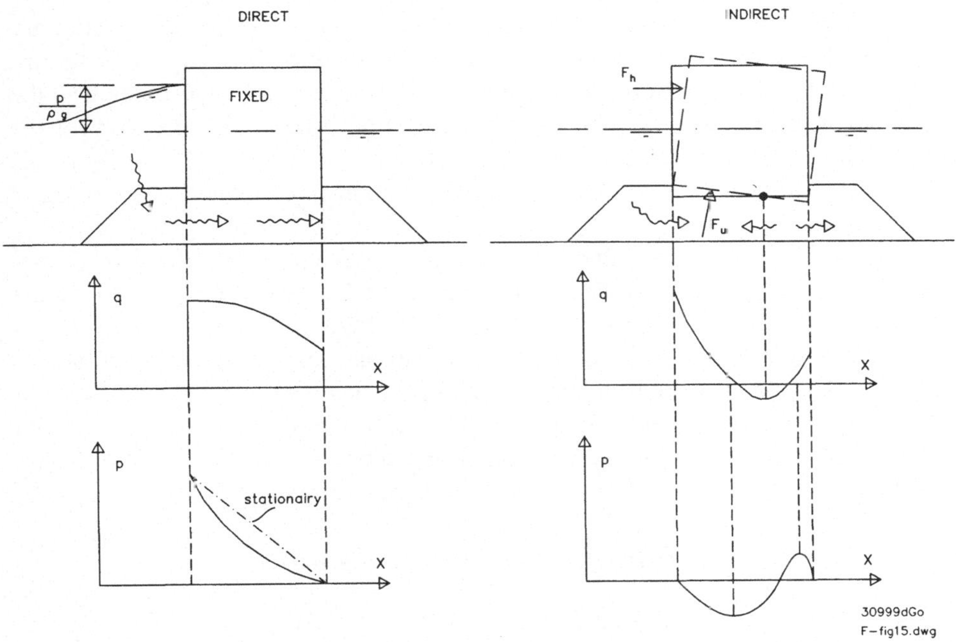

Figure 3-19. Distinction between direct and indirect component with non-stationary flow.

The propagation can be approximated with analytical equations for a sound wave through a one-phase material, if the energy loss by the flow of the pore water through the skeleton can be neglected. Then T_{sound} (subsection 3.5.1, with $A = B_c$) characterises the process. This is the case either if pore water and skeleton move together (in which case $c_p = c_{p1}$) or if the pore water moves freely through the skeleton (in which case $c_p = c_{p2}$).

Usually, however, the grains of the rubble foundation are not fine enough to assume that pore water and skeleton move together and the grains are not coarse enough to assume that the pore water moves freely through the skeleton. Therefore also elastic storage equations for two phase material can be used as an alternative approximation. Now T_{ESP} (with $A = 0.5B_c$) characterises the process. This approximation is also limited, now because the inertia is neglected.

A more refined analysis requires scale tests or a numerical model. Scale tests are faced with the problem of how to model on (Froude) scale the flow at high Reynolds numbers and how to get the correct compressibility of the pore water. Only carefully designed large scale tests may yield reliable estimates of the non-stationary effects. A numerical model must be sophisticated enough to model inertia, compressibility and two-phase flow.

Analyses and tests performed up to now show that the non-stationary effects of the direct component may enlarge the uplift force with up to 30 % compared to the value found with stationary flow. In most cases, however, the non-stationary effects yield a reduction of the uplift force. The effects are the largest if the impact has a very short duration, the width of the wall is large, the gas content in the pore water is rather large (1 % or more) and if rather fine grains are used for the rubble foundation. The effects are not significant if $t_d/T_{sound} > 10$, $t_d/T_{ESP} > 2$, $t_d/T_{ESS} > 10$ and $t_d/T_{ESD} > 10$. More details are found in chapter 4, section 4.2 of Volume IIb.

The indirect effect is the reduction of the pore pressures when the wall is suddenly lifted up and water is forced to flow into the additional room or the increase when the wall falls back. Now T_{ESS} (with $A = 0.5B_c$) characterises the process. Analytical expressions have been derived to estimate the order of magnitude of the effect. The effect is the largest if the impact has a very short duration, the width of the wall is large, the stiffness of the skeleton is limited and if rather fine grains are used for the rubble foundation. Then, more than 30 % reduction of the uplift force can be found at the moment of maximum impact load.

3.5.5 *Instantaneous pore pressures in sandy or silty subsoil*

3.5.5.1 *Relevance of drainage distance*

The flow in the subsoil usually has a quite different character from that in the rubble foundation: T_{ESS}, T_{ESP} and T_{ED} are much larger due to the lower permeability, at least for the larger values of A. Now the drainage distance, A, is the considered depth below the boundary between the subsoil and the rubble foundation or the seabed. "A" is the maximum depth of the rupture surface below this boundary (e.g. 0.1 m for very undeep rupture surfaces and 10 m for very deep rupture surfaces), if rupture surfaces are considered.

Two regions can be distinguished, depending on the considered depth (value of A): the "drained" region where equation (3-28) holds and the "undrained" region where equation (3-29) holds.

3.5.5.2 *Drained region*

This region is characterised by the equation:

$$T/\pi \text{ (pulsating) or } t_d \text{ (impacts)} >> \text{the largest of } T_{ESS}, T_{ESP} \text{ and } T_{ED} \qquad (3\text{-}28)$$

The pore pressure fluctuations for these relatively small values of A are the same as those found at the lower boundary of the rubble foundation or along the seabed. The soil strength is this region can best be expressed by the friction angle φ and the dilation angle ψ. The load by the instantaneous pore pressure must be introduced as a seepage force, F_{hu}, similar as for the rubble foundation (see equations (3-32) and (3-33)).

With dense, coarse sand the maximum of A for which this equation holds, is $A \approx O$ (3 m); for medium dense, silty sand it is $A \approx O$ (0.1 m); for silt and clay it is even smaller. Thus, the direct influence of the water pressure fluctuations above the subsoil on those in the subsoil is usually limited to a thin (top)layer of the subsoil.

3.5.5.3 *Undrained region*

At greater depths the following holds:

$$T/\pi \text{ (pulsating) or } t_d \text{ (impacts)} < \text{ the largest of } T_{ESS}, T_{ESP} \text{ and } T_{ED} \quad (3\text{-}29)$$

The pore pressure fluctuations and the effective stress fluctuations in this part of the subsoil are mainly determined by the total stress fluctuations at the boundary, rather than by the pore pressure fluctuations at this boundary only. In this region, the instantaneous flow velocities of the pore water may be neglected and even a sandy subsoil can best be approximated by considering the soil completely undrained, like clay. Although the phenomena 3, 4 and 5 (section 3.5.1) may be relevant, the approximation to a one-phase material is justified to find the total stress distribution.

The total stress fluctuations are caused by the wave passing over the seabed and the wave induced moment transferred from the wall via the rubble foundation to the subsoil. These fluctuations can be found from stationary calculations with a homogeneous elastic medium with boundary conditions as indicated in Figure 3-20. The non-stationary effect of inertia during wave impacts can be incorporated by application of the dynamic response factor. Such calculations yield two relevant results: the fluctuations of the mean total stress and the fluctuations of the shear stress. Both fluctuations are very strong with high wave attack, especially underneath both edges of the wall.

The fluctuations of the isotropic total stress are partly distributed to the skeleton and partly to the pore water, depending on the ratio of the stiffnesses. If the pore water would not contain any gas, the pore pressure fluctuations would be practically equal to the isotropic total stress fluctuations, which would be nearly equal to the vertical stress variations. However, some gas is probably present and the pore pressure fluctuations are less.

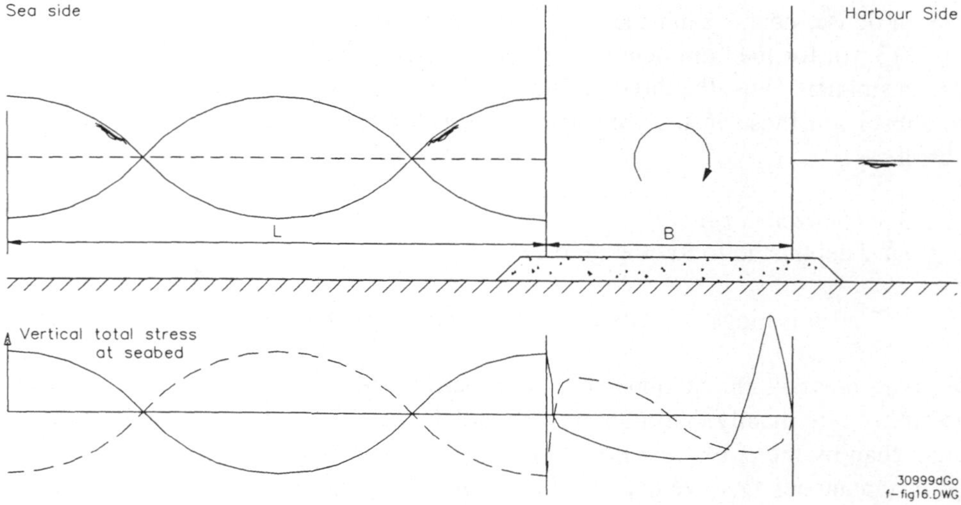

Figure 3-20. Vertical total stress fluctuating along subsoil boundary during wave cycle.

The shear stress fluctuations can be used to find the second component, (negative) excess pore pressures due to instantaneous dilation, provided the relationship between excess pore pressure and shear stress change is available from tests on undrained samples. This relationship is a function of the density, the isotropic stress and the original shear stress. These stresses vary from place to place. Methods to take this variation into account are still to be developed. Just the extreme minimum pore pressure can be calculated, which may be reached before failure if the sand is dense and the gas content of the pore water is low. Unfortunately not much is known about the gas content in nature. The assumption of completely incompressible pore water would yield too optimistic results in this case.

Centrifuge tests (van der Poel, 1997) have shown that very high pore pressure fluctuations may occur, but also that the stability is higher than expected if only the local high pore pressures are taken into account. Also the highly negative pore pressures occurring simultaneously at other locations probably contribute to the stability.

A prediction of the instantaneous pore pressures in a sandy or silty subsoil is rather complicated and associated with several uncertainties. Introduction of these pore pressures, highly varying along any potential rupture surface, is very complicated as well. Therefore it is recommended to express the strength of the sand or silt below the drained region in terms of undrained shear strength at least for feasibility studies and preliminary design. This allows not to make any prediction of the instantaneous pore pressures in the sandy or silty subsoil. The undrained shear strength can be found from undrained tests or can be predicted from the results of drained tests (subsection 3.3.8).

3.6 DEGRADATION AND RESIDUAL PORE PRESSURES

3.6.1 *Relevant phenomena in subsoil*

Cyclic loading of clay, silt or sand may cause a change (usually a reduction) of the effective shear strength and a reduction in the stiffness. This has to do with the development of residual pore pressures.

When undrained sand or silt is loaded by varying shear stress, varying excess pore pressures are observed. Part of these excess pore pressures, usually negative ones, disappear along with the shear stress. These are considered here to be the (second component of the) instantaneous pore pressures. They are caused by an elastic volume change of the skeleton. Each shear stress variation, however, may also yield a non-elastic change in the structure of the skeleton, usually a volume decrease, which brings about a residual increase in excess pore pressure after unloading. This increase adds to the increase from previous shear stress variations, if these excess pore pressures have not dissipated by drainage.

Thus, excess residual pore pressures will develop in the soil, if the duration of the cyclic load history is short compared to the drainage capacity of the soil. The instantaneous pore pressures, if positive, add to the residual ones or compensate residual pore pressures, if they are negative. The resulting excess pore pressures, if positive, cause a reduction in the effective shear stresses in the soil, and the consequence is a reduction in shear strength and shear modulus.

Clays will be undrained during a storm, and possibly also over periods including several storms.

Sands may experience partial drainage during a storm. The amount of drainage depends upon the permeability of the sand and the drainage boundary conditions. The drainage will be significant in most cases, and needs to be accounted for in design.

Experience from laboratory tests has shown that the soil structure and the resistance to further pore pressure generation may be significantly altered when the excess pore pressure due to cyclic loading dissipates (e.g. Bjerrum, 1973, Andersen et al., 1976, Smits et al., 1978). Cyclic loading with subsequent pore pressure dissipation has often been referred to as 'precycling'.

In sands, moderate precycling may lead to significant reduction in pore pressure generation under further cyclic loading, even in very dense sands. Precycling may occur during small storms prior to the design storm, and during the first part of the design storm. The beneficial effect of precycling should therefore be taken into account in cyclic testing of sand in the laboratory by applying some precycling prior to the main cyclic loading.

In clays, experience has shown that normally consolidated and slightly overconsolidated clays will benefit from precycling. Highly overconsolidated clays may, however, experience higher pore pressure generation after precycling than it did before. The beneficial effect for soft clays is normally not accounted for, because the excess pore pressure dissipates relatively slowly, and the design storm may occur relatively soon after construction. For highly overconsolidated clays, the negative precycling effect may be most unfavourable if the design storm occurs after several years with precycling from smaller storms. During this time, however, the clay will also have consolidated under the weight of the breakwater, and the effect of consolidation will in most cases compensate the negative effect of precycling.

Design diagrams for the prediction of shear strength reduction (or increase) and stiffness reduction in (preliminary) design are presented in chapter 5 of Volume IIb. They are based on calculations with finite element programs and measurements in the centrifuge tests.

3.6.2 *Sandy subsoils*

Both the tendency for pore pressure generation and the rate of dissipation of the generated excess pore pressure during the storm will depend on the sand characteristics (e.g. the density of the sand) and the drainage boundary conditions. The generation also depend on the magnitude of the design wave and the design storm composition.

For a feasibility study it may be sufficient in many cases to find out whether the following subsoil conditions are present until a depth of B_c underneath the wall bottom:

- layers of fine, loose or medium dense sand
- layers of loose or medium dense sand interrupted by silt or clay layers.

In these cases significant residual pore pressures may occur and a breakwater may only be feasible after a soil improvement. In other cases no significant pore pressures need to be expected.

This can also be quantified approximately by Equation (3-30):

$$c_u^{cy}/c_u = 1 - 0.5 \cdot \exp(-c_{degra} T/T_{ESS}) \qquad (3\text{-}30)$$

where T is the wave period, T_{ESS} is defined according to subsection 3.5.1 with $A \approx 0.5\ B_c$, $c_{degra} \approx 10$ if rather loose sand is present, $c_{degra} \approx 100$ if medium dense sand is present. There is no need to fear for a reduction of the undrained strength of the sand if only dense sand is present.

The shear modulus reduction may have the order of a factor 2 (reduction to half the original value).

3.6.3 *Clayey subsoils*

The shear strength of clay normally refers to the static (monotonic) shear strength from undrained strain controlled tests with about 1 – 2 hours to failure. In design of vertical breakwaters one should therefore correct this static shear strength for effects of undrained cyclic loading in the design storm and the high rate of loading from the maximum wave.

The cyclic effect can be approximately corrected for by the diagram in Figure 3-21. The input to the diagram is the ratio between the average wave load and the failure load calculated based on monotonic shear strength. The monotonic failure load can be calculated as described in subsection 3.7. The expression along the vertical axis, $P_{f,cy}/P_{f,s}$ can also be substituted by c_u^{cy}/c_u, were c_u^{cy} is the undrained shear strength corrected for the cyclic effects.

The correction for cyclic effect is a function of the clay type and the composition of the design storm. The diagram in Figure 3-21 is based on Drammen Clay, which is a plastic clay with a plasticity index of $I_p = 27\%$, and a typical wave load design storm composition for a vertical breakwater as given in Volume IIb, chapter 5.

The rate effect is partly accounted for in the cyclic correction above, as the diagram in Figure 3-21 assumes a cyclic load period of 10 s. If the load period deviates from 10 s, the additional rate effect can be corrected for by increasing the calculated cyclic failure load by 10% for each tenfold decrease in the load period. This correction is valid for plastic clays with an I_p of more than 20%, and for less plastic clays one should be careful about relying on this additional rate effect. It is therefore recommended to limit the correction factor from the diagram in Figure 3-21 to 1.0 for clays with I_p less than 20%.

The shear modulus reduction may have the order of a factor 2 (reduction to half the original value).

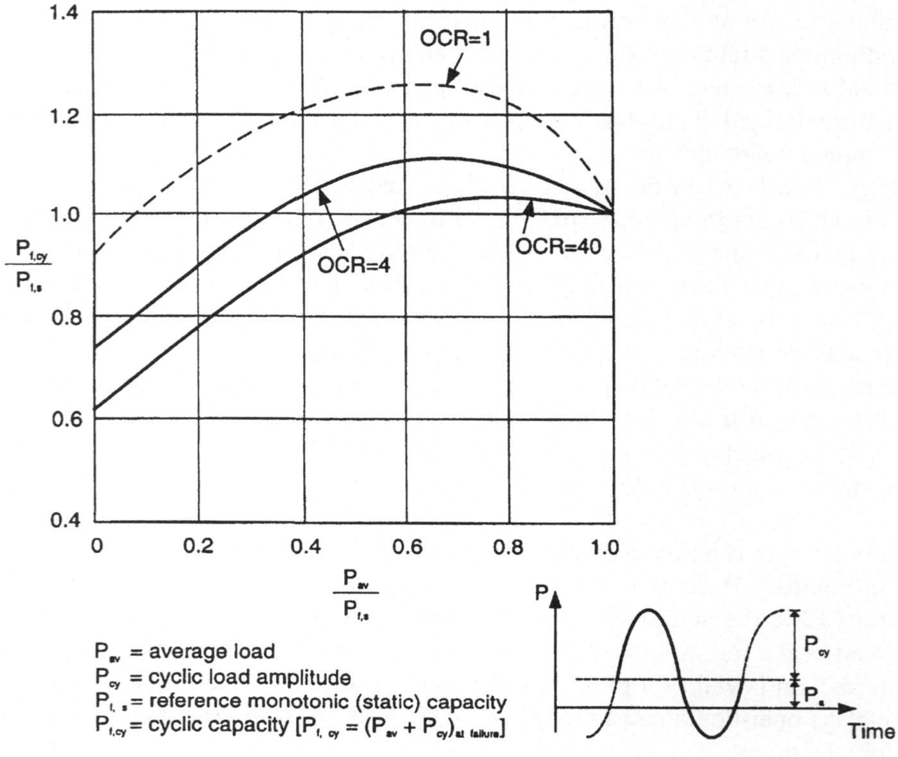

Figure 3-21. Diagram to correct monotonic (static) failure load of clay for effect of cyclic loading.

3.7 LIMIT STATE EQUATION AND OTHER CALCULATION METHODS FOR STABILITY AND DEFORMATION

3.7.1 *Schematisation of loads during wave crest*

Soil consists of a skeleton and of pores filled with water and air. For the prediction of its response to wave loads only the skeleton can be schematised or the whole soil, including the pore fluid. It is recommended to choose the first type of schematisation for the soil that is considered to behave "drained" and the second for the soil that is considered to behave "undrained". The rubble can always be considered to behave "drained", the subsoil, in most cases, to behave "undrained" (see subsection 3.5.5). Thus, if the equilibrium of the wall with part of the rubble and subsoil is considered only the skeleton of the rubble should be schematised and, in most cases, the whole of the subsoil. This has the following consequences (Fig. 3-22):

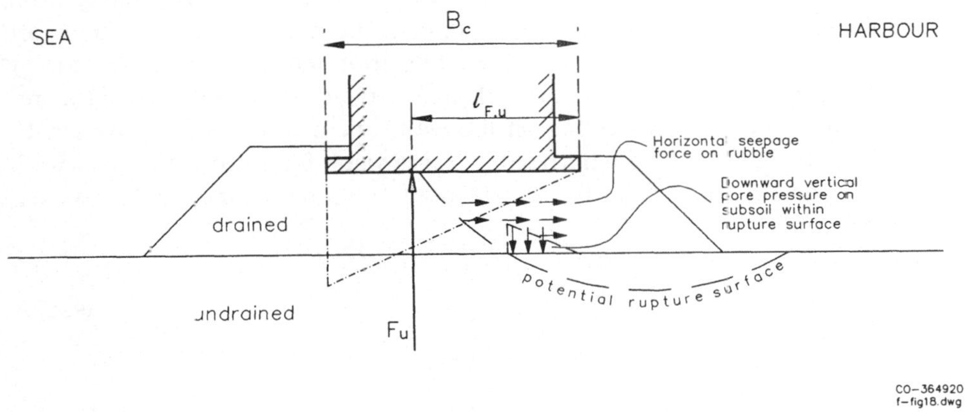

Figure 3-22. Schematisation of pore pressure loads.

- The joint equilibrium is considered of three elements: the wall, the part of the rubble skeleton above/to the harbour side of the rupture surface and the part of the subsoil within the rupture surface.
- The equilibrium follows if all the volume forces acting on these three elements are taken into account and all the surface forces acting along the boundaries of the three elements; the surface forces acting among the elements themselves may be left out as they compensate each other.
- The pore water in the rubble is considered as an external medium, of which the interaction with the rubble must be modelled as external seepage force all over the volume of which the equilibrium is considered.
- Also the surface force from this pore water acting on the bottom of the wall ("uplift force F_u") must be introduced as external force, as well as the surface force from this pore water acting on the boundary of the considered part of the subsoil, a force which is usually a fraction of the uplift force, acting, however, in the opposite direction.
- The pore pressures within the considered part of the subsoil are of no interest: no volume forces.
- The pore pressures along the outer boundaries of the considered part, however, must be modelled in one way or the other as external surface forces. If the subsoil strength is expressed as undrained shear strength, than these pore pressures are, together with the effective normal stresses, along the same boundaries implicitly modelled as total normal stresses. If the subsoil strength is expressed in terms of a friction angle (with cohesion), then these pore pressures

must be modelled explicitly, just like the effective stresses along the same boundaries.

It is often very practical to consider the equilibrium of the wall separately from the equilibrium of the soil (part of the skeleton of the rubble foundation and part of the subsoil). This means that the force acting from the wall to the skeleton of the rubble foundation – the integrated effective stresses - , must be found as resultant from the other forces acting on the wall. These forces are the weight F_g (reduced for the buoyancy), the horizontal excess water force along the front wall, F_h and the vertical force from the excess pore pressures in the rubble foundation, F_u. See Figure 3-23.

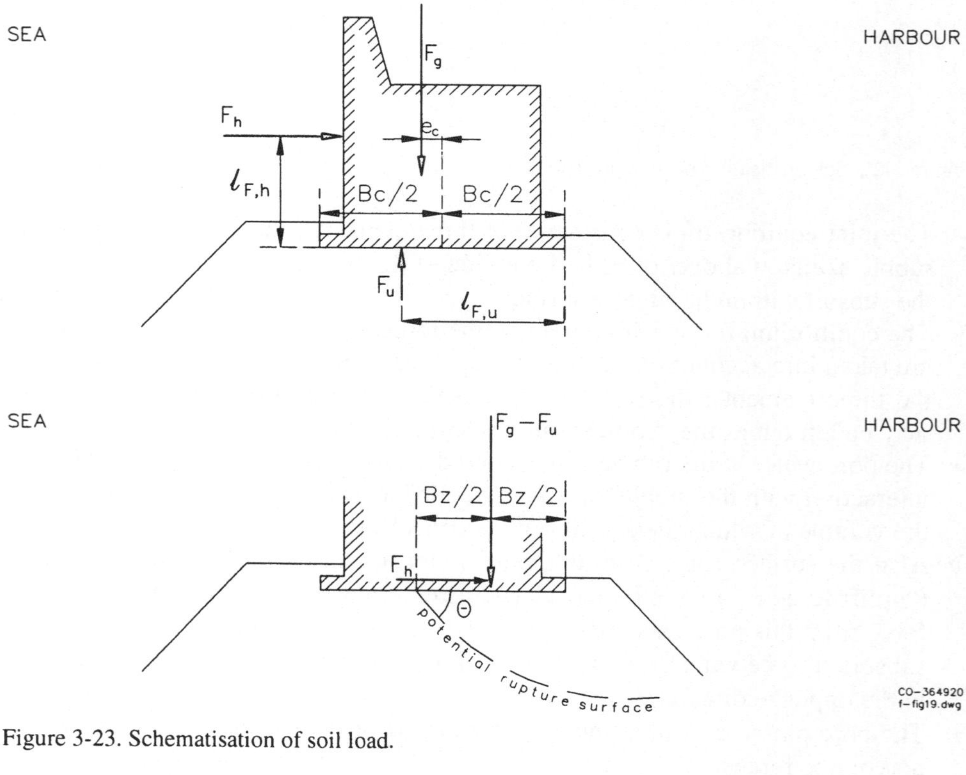

Figure 3-23. Schematisation of soil load.

Taking into account the eccentricity of the weight, e_c, (positive in sea side direction!) and the lever arms l_{Fh} and l_{Fu}, the resultant force acting on the skeleton of the rubble foundation can be expressed with three parameters: the horizontal component F_h, the vertical component (F_g – F_u) and the distance of this force component to the harbour side edge, $B_z/2$. The last parameter follows from the following equation:

$$B_z = 2\frac{F_g(\frac{1}{2}B_c + e_c) - F_h l_{Fh} - F_u l_{Fu}}{F_g - F_u} \tag{3-31}$$

The resulting horizontal seepage force in the rubble foundation, F_{hu}, can be found with the assumptions of triangular pressure distribution in horizontal direction and hydrostatic distribution in vertical direction (subsection 3.5.3). This yields:

$$F_{hu} = \frac{h_r(2B_z - h_r/\tan\theta)}{2B_c}p_u \quad \text{for } B_z \geq h_r/\tan\theta \tag{3-32}$$

$$F_{hu} = \frac{B_z^{\ 2}\tan\theta}{2B_c}p_u \quad \text{for } B_z \leq h_r/\tan\theta \tag{3-33}$$

where θ is the angle between the bottom of the wall and the rupture surface, as indicated in Figure 3-23.

3.7.2 *Limit state equations for main failure (sub)modes during wave crest*

To get a first impression of the limit state, the following rule of the thumb may be applied if the subsoil is at least as solid as medium dense sand or if the rubble foundation is very thick:

$$g_b = B_z - c_{zb}B_c \quad \text{with } c_{zb} \approx 0.4 \tag{3-34}$$

For a better analysis, however, the following five (sub)modes should be considered as the five main failure (sub)modes (see Fig. 3-24 and compare subsection 3.2.1):

a) sliding of the wall over the foundation (failure mode I)
b) rotation failure (bearing capacity failure, submode IIa)
c) rupture surface through rubble only (bearing capacity failure, submode IIb)
d) rupture surface through rubble and along top of subsoil (bearing capacity failure, submode between II and III)
e) rupture surface through rubble and subsoil (bearing capacity failure, mode III)

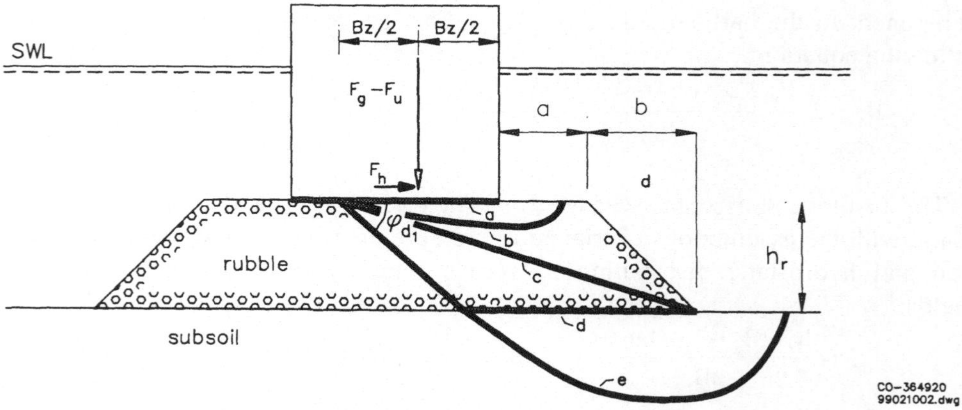

Figure 3-24. Main failure (sub)modes for which limit state functions are available.

The limit state functions, g_a, g_b, g_c, g_d and g_e, presented below are such that $g = 0$ describes the critical condition. The functions g_b and g_e are derived from Brinch Hansen (1968). Part of the background of the other functions is presented in chapter 6 of Volume IIb.

$$g_a = (F_g - F_u) \tan \mu - F_h \qquad (3\text{-}35)$$

$$g_b = B_z^2(\gamma_s - \gamma_w) \tan \varphi \cdot [\tan^2(45° + \tfrac{1}{2}\varphi) \cdot \exp(\pi \tan \varphi) - 1] - (F_g - F_u)\left(\frac{1}{1 - F_h/(F_g - F_u)}\right)^3 \qquad (3\text{-}36)$$

$$g_c = \{F_g - F_u + (\gamma_s - \gamma_w)\, \tfrac{1}{2}\,(B_z + a)\cdot h_r\} - (F_h + F_{hu})\cdot \text{cotan}(\varphi_r - \theta) \qquad (3\text{-}37)$$

where $\theta = \arctan\{h_r/(B_z + a + b)\}$

$$g_d = (B_z + a + b - h_r/\tan\theta)\cdot c_u - (F_h + F_{hu}) \qquad (3\text{-}38)$$

where $\theta = \varphi_r$ ("φ_{d1}" in Fig. 3-24) for undrained subsoil and

$$g_d = \{F_g - F_u + (\gamma_s - \gamma_w)(B_z + a + \tfrac{1}{2}b)\cdot h_r\}\cdot \tan\varphi_s - (F_h + F_{hu}) \qquad (3\text{-}39)$$

for drained subsoil

$$g_e = (B_z + h_r)(2 + \pi)\cdot c_u \cdot \tfrac{1}{2}\left(1 + \sqrt{1 - \frac{F_h + F_{hu}}{(B_z + h_r)c_u}}\right) - \{F_g - F_u + (\gamma_s - \gamma_w)(B_z + \tfrac{1}{2}h_r)h_r\} \qquad (3\text{-}40)$$

where:

μ - the friction angle between caisson bottom and rubble
φ_r - angle of internal friction of rubble
φ_s - angle of internal friction of subsoil material
c_u - undrained shear strength of subsoil material

3.7.3 *Seaward failure during wave trough*

For seaward failure, the same equations and approximations can be used as with harbourside failure.

3.7.4 *More sophisticated methods*

3.7.4.1 *More sophisticated limit state equations*

The above limit state equations have many limitations, among which the limited number of rupture surfaces considered, the prescribed shape of these surfaces and the fact that dilation is not explicitly taken into account. More sophisticated equations, based on the upper bound theory, are presented in Volume IIb, chapter 6. More rupture surface shapes are considered and optimisation procedures are included to find the most unfavourable rupture surface. The influence of dilation is explicitly taken into account.

Nevertheless, the following limitations remain, such as:

- pore pressures and seepage forces are strongly schematised
- subsoil is schematised as homogeneous medium
- variations of undrained shear strength in the subsoil due to variation in stationary stress cannot be modelled.

Numerical models can be used as alternative.

3.7.4.2 *Sliding circle analysis according to Bishop*

One of the most common types of numerical models are those for slip circle analysis according to Bishop. They can be applied to vertical breakwaters, provided the caisson or the load from the caisson is schematised such that sliding circles can pass anywhere through the caisson/rubble interface. The advantages are

- complicated layering can be represented
- complicated pore pressure distributions can be modelled, provided the numerical scheme allows for modelling of separate pressure head functions in each layer
- commonly available, reliable scheme, with which many calculations can be performed in a short time.

Disadvantages, compared to the sophisticated limit state equations:

- only circular rupture surfaces

- the positive effect of dilation cannot be modelled directly.

The reliability of the outcome of sliding circle calculations is probable not greater than that of the sophisticated limit state equations, especially not if the subsoil is rather homogeneous. Comparison of both, however, gives a feeling of the influence of the different assumptions.

3.7.4.3 *Finite element models*

A large spatial variation in soil properties can be modelled with finite elements. Some finite element codes also allow for the modelling of the pore flow or introduction of the pore pressure distribution calculated with a special finite element code for (stationary) pore flow.

The constitutive model of the soil can be made very sophisticated. For the stability analysis of vertical breakwaters, however, a simple constitutive approach, in which the plasticity is modelled by an undrained shear strength or by the combination of cohesion, friction angle and dilation angle, is usually sufficient.

The reliability of the bearing capacity predicted by an finite element model also depends on the type of elements and the size of the elements. The size should at least be small enough to model relevant soil property variations and to model deformations along relative narrow shear bands (rupture surfaces with zero thickness can not be modelled).

3.7.4.4 *Centrifuge model tests*

Physical models in a centrifuge can be a powerful tool to determine the strength of the foundation. All effects, like stress-dependency, contraction, dilation, cyclic mobility, excess pore pressures and drainage can be modelled together. The centrifuge tests performed on a model breakwater on medium dense sand (van der Poel 1997) demonstrated a much larger strength than found from calculations (Zwanenburg and de Groot 1998). Unacceptable deformation was reached as the sum of many minor deformations occurring during several hours of irregular wave loading.

3.7.4.5 *Analysis of unacceptable deformation after several load cycles*

Instability is usually considered to occur if unacceptable deformation of the foundation takes place during one extreme (wave) load. Unacceptable deformation, however, may also be the result of several subsequent loads (Fig. 3-3), as shown during the above mentioned centrifuge tests. This may occur with a vertical breakwater if the subsoil behaves undrained during a wave train of extreme high waves. Adequate description of the "strength" of the soil in such a situation requires more than the single c_u value: the "strength" must be described by the combination of 2 parameters: τ_a and τ_{cy}, where each combination corresponds to a critical combination of average and cyclic shear strain (γ_a and γ_{cy}). It is not immediately clear which combination should be entered in a stability analysis, because the combination that yields the same extreme shear deformation may vary along the rupture surface. How this can be solved is indicated in Volume IIb, section 6.4.

3.7.5 *Three-dimensional rupture surfaces*

If bearing capacity failure occurs, it occurs over a limited length. This may be due to the limitation of the length over which the extreme load occurs (see section 2.5.3) or due to the limitation of the region where the soil is relatively weak. This means that a rupture surface always has a 3-dimensional shape, with the exception of the "rupture surface" of sliding along the base. Nearly all analyses, however, are based on the assumption of 2-dimensional rupture surface. This assumption is always conservative. An idea about the conservatism can be derived from the following observations.

Two factors make that a 3-dimensional shape ("shell shape") causes an increase in foundation strength, compared to the 2-dimensional approach:

- the increase in contact area of the rupture surface
- dilatancy, the strengthening effect which increases with the curvature of the rupture surface.

A very rough estimate of the order of magnitude of the favourable effect can be made, if the depth and the length of the rupture surface can be estimated. If the depth is 10% of the length, the 3-D contact area of the rupture surface becomes ca. 3% larger than the 2-D area. Then, the corresponding increase in strength is at least 3% due to the first factor. It is reasonable to assume another 2% increase in strength due to the second factor, which makes the total increase in strength ca. 5%. If the depth of the rupture surface is 30% of the length, the 3-D contact area becomes ca. 20% larger than the 2-D area and an increase in strength in the order 40% may be expected (20% by each of both factors).

With a caisson breakwater, the effects of a 3-dimensional rupture surface are specially relevant for relatively short caissons which are not interlocked among each other. Indeed, the length of a rupture surface has a minimum of ca. half a caisson length in case of non-connected caissons and a much larger minimum in case of a good interlock between the caissons. Thus, an increase in caisson length or an improved interlock reduce the favourable effects of the 3-dimensional rupture surface. However, these effects are usually more than compensated by the favourable effect of the load distribution: the average wave load is smaller (especially with wave impacts) and the average soil strength is larger, as weak parts are overbridged.

3.8 UNCERTAINTIES

3.8.1 *Survey of uncertainties*

The analysis of the stability of certain breakwater design first requires the selection of the models and input parameters, e.g. the models and input parameters presented above for feasibility studies. To judge the reliability of the outcome of such an analysis, however, a failure analysis is needed, in which the influences of the uncertainties associated with the use of these models and parameters, are quantified.

Any failure analysis requires a clear picture of the input parameters with their uncertainties, the models with their uncertainties and the relationship between input parameters, models and output parameters or limit state functions. Such a picture can be presented in the form of an "operational fault tree", as presented in Figure 3-25. The meaning of the different boxes is as follows:

- In the double circles the input parameters with a stochastic character are indicated.
- In the rectangular boxes the models are indicated. The symbols are explained in subsection 3.8.3. Each model can be represented by an equation or figure discussed before.
- In the single circles the stochastic parameters are indicated which are found with the help of these models.
- The final output consists of the values of the 4 failure functions, also stochastic variables, found with the limit state equations. Failure occurs if one or more are negative.

The uncertainties about the soil parameters will be briefly discussed in subsection 3.8.2; those about the models in subsection 3.8.3.

In a level II or III analysis expected values of the soil parameters must be found and the soil parameter uncertainties will be expressed by a coefficient of variation (CoV = standard deviation divided by the expected value). The model uncertainties will be expressed by (multiplication) model factors m_x with expected values, $E(m_x) = 1$ and with a COV. All probability distributions discussed in this section will be assumed to be log-normal. Remark: with this type of distribution, the standard deviation of the normal distribution of the logarithm of the quotient of the variable and the expected value equals the logarithm of (1 + COV). This means, for instance, that for a parameter "p" with expected value E_p and COV = 0.8 (80%), there is 68% probability that $E_p/1.8 < p < E_p\,1.8$.

The final outcome is a probability for each failure mode that $g_x < 0$.

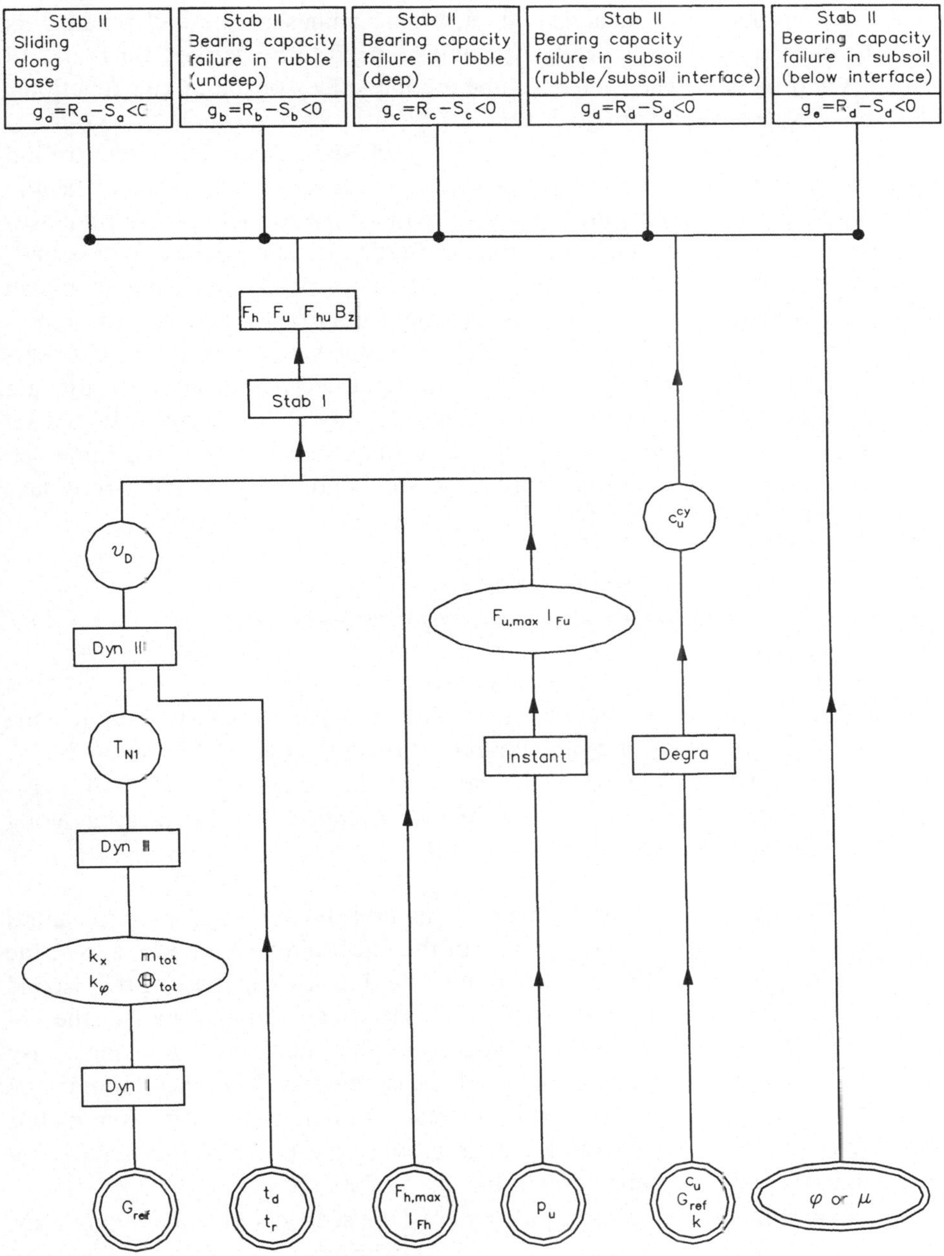

Figure 3-25. "Operational fault tree" (for explanation: see text subsection 3.8.1). Remark: $k_x = K_{11}$ $k_\varphi = K_{22}$ $m_{tot} = M_{tot}$ $\Theta_{tot} = \Theta_{cai} + \Theta_{hyd} + \Theta_{geo}$ and $\Theta_{cai} = M_{22,cai}$ $T_{N1} = T_{N,S}$ $\nu_D = \nu_d \approx \nu_L$

In a level I probabilistic analysis characteristic values of the soil parameters need to be chosen and partial safety coefficients need to be defined for both soil parameters and models. The final outcome are the values of the failure functions. No failure is expected if all are positive.

Several vertical breakwaters have been analysed (see chapter 5) in order to find a good way to define the representative values and a set of partial safety factors which may be used in a feasibility study of more or traditional vertical breakwaters in more or less traditional circumstances (hydraulic and geotechnical boundary conditions). It was found that the expected values of the soil parameters can best be used as characteristic values. One partial safety factor is defined to represent the uncertainty about the soil parameters and one safety factor for the uncertainty associated with the applied models. The suggested partial safety factors are presented in section 5.2.7.3. below. The values are only valid for the presented set of limit state functions and only together with the other values of the same set. The set of values is not valid for complicated soil conditions, such as strong layering, fairly soft clay or liquefiable sand.

3.8.2 *Uncertainties about soil parameters*

Determination of a soil parameter usually requires 3 steps, each of which brings about some uncertainty:

- Measurement, e.g. of the cone resistance q_c
- Translation of the measured parameter into the parameter relevant for the analysis, e.g. translation of q_c into undrained shear strength c_u (Equation 3-1)
- Interpolation or extrapolation of the parameter found at the location of measurement to the same parameter at the relevant location (e.g. the location along the whole breakwater with the weakest soil)

Part of the uncertainties is an inherent (or natural) uncertainty, like the natural spatial variation which is the basic source of the uncertainty of the last step or the change in soil properties after construction of the breakwater (part of the second step) which can never be predicted with 100% accuracy. Another part of the uncertainties is due to the lack of knowledge, like the uncertainty associated also with the last step by the limited number of measurements. The uncertainties due to lack of knowledge can be reduced by increasing the number of measurements, improving the measurement accuracy or improving the physical formulation for the translation in step 2 or the inter/extrapolation of step 3.

Before determination of any soil parameter of the seabed soil, knowledge of the stratification and the types of soil is gained. This brings about uncertainties similar to those associated with the last step, caused by the spatial variation of the soil: the uncertainty about the thickness or even the presence of some layers at the relevant locations. Usually a conservative approach is justified in a feasibility study: the assumption of the weakest layer to be rather thick. This uncertainty and the uncertainties associated with step 3 may be very large or very small, depending on the geological characteristics of the site and depending on the area over which the parameter must be averaged (usually the area of one caisson, because rupture surfaces have the same order of magnitude). Judgement usually requires considerable geological and geotechnical expertise.

Possible uncertainties associated with step 1 and step 2 is presented in Table 3-1 for the main soil parameters needed in a feasibility study. It will be done for a case in which the results of several CPTU's and borings are available and classification tests have been done on all relevant samples. The uncertainties can be reduced considerably (e.g. to half the indicated value) after performance of accurate laboratory tests, like triaxial tests, on representative samples.

Table 3-1. Possible uncertainties of soil parameters (step 1 and step 2).

Soil parameter	**CoV**	**Remarks**
k (sand or silt)	200%	Low value probably (always?) unfavourable
k (clay)	400%	
G	100 %	High G-value probably unfavourable
Γ	10%	limit: $\Gamma \leq 0.5$
μ	10%	
ϕ	5%	
ψ	40%	
c_u (sand or silt, after construction)	20% - 40%	Lower value (20%) to be used if prediction only based on Equation (3-9); higher value if more than 50% of c_u is due to dilatancy
c_u (clay after construction and consolidation)	30%	

CoV = Coefficient of Variation. Lognormal distributions advised.

3.8.3 *Model uncertainties*

The model uncertainties will be expressed by (multiplication) model factors m_x. In the level II and III analyses, these are stochastic parameters with log-normal distributions, with expected values, $E(m_x) = 1$ and with a COV. In a level I analysis these factors are partial safety factors. Suggestions of COV's are presented in Table 3-2.

Table 3-2. Suggestion for model uncertainties.

Model		Model factor	CoV	Remarks
Dyn I	Eq. 3-10 to 3-18	m_{K11} m_{K22}	30-50% 130 - 250%	Smaller values for m_{K11} and m_{K22} for thick homogeneous rubble foundation or very thin bedding layer; otherwise larger values
Dyn III	Fig. 3-12	m_{vD}	5 – 10%	Larger values for $0.5 < t_D/T_{N1} < 1.5$
INSTANT	Eq. 3-26 Eq. 3-27	m_{Fumax} m_{lFu}	20% 10%	
DEGRA	Eq. 3-30 Fig. 3-21	$m_{cucysand}$ $m_{cucyclay}$	50% 15%	
STAB I	Eq. 3-32 or 3-33	m_{Fhu}	30%	
STAB II	Eq. 3-35 Eq. 3-36 Eq. 3-37 Eq. 3-38 or 39 Eq. 3-40	m_a m_b m_c m_d m_e	20% 20% 20% 20% 20%	

CoV = Coefficient of Variation. Lognormal distributions advised.

There are four groups of models:

- **"Dyn"** concerning the dynamics. For a feasibility study these models consist of the equations (3-10) to (3-18) for Dyn I, equations (3-19) and (3-20) for Dyn II and Figure 3-12 for Dyn III. The values of the parameters K_{11}, K_{22} and v_d found by these functions should be multiplied by m_{K11}, m_{K22} and m_{vd} respectively. The estimated uncertainties are based on several hindcasts with large scale tests and prototype tests (Volume IIb, chapter 3 or de Groot et al., 1999).
- **"Instant"** concerning instantaneous pore pressures. For a feasibility study this model consists of the equations (3-26) and (3-27). The values of the parameters $F_{u,max}$, F_{hu} and l_{Fu} found by these functions should be multiplied by m_{umax} and m_{lFu} respectively. Equations (3-28) and (3-29) are meant to find out whether the subsoil should be considered drained or undrained and to select the corresponding stability equation. If the outcome of these equations is not decisive, both type of stability equations should be applied. The estimated uncertainties are based on several hindcasts with large scale tests and prototype tests (Volume IIb, chapter 4).
- **"Degra"** concerning degradation and residual porepressures. For a feasibility study this model consists of equation (3-30) and Figure 3-21. The value of the second term of equation (3-30) should be multiplied by $m_{cucysand}$; the value of c_u^{cy} found by Figure 3-21 should be multiplied by $m_{cucyclay}$.
- **"Stab"** concerning stability. For a feasibility study the models for Stab I consist of the equation 3-10 (i.e. $F_h = v_d F_{h,max}$ and $F_u = v_d F_{u,max}$) and (3-31) to (3-33). No uncertainties are associated with the Stab I equations for F_h, F_u and

B_z. The equation for F_{hu} , however, is a rough approximation. The models for Stab II consist of the limit state equations (3-35) to (3-40) yielding the values of the failure functions. All limit state equations have the form of $g_x = R_x - S_x$. To express the uncertainty, they should be substituted by $g_x = R_x - m_x \cdot S_x$. The estimated uncertainties are based on comparisons between the results of several different models and an incidental hindcast of large-scale tests (Volume IIb, section 7.7).

3.9 INFLUENCE OF DESIGN PARAMETERS

3.9.1 *General*

Design is an iterative process of selecting a breakwater type with provisional dimensions, analysing this provisional design, changing the type or dimensions, analysing once again etc. The previous sections of this chapter mainly concentrate on the analysis of a once selected design. In this section the influence of design parameters on the breakwater stability is discussed and most relevant failure modes are represented as functions of the most important design parameters. This may help during a feasibility study to decide which failure mode should be paid most attention to, as soon as a first provisional design has been made, and to select the dimension to be revised to arrive at a better, though still provisional design. More attention to design improvements will be presented in section 3.10.

In chapter 8 of Volume IIb and in Lesny et al. (2000) the relevance of the two load cases, wave crest and wave trough, and of the relevance of the main failure modes for the stability of a specific breakwater type have been investigated. The type is a vertical breakwater on a thin rubble bedding layer on top of a subsoil of coarse sand or gravel. Many different designs of this breakwater type under many different wave conditions (however, no wave impacts) have been considered. The result consists of the relationship between the most relevant design parameters and the most relevant failure modes and is summarised in subsection 3.9.2.

Section 3.9.3 deals with the question which of the findings of section 3.9.2 would also be valid for other types of vertical breakwaters.

3.9.2 *Vertical breakwater on thin bedding layer and coarse grained subsoil with pulsating wave loads*

3.9.2.1 *Input, analysis and output of performed investigation*

The purpose of the investigation was to get an overview of the most relevant load cases and failure modes in relation to the input parameters. In Table 3-3 findings of this study are summarised in a parameter map of most relevant load-case/failure-mode combinations in relation to the design parameters.

The combinations have been analysed with 4 failure mode equations for wave crest (numbered 2-4, 2-2, 2-5 and 2-1) and 4 equations for wave trough (numbered 3-4, 3-2, 3-5 and 3-1):

- 2-4 and 3-4 correspond to Equation 3-35 (sliding along the base)
- 2-2 and 3-2 correspond to Equation 3-26 (bearing capacity in rubble; undeep rupture surface)
- 2-5 and 3-5 correspond to Equation 3-39 (bearing capacity in subsoil; rupture surface at rubble/subsoil interface)
- 2-1 and 3-1 correspond to Equation 3-40 (bearing capacity in subsoil; rupture surface below rubble/subsoil interface).

Remark: bearing capacity failure in rubble with deep rupture surface is not possible with a thin bedding layer.

An important input parameter (design parameter) is α_c = area of cross section of breakwater/($h_c \cdot B_c$). See Figure 3-26. The design parameter with most influence on the dominant failure mode is the relative weight eccentricity, e_c/B_c (Fig. 3-23). Depending on the measure of the eccentricity and specific soil and load parameters the relevant load-case/failure-mode combination can be identified.

It must be mentioned that the map can not be complete, because it is not possible to describe all influences and dependencies on the caisson stability exhaustively. All results should be used carefully, since it can *not* be guaranteed, that for other parameter combinations than those varied in the study the limits within different load-case/failure-mode combinations become relevant are still valid.

3.9.2.2 *Less relevant load-case/failure-mode combinations*

Load-case/failure-mode combination 3-5, bearing capacity in subsoil with the rupture surface at the rubble/subsoil interface, at wave trough, appeared never to be the dominant combination.

The relevance of load-case/failure-mode combination 2-5, the same mode, but now at wave crest, was found to depend mainly on the ratio of φ'_S / φ'_R or the ratio $\tan\varphi'_S/\mu$, where φ'_S is the friction angle of the subsoil and φ'_R is the friction angle of the rubble. In all calculations $\mu = 2/3 \tan\varphi'_R$ was assumed. This combination may be relevant in the range of $25° \leq \varphi'_S < 30°$ and simultaneously $37.5° < \varphi'_R \leq 45°$ only, if the following inequality is valid:

$$\tan\varphi'_S < \mu \qquad (3\text{-}41)$$

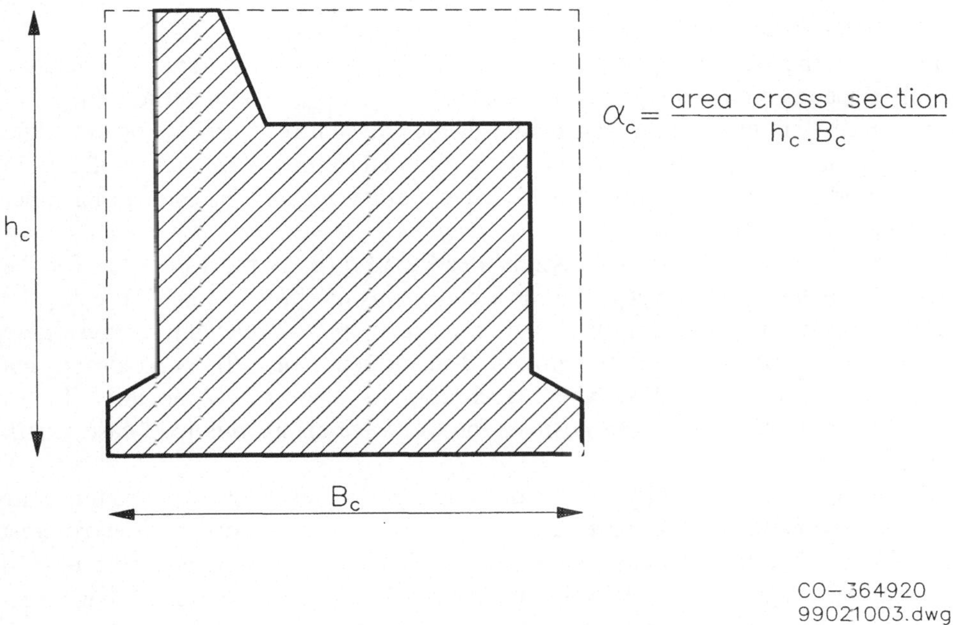

Figure 3-26. Definition of design parameter α_c.

If it is not, sliding along the base (2-4) would dominate over this failure mode. If Equation (3-41) is valid, however, this failure mode (2-5) may be relevant for positive eccentricities within the range of $0.5 \leq e_c/B_c \leq 0.15$, but only if also the following, unusual conditions are met: low weight of caisson: $\alpha_c \cdot \gamma_c/\gamma_W \leq 1.4$ and $h_c < (h_s + h_{crest} - h_r)$ (no overtopping), where h_{crest} is the height of the (standing) wave crest above still water level. For details see Volume IIb, chapter 8.

Sliding along the base at wave trough (3-4) may become the dominant combination only under the following design conditions:

- $0 \leq e_c/B_c \leq 0.05$ (thus, bearing capacity in rubble with undeep rupture surface is not critical)
- and $\varphi'_s \geq 35°$ (no bearing capacity problem) and $\mu \leq \tan 23° = 0.43$
- and $\alpha_c \cdot \gamma_c/\gamma_W \leq 1.7$ and $h_c < 0.9 \cdot (h_s + h_{crest} - h_r)$ (low height of caisson)
- and $H_{si}/h_s \leq 0.2$, $T_p \leq 10$ s (higher wave load at wave trough than at wave crest)

Within the parameter limits given above, other load-case/failure-mode combinations may become relevant as well, but the relevance of this type of bearing capacity failure in the subsoil (rupture surface at the rubble/subsoil interface) at wave crest and sliding along the base at wave trough is restricted to these parameter ranges only.

3.9.2.3 *Important load-case/failure-mode combinations*

Which of the remaining 5 load-case/failure-mode combinations dominates, mainly depends on the design values of e_c/B_c, φ'_S and H_{si}. Within certain ranges of these parameters unit weight and height of the caisson, $\alpha_c \cdot \gamma_c$ and h_c, or the angle of internal friction of rubble mound, φ'_R, could determine the relevance of certain failure modes as it is shown in Tab. 3-3. No *significant* influences of other input parameters have appeared.

The relative eccentricity of caissons' dead load e_c/B_c is most decisive for the question whether wave crest is the most dangerous or wave trough. For $e_c/B_c \leq 0$ (eccentricity to harbourside) only load case wave crest is relevant, because here negative eccentricities are unfavourable. The influence of e_c/B_c is so strong that even for low wave heights, where according to section 2.4.2 negative forces (wave trough) are greater than positive forces (wave crest), this load case is still relevant.

For $e_c/B_c > 0$ (eccentricity to seaside) both, wave crest and wave trough, may become dominant. Positive eccentricities are favourable for load case wave crest but unfavourable for load case wave trough. Here the relevance of a certain load case depends especially on the wave height H_{si}. It has to be mentioned that sometimes very high positive eccentricities may be unfavourable for load case wave crest as well. In these cases the moment of the caissons' dead load is too large in relation to the moment of the wave load, especially at low waves. For high waves this effect is not so obvious.

The eccentricity of caissons' dead load also largely determines which of the failure modes dominates. If $e_c/B_c < -0.05$ only combination (2-2), bearing capacity failure in rubble with undeep rupture surface and combination (2-1), bearing capacity failure in subsoil with rupture surface below the rubble/subsoil interface, are relevant because of the large resultant moment, which reduces the effective width of the caisson and enlarges the resultant eccentricity. If $e_c/B_c \geq -0.05$ every failure mode can be relevant. Here their relevance is determined by other parameters. For $e_c/B_c \geq 0.05$ combination (2-1) is no longer decisive, because positive eccentricities act favourably within this failure mode.

Apart from the influence of e_c/B_c, especially the angle of internal friction of subsoil, φ'_S, and the wave height, H_{si}, determine the relevance of the combinations. Parameter φ'_S is decisive for the relevance of bearing capacity failure in subsoil (combinations 2-1 and 3-1). Generally, its importance decreases with increasing φ'_S, whereas sliding along the base (2-4) and bearing capacity failure in rubble (2-2 and 3-2) become more and more relevant. The value of parameter H_{si} decides whether wave crest or wave trough is the relevant load case.

Table 3-3. Parameter map of relevant load-case/failure-mode combinations after Lesny et al. (2000) (for legend see Table 3-3 continued).

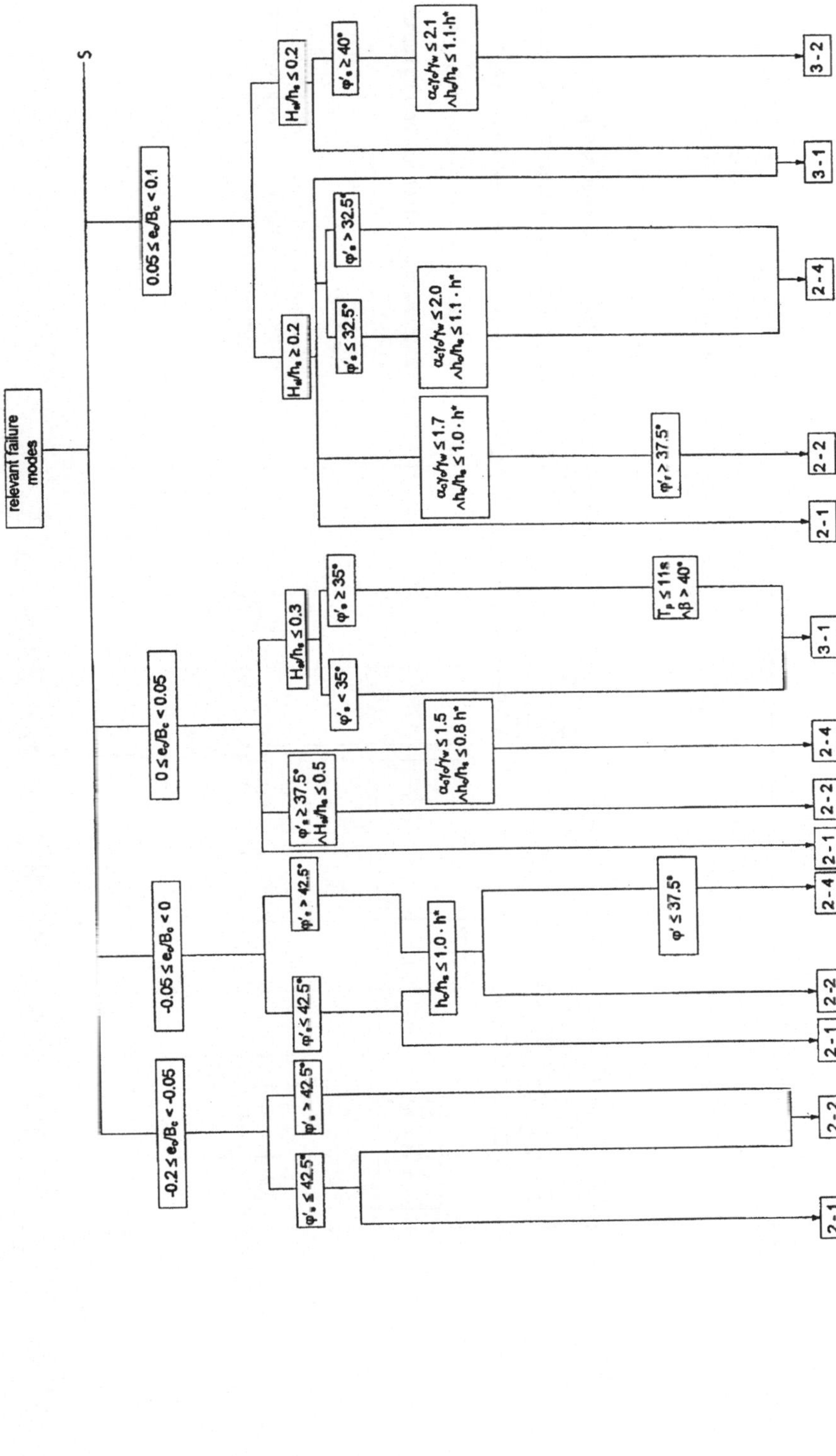

Table 3-3, continued. Parameter map of relevant failure modes after Lesny et al. (2000).

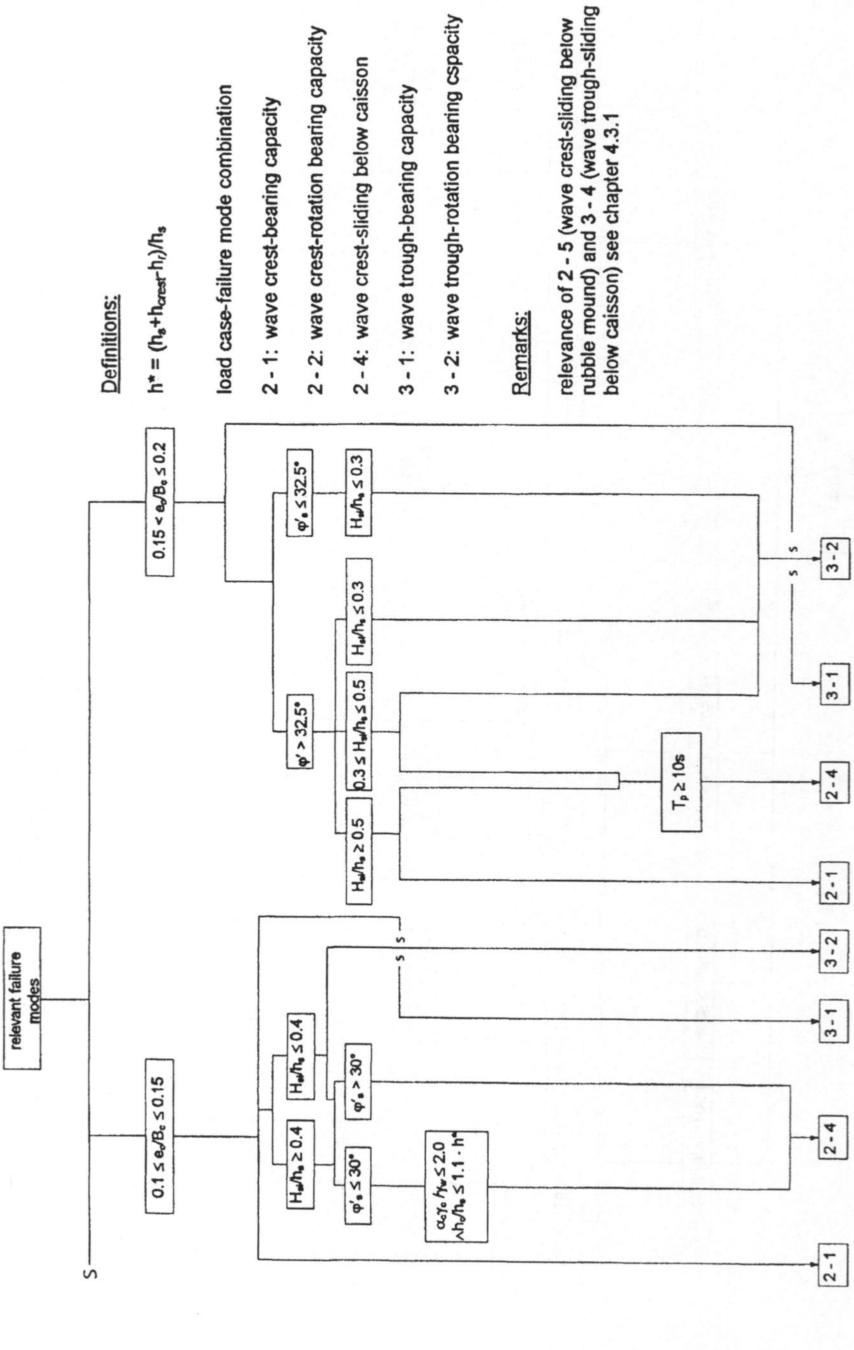

3.9.3 *Effects with other breakwater types*

3.9.3.1 *Effect of a high rubble foundation*

Although many simplifications have been assumed for the investigation summarised in subsection 3.9.2, most of the general tendencies found will probably also be valid for vertical breakwaters on high rubble foundations, provided the foundation is rather wide and the subsoil consists of rather solid material. The main difference will be that the role of bearing capacity failure in the subsoil (load-case/failure-load combinations 3-1 and 2-1) will be taken over by bearing capacity failure in the rubble with relatively deep rupture surface (compare equation 3-37 for failure function g_c).

3.9.3.2 *The effect of wave impacts*

Wave impact loads, if they occur, are usually much higher than pulsating wave loads. Sometimes certain constructional measures may help to reduce the wave impact load to the level of the pulsating load or lower. If this is not possible, the dominant failure modes will be those which occur with wave crest, even where wave trough dominates according to the findings of subsection 3.9.2. Then, an increased eccentricity may be helpful to arrive at the optimal design.

3.9.3.3 *The effect of fine grained subsoil*

Fine sand, silt or clay as subsoil material will behave nearly completely undrained under wave loading. The effective shear strength is usually smaller than of a coarse grained soil, due to instantaneous and/or residual pore pressures or degradation, unless the subsoil consists of sand/silt with a high relative density or of highly overconsolidated clay. Even with high rubble foundations (see „Genoa Voltri V1“ in section 5.8), bearing capacity failure in subsoil (combinations 2-1, 3-1, but also 2-5 or 3-5) may become the dominant failure modes instead of sliding along the base or bearing capacity failure in rubble.

3.10 POSSIBILITIES FOR DESIGN IMPROVEMENTS

3.10.1 *Variation of design parameters if rubble foundation is present*

3.10.1.1 *Increase the mass of the wall*

This helps for nearly all failure modes. If bearing capacity failure is critical also the foundation width B_c must be increased. The other measures discussed below are based on the assumption of keeping the mass constant. The mass of perforated wall caissons is relatively small. Extra mass in the superstructure may be needed for compensation. The presence of this mass above the water level is favourable (see point 3.10.1.3).

3.10.1.2 *Increase or decrease weight eccentricity e_c*

For conditions without wave impact it was found that $0 \leq e_c/B_c \leq 0.1$ yields the optimal balance between seaward failure and harbour side failure. In case of significant wave impacts, a larger value (weight eccentricity more seaward) is likely to be more economical. Perforated wall caissons usually have a negative weight eccentricity and it may be very helpful if this is (partly) compensated by a superstructure with a largely positive weight eccentricity.

3.10.1.3 *Reduction of wall volume below still water level*

Favourable for the stability is the increase in effective weight of the wall. Given the mass of the wall, this can only be reached by reducing the volume below still water level, e.g. by reducing the width of the caisson (while keeping B_c constant to avoid loss of stability). Perforated wall caissons usually have "automatically" a relatively small volume below the still water level.

This loss of volume must be compensated either by increasing the unit mass or by increasing the volume above the still water, i.e. increasing the volume of the superstructure. This means also increasing α_c and/or h_c. For conditions without wave impacts the following values of these parameters appeared to be favourable:

- $h_c > h_{crest} + h_s - h_r$ where h_{crest} = height of the wave crest above SWL
- $1.7 \leq \alpha_c \cdot \gamma_c/\gamma_w \leq 2.0$ for $H_{si}/h_s < 0.5$
- $2.0 \leq \alpha_c \cdot \gamma_c/\gamma_w \leq 2.3$ for $H_{si}/h_s \geq 0.5$

3.10.1.4 *Enlargement of B_c*

Enlargement of B_c, while keeping the mass constant, may help the stability if one of the bearing capacity failure modes, especially bearing capacity failure in the rubble.

3.10.1.5 *Enlarging the rubble foundation*

Placing more rubble adjacent to the wall at the harbour side increases the stability, especially if sliding along the base at wave crest determines the safety. The increase in horizontal sliding resistance is roughly equal to the under water weight of the added rubble. If bearing capacity failure in the subsoil dominates, it would be more efficient to place the rubble on top of the subsoil adjacent to the rubble foundation, thus widening the rubble foundation.

Another way to enlarge the rubble foundation is increasing its height, thus lifting up the wall. If the hydraulic load does not increase, this helps the stability in two ways:

- increase of the effective weight of the wall (compare measure c)

- increase of the stationary vertical effective stress in the soil, yielding increase of the shear strength.

The enlargement of the rubble foundation may increase the settlement of the subsoil.

3.10.1.6 *Connecting caissons to each other*

This measure is of particular interest if wave impacts determine the stability. The connections should be such that the total length of caissons working together is at least two or three times the characteristic length of the load (section 2.5.3 and chapter 4). The connections must be able to transfer forces in the x-direction and moments around the y-axis (longitudinal axis). If unequal settlements are to be expected, the connections must not transfer forces in the z-direction and moments around the x-axis.

3.10.1.7 *Soil replacement or soil improvement*

If the subsoil consist of (too) soft clay or silty sand with the risk of the generation of considerable residual pore pressures, bearing capacity failure with rupture surfaces along or in the subsoil may determine the stability. An effective measure is to dredge the upper 5 to 10 m of the soft soil and replace it by clean sand (content of grains finer than 0.063 mm smaller than ca 5%). Subsequent densification of the sand is also often needed to avoid considerable settlement after construction caused by the wave induced cyclic loading and/or to avoid the risk of too high residual pore pressures during extreme cyclic loading. With soft clay an alternative could be pre-consolidation by a considerable weight of rubble, before the caissons are placed. Pre-consolidation, however, often requires several years. Several other soil improvement techniques (mixing with cement, stone columns etc) could also be considered. The construction at sea, however, makes these techniques usually extremely expensive.

3.10.1.8 *Prolongation of seepage path in rubble foundation*

The seepage path of the water flowing through the rubble foundation during wave crest, can be made longer, in order to reduce the uplift force. This would be favourable for all failure modes, although the effect is only significant with relatively light structures. The path can be made longer by:

- wide apron slabs which are placed tightly against the caisson footing
- reduction of permeability of the rubble immediately underneath the seaward toe of the wall (fine grains, or injection)
- increasing the permeability directly underneath the middle and the harbourside toe of the wall.

These measured cause a reduction of the uplift during wave crest. During wave trough, however, they cause an opposite, unfavourable effect. Therefore it is only useful if the wave crest load dominates failure.

3.10.2 *Caisson foundation directly on sand*

Under certain conditions it may be wise to consider a foundation of a caisson directly on the sandy seabed. In such a design no rubble is placed on the seabed to raise and level the foundation bed. No rubble is present either to take care for drainage of excessive pore pressures and to act as filter against the loss of sand particles, at least not underneath the caisson floor. Ribs and/or a bed protection directly adjacent to the wall may be necessary to avoid extreme excess pore pressure (gradients) and the loss of fine material. Experience with this type of structures is gained with offshore platforms and may be useful for vertical breakwaters.

3.10.3 *Skirts to improve foundation capacity in clayey soils*

At locations where the natural soil consist of soft clays of clayey silty sands, the bearing capacity will in many cases not be sufficient to support a rubble mound and the breakwater structure. This may require a staged construction where the rubble mound is placed first. The soil will then consolidate under the weight of the mound. The time required for consolidation can be several months and even years before the superstructure can be installed, and the settlements will continue during the operational life of the breakwater. See Figure 3-27a.

Skirted foundations have been used with great success for offshore gravity platforms. The base is equipped with skirts along the periphery and also compartmented in accordance with the internal walls of the superstructure, see Figure 3-27b. The structure can be built as a self-floater, partly or fully completed in a dry dock prior to float out and installation. The breakwater sections are towed to location, positioned accurately and ballasted. The skirts will penetrate under gravity loads, and by applying suction full penetration to design depth can be achieved. By varying the underpressure in the different skirt compartments full control of tilt can be achieved. From a foundation point of view skirted concrete caissons have a number of advantages.

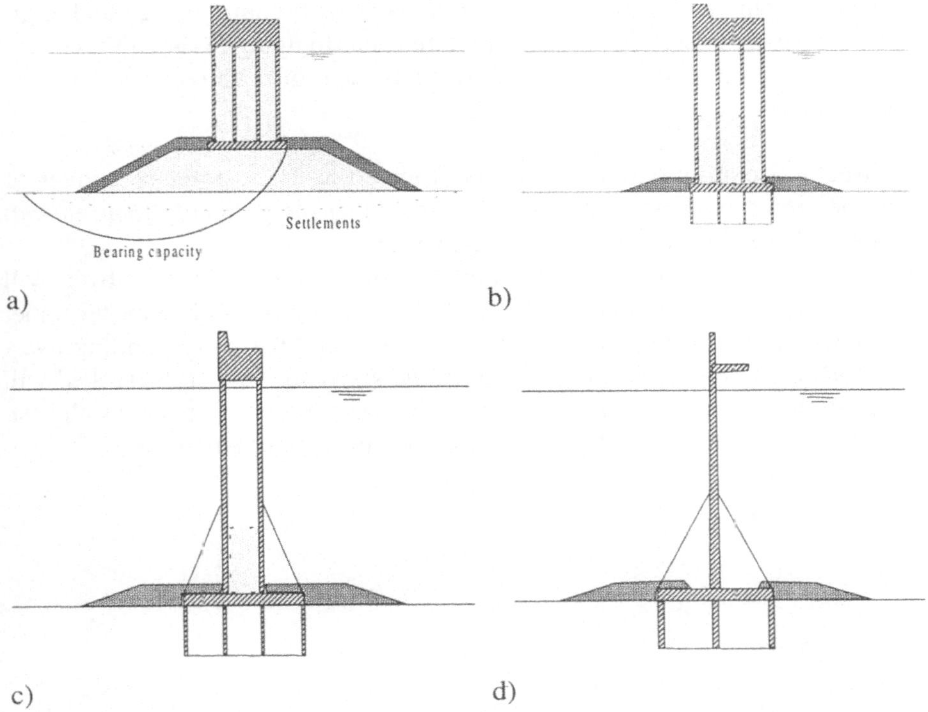

Figure 3-27. Skirted breakwaters compared with traditional design. Possible configurations for skirted breakwater structures on clayey soils.

- The sliding and the bearing capacity on soft ground can be improved considerably compared to a flat based structure
- Staged construction can be avoided, i.e. short completion time for breakwater
- Underbase erosion (scour) is prevented
- The underbase water pressure and corresponding overturning moment acting on traditional breakwater structures is avoided
- A caisson cantilever base can reduce the overturning moment
- A considerable weight reduction can be achieved with increased safety against bearing capacity failure and with strongly reduced settlements
- Quick and proven installation method.

There are certainly also limitations and disadvantages:

- The additional concrete work and cost compared with rubble mounds made of rockfill and gravel
- Availability of construction sites and skirt depth restrictions during towing out

➢ The water depth variations along the breakwater will require design and engineering of several sets of concrete structures and dredging (although a certain water depth variation can be covered by adjustment of penetration and under-base grouting).

The structural configurations may vary from a box type structure similar to many of the representative structures with skirts under the periphery to a sea wall with a skirted base plate as outlined in Figure 3-27b, c and d.

The lack of a rubble mound and the presence of skirts in a clayey subsoil will change the possible failure modes. The skirts prevent intrusion of seawater under the base under wave loading. The direct uplift force and the corresponding over-turning moment is thus eliminated. However, the wave pressure at the seabed will tend to push the soft soil down and in under the base. Figure 3-28 shows alternative failure modes that have to be investigated for this type of structure.

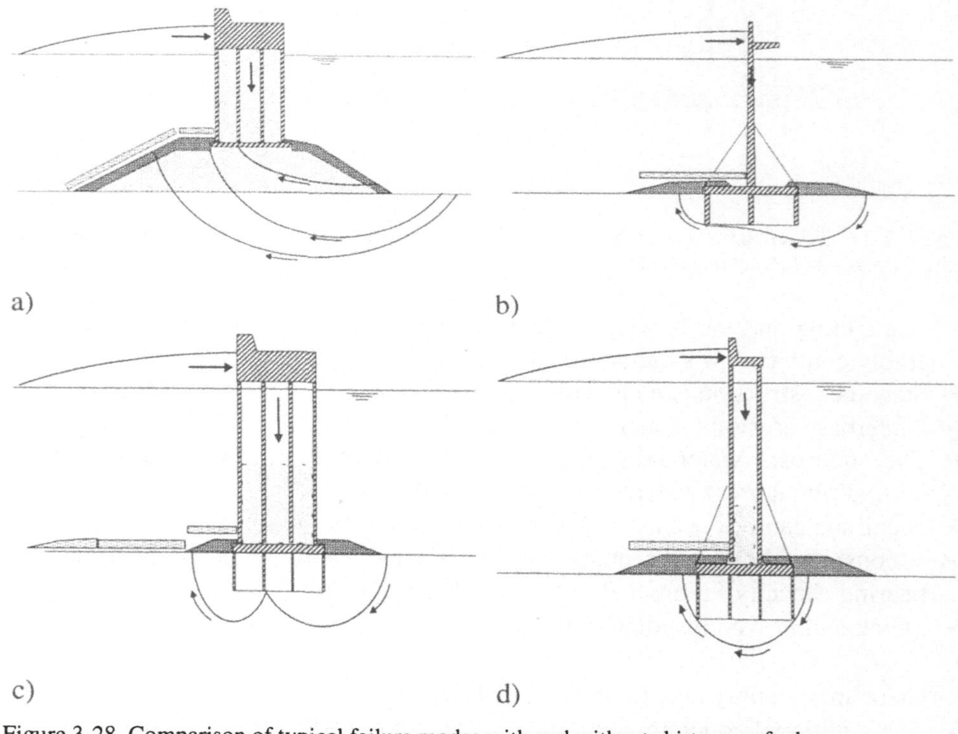

a) b) c) d)

Figure 3-28. Comparison of typical failure modes with and without skirts on soft clay.

Sliding at base level and possible bearing capacity failure under the most loaded part (Fig. 3-28a) are typical failure modes for a gravity type breakwater structure. The skirts allow utilisation of the suction capacity under the heel of the base and mobilisation of passive earth pressure in front of the skirt. The weight can thus be reduced compared with a traditional structure. This change in proportion between horizontal and vertical force and will influence the shape of the critical failure surfaces. With increasing water depth and increasing overturning moment the critical failure surface will change from bearing capacity type failure (Fig. 3-28b) toward combined suction/bearing capacity failure (Fig. 3-28c) towards a pure rotational failure with a circular failure surface (Fig. 3-28d).

ACKNOWLEDGEMENTS

Much of the work described here is based on studies under the PROVERBS project of the EU Mast programme under contract MAS3-CT95-0041, with some information produced by the MCS-project under contract MAS2-CT92-0047.

Substantial additional work has been completed by the PROVERBS partners under their own funding, or matching funds provided by other organisations. At GeoDelft additional support to the research work presented here has been afforded by the Ministerie van Verkeer en Waterstaat, Rijkswaterstaat (Public Works Department), Dienst Weg- en Waterbouwkunde.

The content of this chapter is mainly based on the work by:

Aalborg University (AU: L.B. Ibsen, K.P. Jakobsen, J.D. Sørensen)
GeoDelft (DG: M.A. Allard, M.B. de Groot, P. Hölscher, A. van Hoven, J.T. van der Poel, C. Zwanenburg)
Leichtweiss Institute (LWI: A. Kortenhaus)
Norwegian Geotechnical Institute (NGI: R. Dyvik, T.J. Kvalstad)
University of Bologna (UoB: L. de Angelis, A. Lamberti, L. Martinelli)
University of Essen (UGE: K. Lesny, E. Perau, W. Richwien)

Particular acknowledgement is given for the comments by A.H. Nooy van der Kolff and J. van 't Verlaat.

REFERENCES TO VOLUME IIb

Chapter 1: DE GROOT, M.B. (1999): Introduction.

Chapter 2: KVALSTAD, T.J. (1999): Soil investigations and soil parameters.

Chapter 3: LAMBERTI, A.; MARTINELLI, L.; DE GROOT, M.B. (1999): Dynamics.

Chapter 4: DE GROOT, M.B. (1999): Instantaneous pore pressures and uplift forces.

Chapter 5: KVALSTAD, T.J. ; DE GROOT, M.B. (1999): degradation and residual pore pressures.

Chapter 6: IBSEN, L.B.; JAKOBSEN, K.P. (1999): Limit state equations for stability and deformation.

Chapter 7: KVALSTAD, T.J. ; DE GROOT, M.B. (1999): Uncertainties.

Chapter 8: LESNY, K.; PERAU, E.; RICHWIEN, W. (1999): Influence of design parameters – stability analysis on feasibility level.

Chapter 9: KVALSTAD, T.J. (1999): Alternative foundations.

REFERENCES

Aas, G.; Lacasse, S.; Lunne, T.; Hoeg, K. 1986. Use of in-situ tests for foundation design of clay. *Proceedings. In Situ '86*, ASCE, Blacksburg, Virginia, USA, June 1986, pp. 1-30.

Andersen, K.H.; Brown, S.F.; Foss, I.; Pool, J.H.; Rosenbrand, W.F. (1976): Effect of cyclic loading on clay behaviour. *Proc. Conf. Design and Construction of Offshore Structures, London.*

Andresen, A.; Lunne, T. 1986. Soil investigation for strait crossing. *Proceedings International Conference on Strait Crossings.*, Stavanger, Norway, October 1986, Volume 2, pp. 783-798.

Baldi, G.; Bellotti, R.; Ghionna, V.; Jamiolkowski, M.; Pasqualini, E. 1986. Interpretation of CPT's and CPTU's; 2nd part: drained penetration of sands. *Proceedings 4th International Geotechnical Seminar*, Singapore, pp. 143-156.

Barton, N.; Kjaernsli, B. 1981. Shear strength of rockfill. *Journal of Geotechnical Engineering Division*, ASCE, vol. 107, no. GT7, pp. 873-891.

Bjerrum, L. (1973): Geotechnical problems involved in foundations of structures in the North Sea. *Geotechnique, Vol. 23 – 3,* pp 319-358.

Brinch Hansen, J. (1968): A Revised and extended Formula fo Bearing Capacity. *Reprint of Lecture in Japan*, October 1968.

CIRIA/CUR 1991. Manual on the use of rock in coastal and shoreline engineering. Rotterdam, The Netherlands / Brookfield, USA: A.A. Balkema, 607 pp.

CUR/RWS 1995. Manual on the use of Rock in Hydraulic Engineering. CUR report 169, Gouda, The Netherlands, 902 pp.

De Groot, M.B.; Andersen, K.H.; Burcharth, H.F.; Ibsen, L.B.; Kortenhaus, A.; Lundgren, H.; Magda, W.; Oumeraci, H.; Richwien, W. 1996. Foundation design of caisson breakwaters. *Norwegian Geotechnical Institute*, no. 198, 2 volumes, Oslo, Norway, 126 pp; 9 Appendices.

De Groot, M.B.; Hölscher, P; Meijers, P.; Kortenhaus, A. 1999. Stability of caisson breakwaters under wave impact loading. *Geotechnical Engineering for Transportation Infrastructure, Barends et al (eds)* Balkema, Rotterdam, ISBN 90 5809 0477, Vol 2, pp 747-754.

Flügge, W. (ed.). 1962. Handbook of engineering mechanics. New York, USA: McGraw-Hill, 1st edition, 1000 pp.

Hölscher, P.; Zwanenburg, C.; De Groot, M.B. 1998. Hindcast Hannover breakwater. *Research Report, MAST III, PROVERBS-Project, GeoDelft (formerly "Delft Geotechnics")*, CO-364920/127, Delft, The Netherlands, ca 500 pp.

Lesny, K.; Perau, E.; Richwien, W.; Wang, Z. (2000): Some aspects on subsoil failure of vertical breakwaters. *Forschungsberichte aus dem Fachbereich Bauwesen*, Heft 83, Universität GH Essen: Der Dekan des Fachbereichs 10, to appear 2000.

Lunne, T.; Powell, H.P. 1992. Recent development in in-situ testing in offshore soil investigations. *Proceedings SUT Conference on Offshore Soil Investigations and Foundation Engineering*, London, U.K., September 1992, pp. 147-180.

Martinelli, L. 1998. Risk analysis of Genoa Voltri breakwater: the case of Genoa Voltri. Ph.D. Thesis, Università di Bologna, 15 December 1998, Bologna, Italy.

Meijers, P. 1994. Spring and dashpot parameters from large scale tests. *Research Report. "Geotechnical Group" of the MAST II / MCS-Project (MAS2-CT92-0047)*, CO-330040/49, Delft, The Netherlands, 32 pp.

Norwegian Geotechnical Institute 1998. Data base of laboratory test results on selected sands and silts. *NGI Report. MAST III, PROVERBS-Project: Probabilistic Design Tools for Vertical Breakwaters*, no. 524094-1, Oslo, Norway, 39 pp., 22 Annexes.

Oumeraci, H.; Kortenhaus, A. 1994. Analysis of dynamic response of caisson breakwaters. *Coastal Engineering, Special Issue on 'Vertical Breakwaters'*, Eds.: Oumeraci, H. et al., Amsterdam, The Netherlands: Elsevier Science Publishers B.V., vol. 22, nos. 1/2, pp. 159-183.

Robertson, P.K. 1990. Soil classification using the cone penetration test. *Canadian Geotechnical Journal*, vol. 27, no. 1, pp. 151-158.

Sekigushi, H.; Ohmaki, S. 1992. Overturning of caissons by storm waves. *Soils and Foundations*, vol. 32, no. 3, pp. 144-155.

Smits, F.P.; Andersen, K.H.; Gudehus, G. (1978): Pore pressure generation. *Proc. Int. Symp. on Soil Mech. Research and Found. Design for the Oosterschelde Barrier, Delft, the Netherlands,* Vol 1 – II 3. 16 pp.

Van der Poel, J.T.; De Groot, M.B. 1997. Results of centrifuge tests on a caisson breakwater. *MAST III, PROVERBS-project, GeoDelft (formerly "Delft Geotechnics"),* CO-364920/89, Delft, The Netherlands, ca 1000 pp.

Van der Poel, J.T.; De Groot, M.B. 1998. Cyclic load tests on a caisson breakwater placed on sand. *Proc Int. Conf. "Centrifuge 98", Tokyo*, Balkema, Rotterdam, Vol 1, pp. 403-408.

Van Gent, M.R.A. 1993. Stationary and oscillatory flow through coarse porous media. *Communications on Hydraulic and Geotechnical Engineering*, TU Delft, Delft, The Netherlands, Report No. 93-9, 62 pp.; 4 Appendices.

Vermeer, P.A. 1990. The orientation of shear bands in biaxial tests. *Géotechnique*, Vol 40, No. 2, pp. 223-236.

Zwanenburg, C; De Groot, M.B.; Bakker, H.L. 2000. Foundation of the Genoa Voltri Breakwater. *Coastal Structures, Losada (ed)*, Balkema, Rotterdam, ISBN 90 5809 0922, Vol 2, 623-629.

CHAPTER 4

Structural aspects

Roger Crouch

4.1 INTRODUCTION

4.1.1 *Background*

Vertical breakwaters constructed from cellular reinforced concrete caissons can provide excellent performance and long service as part of a coastal structure, provided care is taken not only in the design and construction phases but also in the development of a properly managed maintenance plan. This chapter offers a synthesis of observations on the structural design of the reinforced concrete caissons. A significant part of the text deals with the use of the Finite element method to analyse these structures. The chapter begins with a discussion on the phases typically followed when structurally designing a caisson breakwater. It then goes on to identify the common caisson types (section 4.2) and list some relevant geotechnical and hydraulic parameters needed by structural engineers to complete the design (sections 4.4 and 4.5).

The likely modes of, and consequences of, structural damage (appropriate to the transportation, placing and in-situ situations) are then discussed in section 4.6 together with a summary of the structural roles played by the key components in a typical cellular caisson. The floating stability and towing of the caisson are also briefly covered in this section. No single code of practice currently exists which explicitly covers the complete design of the reinforced concrete elements forming a caisson breakwater. Therefore, in section 4.7 a summary of some of the more relevant standards is made and some tentative suggestions offered for a new design code for caisson breakwaters.

In section 4.8 a discussion on the basic load transfer mechanisms is given which lead to a series of simplified limit state equations (LSE). The LSEs address the flexural and shear failure of beam/slab members as well as basic cracking and chloride penetration models.

The justification for the simplified structural idealisations is examined in section 4.9 where a hierarchy of increasing complex structural analysis methods is presented. The first approach is based on a 3 degree-of-freedom, lumped parameter, dynamic model. This class of model can help the engineer assess whether a dynamic analysis is warranted for the front wall. Moving on to a more sophisticated approach, the use of a non-linear dynamic finite element shell analysis code is covered next. This section describes the manner in which yielding in the reinforcement and cracking in the concrete may be simulated. Some results (where large deformation effects are also included) are given for a typical rectangular caisson under severe wave slamming.

The benefits of a full 3-dimensional continuum analysis is then discussed and the results from a series of analysis at two stages of construction presented. The current state of the art of non-linear dynamic FE modelling is reviewed and the section concludes with how dynamic fluid-structure-soil interaction effects may be introduced into the analysis.

This chapter concludes (in section 4.10) with a series of practical points, relevant to the design of cellular caissons.

4.1.2 *Design sequence*

Prior to presenting some typical concrete caisson types, it is appropriate to briefly review the three basic phases of structural design. The first stage consists of devising an overall structural scheme which meets the intended use, is safe, constructible and economically viable. The caisson arrangement will be influenced by the placing and transportation method adopted (for example, lowered by cranes from barges, carried by an overhead crane running on the existing caissons, floated out, formed partly in-situ). This, in turn, may be controlled by the availability of local materials (for example, aggregate), skills and a suitable pre-casting site. The second stage comprises performing the initial calculations to determine the approximate sizes of the structural components and estimate the cost of the materials, temporary works and develop a more detailed construction method and sequence. Rules of thumb are often used to quantify loadings and idealised models are used to approximate the manner in which the structure carries the loads. In the final stage, the adequacy of each structural member is assessed for a full suite of possible load cases. Detailed checks on internal resistance, structural deformations, reinforcing arrangements and materials specification form part of the third phase.

In the case of a caisson breakwater, the conceptual design will identify the constrains operating at the particular site. The overall dimensions and geometric form of the structure will typically be dictated by geomechanical and hydraulic conditions. For example, the frictional resistance which can be mobilised at the foundation-base interface (to prevent rigid body sliding) will control the width of the base slab. The height of the structure will be governed by the structure's intended purpose, the tidal range and the maximum wave height to be resisted without over-topping.

4.2 GENERIC TYPES OF REINFORCED CONCRETE CAISSONS

4.2.1 *Planar rectangular multi-celled caissons*

A typical structural arrangement for a cellular, rectangular, reinforced concrete caisson with a vertical face is illustrated in Figure 4-1. The diagram shows an isometric view of one half of a 7 by 4-celled caisson.

This form of caisson typically comprises 8 different types of load-bearing elements (i) the front wall, (ii) rear wall, (iii) side walls (not shown), (iv) internal walls, (v) base slab, (vi) top slab, (vii) crown wall and (viii) shear keys (not always present).

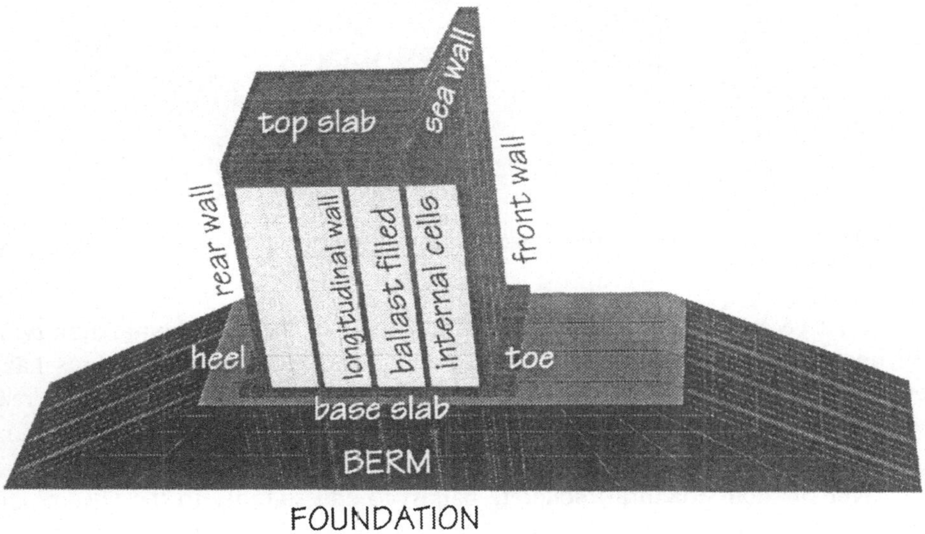

Figure 4-1. Isometric view of one-half of a caisson, Genoa Voltri, Italy.

The planar front wall in this class of structure reflects the incident wave. Internal cell sizes are usually of the order of 4 to 5m square, although circular (cylindrical) internal cells have been used successfully on some projects (for example, the extension to the Reina Sofia breakwater at Las Palmas, Spain). Use of the latter can result in a smaller reinforcement requirement as the loads are transferred through the caisson more by compressive arching action, than flexure. The walls are usually slip-formed either continuously, or in distinct lifts.

4.2.2 *Perforated rectangular multi-celled caissons*

Perforated caissons are becoming increasingly popular because they can create a more tranquil sea state in front of the structure (due to reduced reflections) and also lead to reduced material costs although the latter may be offset by increased formwork costs.

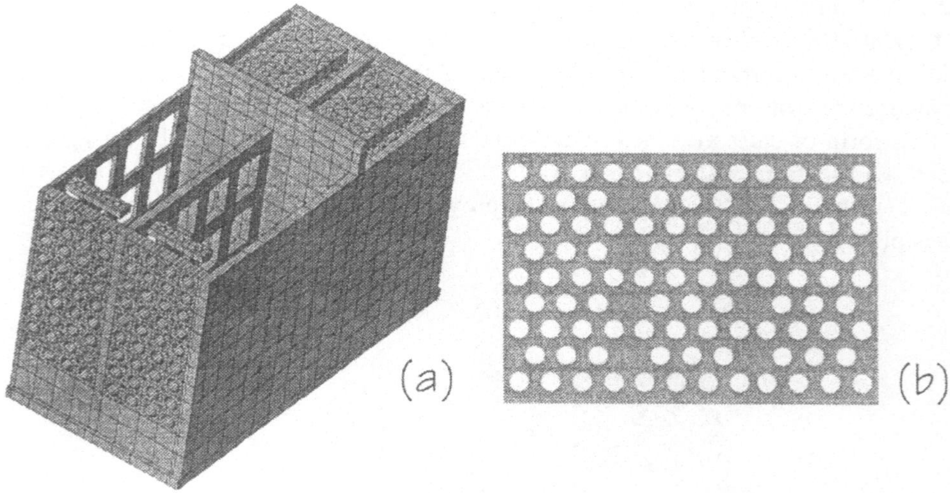

Figure 4-2. (a) Part of a perforated rectangular caisson (half-structure, Dieppe, France); (b) Typical perforation arrangement

The relative area of the perforations with respect to the total frontal area typically lies in the range 25 to 40%. Both circular and rectangular apertures have been used. In the case of the breakwater at Dieppe (France), the front and rear faces (as well as the top slab) have circular holes, whereas the internal walls have large rectangular perforations. Internal cells in a perforated caisson often have a thick layer of (non-structural) concrete ballast to add stability to the base of the structure.

4.2.3 *Circular-fronted caissons*

Circular fronted caissons do not require such large wall thicknesses as rectangular caissons because the external wave pressure is transmitted to the foundation by in-plane compression (that is, through compressive membrane stresses) rather than flexure. As examples, the Hanstholm (Denmark) and Brighton (UK) breakwaters have similar circular forms, whereas the Duca degli Abruzzi breakwater in Naples (Italy) exhibits a hybrid rectangular-circular footprint (Van der Meer & Benassai, 1984). Whilst total impact forces may be reduced on circular caissons, care is needed to avoid wave trapping and local high pressures in the clutches where two neighbouring caissons meet. In the case of the Hanstholm and Brighton caissons, the units were lowered into position by means of a rail mounted gantry crane straddling adjacent caissons.

4.2.4 *Alternative designs*

There is a growing tendency to adopt hybrid caisson forms in new breakwater designs to optimise the solution. Thus, perforated circular front walls with an open structure to the front cells can be combined with a planar rear wall which has smaller perforations. One problem with adopting a perforated structure lies with difficulty in obtaining reliable design pressure intensities as a result of the highly turbulent flow within the cells.

In deeper waters, some novel solutions which build upon experience gained from off-shore platform designs are emerging. Chapter 3 of this document illustrates some of these structures; where the arrangement is driven largely by geotechnical conditions (see also Volume IIb, chapter 9).

4.3 LOADS ACTING ON THE CAISSON

Possible loading during the in-service life include (i) *permanent* loads resulting from the dead weight of the structure (using submerged densities, where appropriate) and the superstructure as well as the horizontal soil pressure from the fill inside the cells and from the foundation reaction (ii) *variable* loads arising from changes in the water level, from pulsating and impact loads (including uplift effects under the base slab) and over-topping wave loads as well as superimposed harbour traffic loads (iii) *accidental* loads resulting from vessel impacts during mooring and falling masses during cargo loading/unloading operations. Clearly, in regions where seismic activity occurs, the earthquake induced ground motions can lead to structural distress.

4.4 GEOMECHANICAL FACTORS RELEVANT TO THE STRUCUTRAL RESPONSE

As noted earlier, the stability of a caisson breakwater depends largely on its resistance against rigid-body sliding and the strength reserve in the foundation. *Failure* analyses based on these mechanisms are typically carried out by geotechnical engineers (chapter 3). However, structural engineers are also interested in some geotechical parameters when designing the reinforced concrete elements. For relatively thin walled caissons under short duration wave impact, the characteristics of the granular fill in the cells and elasto-plastic properties of the foundation can have an influence on the structural dimensions and dynamic[3] response of the front wall.

4.4.1 *Characteristics of the ballast fill in caisson cells*

A horizontal thrust from the granular fill in the cells acts outwards on the external walls. A linear increase in horizontal pressure with depth, using the coefficient of earth pressure at rest, provides the following upper limit to these pressures $\sigma_h=0.6\rho_s\, g\, z$ where ρ_s is the soil density, g is acceleration due to gravity and z is the height of the fill. Note that the full saturated soil density should be used if no drainage is present in the cell[4].

Ballast density	ρ_s	$kg.m^{-3}$

4.4.2 *Characteristics of rubble foundation and sub-soil*

A complete analysis of the dynamic soil-structure-fluid interaction problem would necessitate inclusion of a realistic elasto-pastic constitutive model for the soil (Crouch & Wolf, 1994), a capability to simulate sliding and loss of contact at the caisson base/foundation interface and the transport of pore fluid within the soil skeleton (see chapter 3). Such a detailed analysis, whilst enlightening, is not normally required. A simplified approach would attempt to choose reasonable values for the rubble mound and foundation shear moduli and Poisson's ratios to enable isotropic, linear elastic soil models to be used (Wolf, 1994). Although these elastic constants are rather difficult to estimate, the following relationship for the rubble mound shear modulus may be assumed $G \approx G_0 \sqrt{\sigma_v'/0.1}$ MPa where G_0 lies in the range 20-100MPa and σ_v' is the effective vertical stress (in MPa).

Rubble mound density	ρ_{rm}	$kg.m^{-3}$
Foundation soil density	ρ_f	$kg.m^{-3}$
Rubble mound shear modulus	G_{rm}	$N.m^{-2}$
Foundation shear modulus	G_f	$N.m^{-2}$
Rubble mound Poisson's ratio	ν_{rm}	1
Foundation Poisson's ratio	ν_f	1

[3] For stiffer, thick-walled caissons the bending and shear forces generally may be determined from static analyses as the fundamental period of the walls can be considerably shorter than the loading duration.

[4] Through-holes linking the internal walls are not recommended as this will reduce the righting moment during floating due to the internal water ballast acting as a single volume.

The effective densities of the rubble mound and foundation are required to estimate not only the in-situ stresses but also the mass contribution in a full dynamic analysis (the latter is idealised as a geodynamic mass in multi-degree-of-freedom lumped parameter models).

4.4.3 *Unevenness of the foundation*

In order to achieve the required vertical alignment of the caisson walls, the top of the rubble mound should typically be levelled to achieve a mean surface profile to within 2% of the horizontal, over a 5m length, with no local depression being larger than about 200mm deep. In certain cases, especially for larger caissons, the effort expended by divers (or robots) preparing the foundation surface to achieve a pre-defined degree of flatness (and horizontality) may be offset by relaxing these tolerance a little and stiffening-up the caisson base slab such that it could safely span over local surface irregularities. In this case, information on the degree of unevenness is required in order to carry out a check on the likely distress caused to the base slab. However, in general the problem with uneven foundations is not so much one of potential rupture of the base slab, but rather one of miss-alignment between caissons.

4.5 HYDRAULIC DATA REQUIRED TO DESIGN A REINFORCED CONCRETE CAISSON

In order to design the wall thickness and percentage of reinforcement required for the front face of a concrete caisson, a realistic assessment is required of the distribution, magnitude and duration of the pressure loading resulting from a wave impact associated with a particular return period.

4.5.1 *Pressure distribution on front face*

Chapter 2 describes the state of the art as far as wave pressures on vertical breakwaters are concerned. In a preliminary analysis, the horizontal pressures should be assumed to act along the entire length of one caisson, assuming a normal wave attack on the front face. On the basis of numerous experimental trials and field measurements, the pressure intensity appears greatest near the mean water level. Using Goda's notation and bilinear vertical distribution, p_1, p_3 and p_4 characterise the pressure intensities at the mean water level, base of the caisson and top, respectively. In the case of very short duration impact, the pressure distribution does not appear linear over the height as the peak intensity is localised over a small region. The magnitudes of pressure may be estimated from the formulae in chapter 2.

Maximum pressure at mean sea level	p_1	N.m^{-2}
Maximum pressure at top of crown wall	p_3	N.m^{-2}
Maximum pressure at base	p_4	N.m^{-2}
Rise-time for impulsive load	t_r	S
Duration of impulsive load	t_d	S
Duration of impulsive and quasi-static phases	t_q	S
Total duration of pressure loading	t_t	S

The shape of the pressure-time history depends on the type of wave striking the caisson. Recent work has suggested that an impact pressure-time signature may be decomposed into a quasi-static component, an oscillatory component and an impact component. The latter can account for very high intensity loads occurring over very short durations. *Church-roof*, or simplified triangular, idealisations are considered as adequate for most dynamic analysis exercises.

Wave pressures act not only inwards on the front face of a perforated caisson, but also outwards and on the internal walls. Chapter 2 gives some guidance on likely pressure magnitudes in perforated caissons, compared to equivalent planar caissons. However, the phasing of the pressure pulses can be quite complex. In this respect, care is needed to ensure that the worst probable case is identified.

The phasing of the peak pressure acting around the caisson wall could be taken into account when analysing a circular-fronted breakwaters, although this extra detail might have a very small effect on the maximum bending moments and shear stresses experienced in the wall.

4.5.2 *Uplift pressure distribution on base slab*

Much work has recently been done on determining the distribution, intensity and duration/phasing of uplift pressures acting vertically on the underside of the caisson. Consideration of these pressures becomes important in global stability analyses. However, the consequences of these pressures on the bending moments experienced in the base slab are probably small, provided they do not lead to loss of contact of the base during rocking.

4.5.3 *Over-pressure on top slab and super-structure*

Superimposed loading on the top slab (in addition to the dead weight) may include not only the traffic loads and cargo handling/crane loads but also loads resulting from over-topping and water slamming down onto the upper surface. Estimates of likely water volumes and possible projection heights would enable a simple calculation of the probable vertical impulse.

4.6 FAILURE MODES ASSOCIATED WITH PRE-SERVICE AND IN-SERVICE CONDITIONS

Before detailed design calculations are undertaken, a series of plausible *failure* scenarios need to be identified and eliminated. It is important to note that the transportation and placing of cellular caisson often represent the most critical loading conditions to which a caisson is subjected, yet little design advice exists on this subject. By considering some of these issues, this section aims to offer new insights for coastal engineers.

4.6.1 *Pre-service states*

Pre-cast concrete caissons are either floated and towed out to their final location or hoisted into position by large cranes (mounted on barges or the existing caisson units). The act of floating and towing the caisson introduces a different set of forces onto the structure compared to those acting whilst in-service.

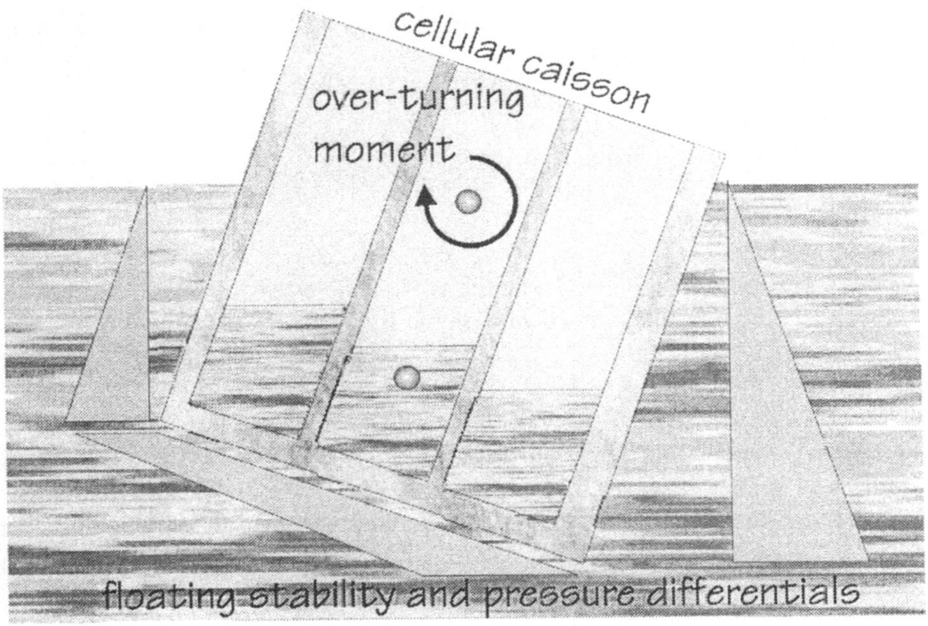

Figure 4-3. Cross-section through floating caisson (stability check).

If the unit is launched and transported in too rough a sea, then damage may occur due to either accidental sinking (this can be avoided by closing the top surface, or providing sufficient freeboard) or the slamming of adjacent caissons into each other. If the freeboard is too high, the floating caisson becomes unstable;

causing the unit to roll very easily. If the freeboard is too low, then over-topping may occur. Key calculations include a check on the availability of sufficient righting moment when the floating caisson is heeling over at an angle at which water may enter the cells (see Fig. 4-3 and Volume IIc, chapter 3 for further details). This calculation requires the meta-centric height to be determined. When floating, the base slab is required to resist a buoyancy pressure on its underside. Even a mild sea swell can induce global torsion and flexure in the unit which may lead to premature damage in the walls. Note that the use of circular internal cells can provide a higher torsional rigidity than rectangular cells.

Another situation which needs to be examined is the effect of drag forces on the caisson walls during the towing operation. The magnitude of the towing forces (which will be a function of the towing speed) will control the degree of local stiffening required in the wall, near the tow anchorage points.

As noted in section 4.4.3, uneven preparation of the foundation surface will cause a non-uniform pressure distribution on the base slab. In extreme cases (especially for a very large caisson) this may result in fracturing of the slab and rupture of the cross-walls. The ultimate consequences of this are the loss of fill and/or lack of geometric alignment.

Summarising, the following pre-service limit states may be considered:

- Uncontrolled sinkage during transport;
- Cracking or collapse of bottom slab or walls during floating stage;
- Cracking or collapse of bottom slab due to placement on an uneven bottom.

The pre-service states are not quantitatively dealt with in this chapter. A detailed analysis is given in chapter 3 of Volume IIc.

4.6.2 *In-service states*

In principle, once in service, the *failure* of concrete caisson breakwaters to provide tranquil water within a harbour by breaching the sea wall may be the result of both large-scale rigid body translation of the structure (due to global sliding at the base-foundation interface or rotational collapse of the foundation) **and** local rupture in the structural elements. The latter requires a sequence of damaging events to lead to a failure state.

The progressive loss in structural integrity may start by chloride ingress in the splash zone of the face of the breakwater. Small cracks may be present in the front wall, near the transverse cross walls, as a result of earlier wave impacts. If unheeded (and if exacerbated by thermal cycling), the chlorides may penetrate to the reinforcing steel, building-up sufficient concentration to provoke the onset of corrosion. Continued corrosion can result in a loss of bond, reduction in steel cross-sectional area, weakening of anchorage and bursting-off of the cover concrete. All these mechanisms can further weaken the reinforced concrete cross-section. If no significant reserve of strength exists at that section, the wall may rupture under repeated storm loading. Without a regular programme of inspection, diagnosis and repair, progressive deterioration of a wall panel may occur. Should sufficiently large cracks be induced in the front face, then this may lead to a washing-out of ballast in the cells. The ultimate consequence of loosing ballast, will be to reduce the frictional resistance at the foundation-base interface resulting in an increased risk of sliding failure. In order to prevent such a chain of events, coastal/structural engineers need some guidance on how to assess the likelihood of each mechanism. Note that partial collapse of the front wall or even minor shearing dislocation between caissons could also lead to loss of support and serviceability of the top slab, damaging crane rail-tracks, service ducts and/or vehicle access. The consequent reduction, or loss, of access to the structure may significantly restrict harbour operations without actually resulting in a breech of the sea wall.

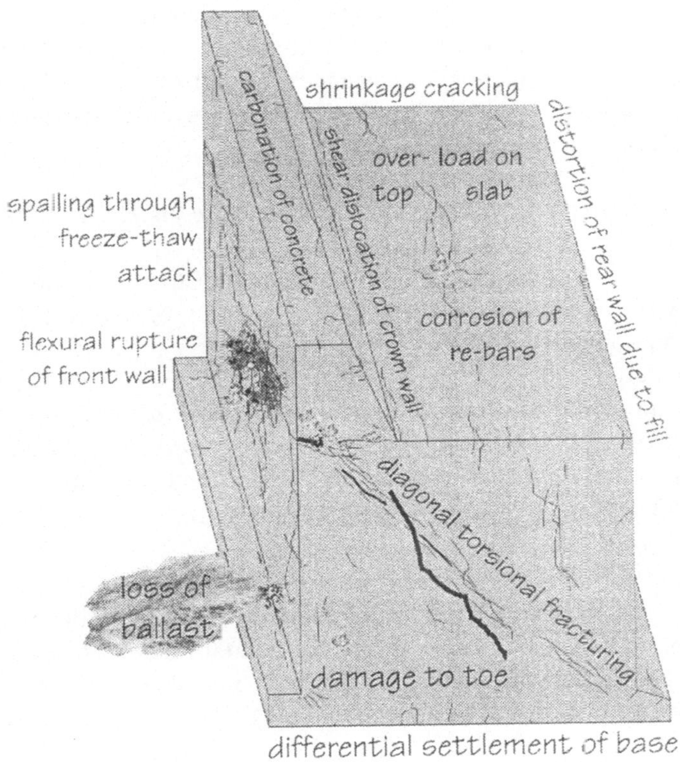

Figure 4-4. Sketch of some possible damaging mechanisms in a reinforced concrete caisson.

Multi-celled reinforced concrete caissons are generally highly redundant structures with several alternative load paths available. Local damage to the sea wall, in the form of the bursting or spalling of concrete will not immediately lead to a critical *failure* situation. Reasonable engineering judgement must therefore be exercised before structures are condemned just on the basis of *unsightly* corrosion stains or local loss of cover. In many cases, the structure may go on to provide years of active service before a collapse state is approached.

4.7 THE NEED FOR A NEW INTEGRATED DESIGN CODE

A number of codes of practice exist world-wide for the design of reinforced concrete structures; a few codes have been developed for the design of maritime structures. However, no single code is available which gives detailed guidance on the sizing of structural elements in a maritime environment (for example, within a reinforced concrete caisson). Designers are therefore currently forced to gather information from a variety of sources. Such an approach can lead to inconsistencies if carried out by inexperienced engineers, particularly when it comes to the specification (and factoring) of the loads acting on vertical breakwaters, the material parameters and the sectional resistance. As part of PROVERBS, five codes have been examined in some detail to assess their relevance, to make comparisons and to identify any significant omissions.

4.7.1 *Design standards relevant to reinforced concrete caissons*

The following codes have been examined

- (i) ACI 318-5 Building Code Requirements for Structural Concrete, USA
- (ii) CEB-FIP 1978 *Model Code for Concrete Structures*, Comite Euro-International du Beton
- (iii) EC1 Part 1 and EC2 Parts 1 and 2 *ENV 1991, Eurocode 1, Basis of Design and Actions on Structures, Part 1 Basis of Design* and *ENV 1992 Eurocode 2, Design of Concrete Structures, Part 1-1 General Rules and Rules for Buildings*, European Standardization Committee
- (iv) BS 6349 Parts 1, 2 and 7 *British Standard Code of Practice for Maritime Structures*, UK
- (v) ROM 0.2 90 *Maritime Works Recommendations: Actions in the Design of Maritime and Harbour Works*, Ministerio de Obras Publicas, Spain

This is not an exhaustive list, but one considered relevant to European coastal engineers[5]. The American ACI code has been included because it is often the required standard in overseas contracts.

---

[5] Other relevant codes could include the Norwegian NS 3473E and the French BAEL-91.

4.7.2 *Scope of selected codes*

The first three codes listed above, provide detailed guidance on the design of reinforced concrete elements such as beams, slabs, walls and columns irrespective of the use of the construction. These codes focus on the material and resistance parameters and offer a strategy for safely designing simplified structural components. Each of these three codes specify different load factors which should be used to multiply the various *characteristic load* intensities in order to finally arrive at the *design load.* The design effect of the loads results from the combination of the design values of the loads considered to act simultaneously for the particular situation[6] being studied. Both the EC2 and CEB-FIP codes are based on a semi-probabilistic framework whereby a series of limit states are identified and the concrete section designed such that these states have a notional probability of non-exceedence. None of the first three codes provide any information on how to calculate the magnitude and duration of wave loads acting on a vertical breakwater.

The last two codes give very general guidance on the planning and design of maritime structures, without specific reference to the design of reinforced concrete sections or structural modelling techniques. Both the British and Spanish codes cover buoyant and fixed structures. No advice is given on the determination of the characteristic and design wave load.

4.7.3 *Comparisons between design codes*

The five codes have been examined to highlight their differences (Volume IIc, chapter 2). The following section represents a brief summary of these findings. By way of example, comparisons are made between (i) the way in which individual load cases are classified (for example, transient or variable, permanent or accidental), (ii) whether maritime aspects are addressed, (iii) if a target reliability concept is used (and the return period and structure lifetime are addressed), (iv) the type of structural models to be used in the analysis, (v) the local structural models (for example, flexural bending and axial loading) and (vi) the cover requirements for the reinforcement.

ACI The ACI code gives no consideration of the *target reliability* nor does it address the different design *situations.* All situations are effectively treated as permanent, apart from seismic loads. Maritime aspects are not considered at all and there is no mention of a structure's design life or the relevant return period for a load. The ACI code considers both elastic and plastic behaviour and discusses both static and dynamic loads on beams, rafts, walls, footings and shells. The

[6] A situation corresponds to a set of physical conditions representing a certain time interval for which the design will demonstrate that relevant limit states are not exceeded and for which the stochastic safety aspects are considered to be stationary.

code does not explicitly employ a limit state philosophy, but local failure mechanisms are examined for bending, axial loading, shear, torsion, cracking and deflection. The concrete cover required over the reinforcement is specified as a function of the method of casting, the type of exposure, the size of the rebars and the type of structural element (a maximum cover of 75mm is suggested for the most extreme environments).

CEB-FIP The CEB-FIP model code[7] provides target probabilities of occurrence of the limit states in relation to the average number of people endangered together with the economic consequences. The code identifies permanent, transient and accidental design situations. The CEB-FIP code suggests a design life of 5 years for temporary works, 50 years for normal construction and 500 years for monumental construction. It gives an outline methodology for assessing the return period where a figure of between 125 to 200 years represents a typical value. The code encompasses elastic analysis, elastic with re-distribution, plastic analysis and second-order effects. Both static and dynamic actions are considered. A limit state approach is used, addressing local failure modes such as axial compression, shear in beams and punching shear in slabs, torsion and buckling. Concrete cover requirements depend on the exposure class, with a maximum of 40mm. The code indicates that the cover can be higher for sea structures, although no figures are suggested.

EC1/EC2 Eurocode 1 Part 1 (*Basis of Design*) describes the principles and requirements for safety, serviceability and the durability of structures. It is based on the limit state concept used in conjunction with the partial factor method. Ultimate limit states include loss of equilibrium, lack of mechanical strength and geometric instability. Cracking and excessive vibration and deformation are treated as serviceability limit states. Eurocode 2 gives detailed guidance on the design of reinforced concrete civil engineering works. EC2 was largely derived from the CEB-FIP model code. The target safety indices, β, are given as follows

Ultimate limit state for the design working life	3.8
Ultimate limit state for an annual event	4.7
Serviceability limit state for the design working life	1.5
Serviceability limit state for an annual event	3
Fatigue limit state for the design working life	1.5-3

The Eurocodes deal with permanent, transient and accidental (which includes seismic) situations. EC1 suggests a design working life of 1-5 years for temporary structures, 25 years for replaceable structures, 50 years for buildings and 100 years for monumental structures such as bridges and other civil engineering

[7] Note that a later version of the CEB-FIP code is available (CEB-FIP 90).

structures. EC2 gives only limited guidance on the type of structural analysis method to be used, however the adopted approach must be based on established engineering theory. For example, EC2 provides some basic rules for the analysis of simple structural elements and notes that, where necessary, the method used should be verified experimentally.

Partial safety factors for permanent, transient and accidental loads at the ultimate and serviceability limit states are defined depending on whether the action is favourable, or unfavourable. Different partial safety factors also apply to the material strengths for the two limit states.

Eurocode 2 specifies a minimum concrete cover of 40mm in the splash zone of a coastal structure where freeze-thaw action is possible.

BS 6349 The British code does not explicitly address target reliabilities, but the framework of the code makes reference to risk analyses. Design situations are not discussed in BS 6349, whereas maritime conditions are covered in considerable depth (including guidance on determining the wave climate). A design life of 100 years is recommended for flood protection works and 60 years for shore protection works, breakwaters and quay walls. Goda's wave pressure formulae are referred to in BS 6349. No information is given on the local structural detailing, although some typical construction arrangements are presented. There is no information in BS 6349 on the limit states, failure mechanisms or concrete and reinforcement strengths. Similarly, the required concrete cover is not discussed.

ROM 0.2 90 The Spanish code provides a list of maximum admissible quantified risks associated with three levels of danger and two levels of potential human loss. A distinction is made in the code between the construction and service phases. ROM 0.2 90 provides the partial load factors for various maritime actions but does not describe how to calculate the *characteristic* values. The design life of a marine structure is considered to be dependent on the safety level required of the construction and the type of installation. The Spanish code gives no real guidance on the types of structural models which should be used to analyse a caisson breakwater. No mention of the concrete cover requirements are given in ROM 0.2 90.

4.7.4 *Suggested features for a possible new unified design code*

From the previous review, it emerges that three of the codes are worthy of further study: the **CEB-FIP** model code for the basic principles of a semi-probabilistic approach, **BS6349** for dealing with the maritime aspects (despite being formulated essentially from a deterministic view point) and the **Eurocodes**, not only because they are the most recent application of the CEB-FIP semi-probabilistic principles, but also because they will become the European Community's unified codes.

There is a real need to formulate recommendations for maritime structures which harmonise the safety approaches. Simply assembling sections from each of the different codes would be inappropriate because of the lack of homogeneity. In fact, such a piecemeal approach may lead to quite unreasonable results. For an illustration of this, see Volume IIc, chapter 2 where the wave loadings adopted for the ultimate limit state design calculations lead to a lower reinforcement requirement than that needed to satisfy a serviceability limit state.

It is the lack of guidance on the appropriate wave loading which is of most concern for structural designers. The information currently available on the time and spatial distribution of wave loads (pulsating or impact) is not easy to apply within a limit state structural calculation. When examining the Eurocodes, a given action may have many different values: characteristic, design and combined values, each of which may be associated with frequent, quasi-permanent or accidental conditions.

- The *characteristic* value is generally determined from wave data. The questions are: (i) which wave parameter should be used (the pressure calculated on the basis of a significant wave height, or a maximum wave height)? (ii) which return period should be used? Is the 50 year return period (generally adopted for the characteristic value of a variable action) also relevant for wave loads? When determining the characteristic value of the wave parameter, consideration must be taken of the size and quality of the data describing the wave climate.
- The *design* value can be determined directly from the wave data, or by applying a partial factor to the characteristic value. In the latter case, should the partial factor be applied to the wave height or the wave force (for example, wave pressure or overturning moment)?
- The *combined* value should be consistent with the representative values of other parameters such as: water levels, wave period, wave direction and current velocity. For example, the combination of a 100 year wave height and a 100 year surge is likely to be have a probability far less than 1/100 to be exceeded in one year. What are the appropriate water level conditions, wave periods and current velocities to be used in combined situations?
- The *frequent* and *quasi-permanent* values can be determined either directly from the wave data, or with a ψ_i factor applied to the characteristic value.

It is argued that the design life of a caisson structure forming part of a *permanent* harbour wall should be at least 100 years. Some components of the breakwater might require renewal during this lifetime (such as parts of the superstructure) but the main structural elements should remain durable without requiring major maintenance throughout this period.

The behaviour (and hence the design) of a caisson involves significant soil-fluid-structure interaction, yet the existing codes do not offer a consistent set of partial safety factors for each of the phases soil-fluid-structure). One way to achieve consistency between the different approaches (without significantly re-writing the existing codes to include maritime aspects) would be to separate the treatment of the uncertainties. It is suggested to make some minor adaptations to the semi-probabilistic way of thinking. This approach would consider the uncertainties comprehensively for a given limit state function. Partial factors operate on the actions and materials; their values could be taken from existing codes or regulations. For example, rules for determining representative concrete values are already given in EC2 (with a partial factor of $\gamma_c = 1.5$ for concrete, and γ_s=1.15 for steel reinforcement). Recommendations for determining representative soil parameter values are already given in EC7. Partial factors should only account for the intrinsic uncertainties of the parameters, without consideration of the particular limit state function.

Model factors are introduced in the limit state function at the last stage of the design process. They differentiate between safety levels according to the limit state and allow for (i) the discrepancy between the model (for example, simplified equivalent static beam analogy, or full non-linear dynamic FE study) and reality, (ii) the required safety level and (iii) the design working life.

The model factors should be calibrated using probabilistic procedures. Their values depend on the pre-determined safety levels specified by the National Regulation Authorities. According to the Eurocodes, the model factor can be split into an *action* model factor γ_{sd} and a *resistance* model factor γ_{rd}. The canonical form of the limit state function could then be written as

$$\gamma_{sd}\, E(\gamma_g\, G_k + \gamma_h\, H_k) < \gamma_{rd}\, R(X_k / \gamma_m) \qquad (4\text{-}1)$$

where γ_g are the pre-determined factors for permanent actions, γ_h is the pre-determined factor on the wave and γ_m are the pre-determined material factors. G_k represents the characteristic value of the permanent action, H_k the characteristic value of the wave, X_k the characteristic value of a material parameter, E( ) signifies the effect of an action and R( ) indicates the structural (or foundation) resistance.

4.8 SIMPLIFIED LIMIT STATE EQUATIONS

4.8.1 *Identification of structural idealisations*

Before individual structural members are designed, the load paths must be identified and the basic global structural action understood. It is not possible to treat the structure as an equivalent 2-d plane strain problem because of the arrangement of cross walls which stiffen a rectangular caisson (this also holds for circular caissons, for obvious reasons). The three-dimensionality of the structure therefore needs to be taken into account in order to appreciate the manner in which the forces are transmitted through the walls to the base slab, and down through the foundation.

4.8.1.1 *Simplified beam and slab analogies and associated limit state equations*

Consider a rectangular, planar-fronted multi-cell caisson. Each of the main structural elements will now be examined and the simplest possible idealisation identified.

The **front wall** will be subjected to horizontal pressures acting outwards (due to the ballast fill in the cells) at low tide[8] and horizontal pressures acting inwards when struck by a storm wave. This loading will cause a rectangular panel to act rather like a one-way horizontally spanning slab supported along its length by the vertical cross-walls (Fig. 4-5).

This one-way action only holds for horizontal strips remote from the top and bottom slabs (that is, at least one span width above the base slab and below the top slab). Close to the base and top slabs, the action is essentially two-way and the deflections will be reduced. It is worth pointing out that the maximum pressure from the internal ballast will occur near the base slab whereas the maximum pressure from the wave loading will typically occur near the top slab. For the purpose of a preliminary sizing of the front wall, the peak wave pressure[9] may be considered to be acting uniformly over a horizontal strip across the caisson face. A unit width beam continuously supported over the internal cross-walls may be analysed to determine the wall thickness and maximum percentage of reinforcement required. Note that the ability of the wall to resist the outward pressure from the ballast alone must be considered as an important load case.

The stiffening effect in the wall due to the fill material behind it, as a wave strikes the front face, is quite difficult to accurately assess; although simple calculations indicate that this effect will be small and so it is generally neglected. It must be remembered that the front wall may also carry a moderately high vertical axial load and bending moments from the top slab. The compression loading will come from the weight of the crown wall and top slab and self weight of the front face, in addition to some proportion of any live load acting on the top slab.

The **rear wall** will be subjected to a similar loading regime as the front wall except that the wave pressures will be very much reduced. Berthing forces could, however, be significant for a harbour quay.

[8] Or when a wave trough occurs in front of the wall.

[9] Minus the active ballast pressure, if the fill is in full contact with the wall near the top.

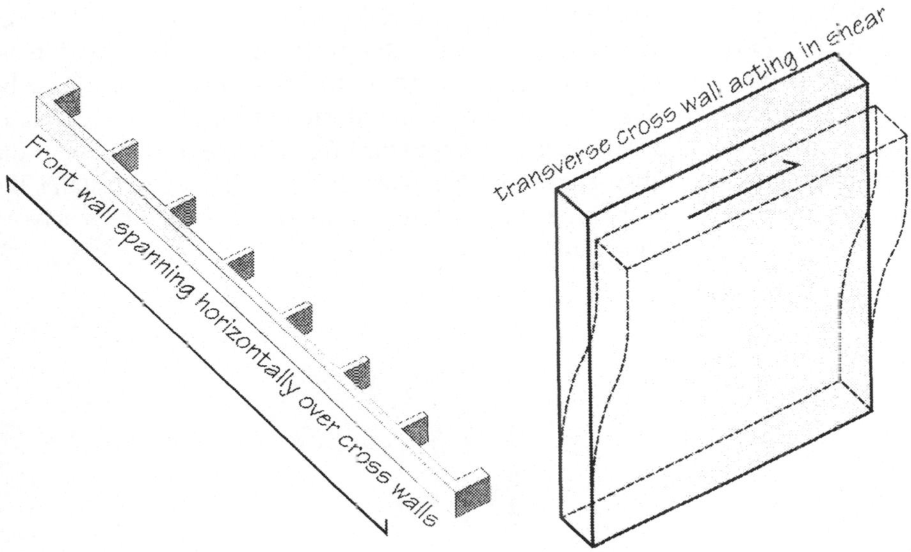

Figure 4-5. Isometric view of front wall (1m strip) and a cross wall panel.

The **side walls** must be designed to retain the ballast fill and resist in-plane shear stresses in order to transfer the horizontal loads from the front face to the base slab. The in-plane shear stiffness will generally be so high as to render these stresses very small. Depending on the degree of inter-connectivity between adjacent caissons, the side walls may also be required to resist the local horizontal forces carried by the vertical shear keys and the (relatively minor) wave impacts in the clutches.

The **internal cross-walls** will carry the vertical loads from the top slab to the foundation and contribute to the transverse stiffness of the caisson box by transferring the horizontal forces (mobilising the transverse, front-to-back, wall's in-plane stiffness) from the external walls to the base slab. These walls should be designed to support the ballast fill pressures assuming no fill in the neighbouring cell[10]. The presence of the internal transverse and longitudinal walls add greatly to the torsional rigidity of the caisson, an important consideration during the float-out and towing phases.

The **base slab** will be subjected to vertical pressures acting upwards from the supporting foundation and uplift water pressure during a wave impact. These loads will be in equilibrium with the downward forces arising from the weight of the caisson. The vertically downward loads will be transmitted to the base slab via the walls and ballast. Bending moments resulting from horizontal pressures acting

[10] This condition could occur during the placing stage.

on the walls will also be carried into the base slab. A *Beam-on-an-Elastic Foundation* model could be used to determine the design moments and shear forces, however a simpler one-way spanning (front-to-back) beam model could also be used for the preliminary design. Moments acting at each of the side-to-side internal wall locations could be shared according to the effective lateral stiffness of the front-to-back internal walls. The base slab must also be able to withstand the bending moments and shear forces induced as a result of the structure receiving only partial support from the foundation.

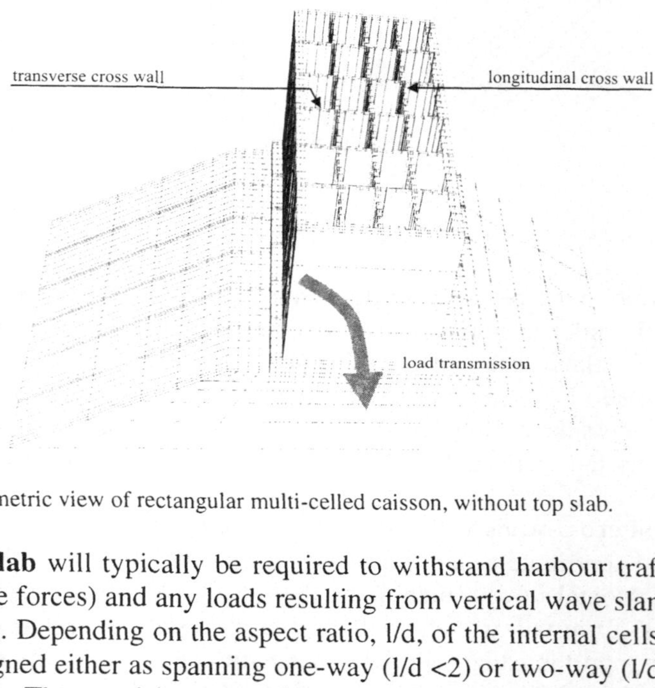

Figure 4-6. Isometric view of rectangular multi-celled caisson, without top slab.

The **top slab** will typically be required to withstand harbour traffic loads (including crane forces) and any loads resulting from vertical wave slamming during over-topping. Depending on the aspect ratio, l/d, of the internal cells, the top slab may be designed either as spanning one-way (l/d <2) or two-way (l/d >2) over the internal walls. The top slab can provide very considerable transverse stability to a cellular caisson by virtue of its high in-plane stiffness. This membrane action contributes to the distribution of the horizontal forces (acting on the front face) out to other internal and external walls. The top slab may be cast directly on the ballast fill, or formed by casting a thinner in-situ reinforced layer over a series of pre-cast slabs or beams. The latter construction technique, although quicker, will leave a void underneath the top slab.

The **crown** (or sea) **wall** and associated super-structure must be designed to resist a severe storm wave crashing onto its vertical face without inducing significant damage. This element is subjected to the largest temperature variations. Depending on the location of the breakwater, the concrete may be exposed to temperatures below freezing, or temperatures up to 40°C. Structurally, the crown wall may be treated as either a simple, monolithic gravity element, or a vertical cantilever depending on its relative slenderness. In either case, the horizontal load may be idealised as being transferred by the shear resistance acting at the horizontal interface between the base of the crown wall and the top slab.

If present, **shear keys** form a mechanical interlock which is designed to share the load between adjacent caissons. These are considered to be highly desirable. In one approach, transfer of the horizontal wave loads is achieved by relying on the concrete's shear resistance (in a vertical plane) over the full height of the key. A second, preferred approach, is to introduce a granular fill into the gap between caissons (over the full height and most of the width) to mobilise the frictional resistance of the confined material. This technique places fewer restrictions on the precision of the geometric alignment needed between neighbouring caisson units.

4.8.2 *Limit state equations*

No single simple analytical model is relevant for all caisson structures. Because of the relatively complex load sharing which takes place within a cellular caisson, most design engineers undertake linear finite element analyses to determine maximum bending moment and shear force envelopes. This technique has transformed the way in which structures have been designed during the past 35 years. Today, even the smallest design office can gain access to a general purpose linear analysis FE programme. However, despite enormous increases in the processing power of modern computers, full three-dimensional dynamic analyses for impact problems demand significant computer resources. For this reason, simplified approaches based on the assumed behaviour of individual elements, are still used in the preliminary design stage. The following section identifies some elementary, but useful, limit state equations. Where available, and considered appropriate, the EC2 expressions have been used. The following notation applies: D denotes a deterministic variable, N(m, σ) a normal distribution with mean value of m and standard deviation σ and LN implies a log-normal distribution.

4.8.2.1 *ULS for flexural failure of a reinforced concrete member*

Considering an equivalent unit width beam (with no compressive reinforcement), spanning one-way continuously over at least 6 equal-span cells, the LSE for flexural failure in an under-reinforced section is given by (see, for example, O'Brien & Dixon, 1995)

$$g_1 = g_1(\rho_r, d, f_y, \alpha, f_{ck}, p, L) = \rho_r\, d^2\, f_y\, (1-(0.4\rho_r f_{ck} / \alpha\, f_{ck})) - 0.08pL^2 \qquad (4\text{-}2)$$

where ρ_r is the area ratio of steel reinforcement with respect to the concrete cross-sectional area D(0.015-0.04), d (see Fig. 4-7) is the depth of the section from the compression face to the centre of the tensile steel reinforcement D(0.25-1.5m), f_y is the characteristic yield strength of steel reinforcement LN(460MPa, 10MPa), α is a coefficient which takes account of the long-term affects on the compressive strength and of the unfavourable effects resulting from the way in which the load is applied (adopt α=0.85 as a default value), f_{ck} is the characteristic compressive strength of concrete LN(40-60MPa, 4-8MPa). EC2 denotes a concrete with a characteristic cylinder strength of 30MPa and a characteristic cube strength of 37MPa, as grade C30/37 concrete. Other grades include C35/45, C40/50, C45/55 and C50/60. Concrete of grade at least C40/50 should generally by used in a marine environment to limit the chloride diffusion. p is the net uniformly distributed pressure acting on the member (in the case of the front wall, p is the arithmetic sum of the applied wave loading and the internal cell pressure. Finally, L is the effective span distance between the supports. The factor 0.08 is chosen as a representative value for the maximum (mid-span) bending moment occurring in the middle of the outer-most span of a caisson with 6, or more, cells.

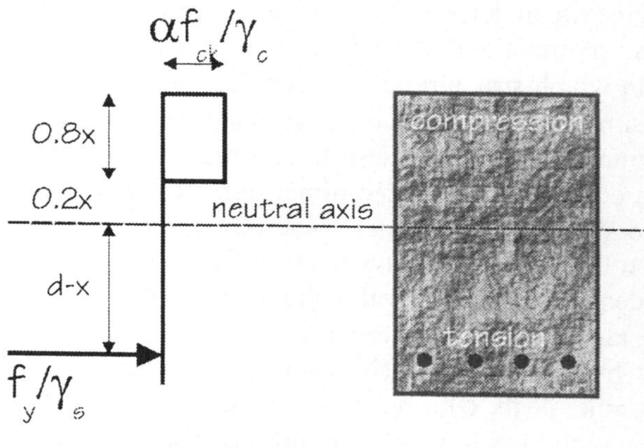

Figure 4-7. Idealised stress block for a reinforced concrete beam under flexure.

Note that if the EC2 partial safety factors for strength (γ_c and γ_s) and model factor (γ_d) are included in the above expression, the LSE becomes

$$g_1 = (\rho_r\, d^2\, f_y /\gamma_s)(1-(0.4\rho_r f_{ck}\gamma_c /\alpha\, f_{ck}\gamma_s))-(0.08pL^2)/\gamma_d \qquad (4\text{-}3)$$

4.8.2.2 *ULS for shear failure of a reinforced concrete member*
The following expression applies to the shear state in a beam at a distance d from the edge of the support wall. The factor 0.6pL corresponds to the maximum shear force experienced in the outer-most span, nearest the internal support.

$$g_2 = g_2(f_c, d, \rho_l, \sigma_{cp}, p, L) = (0.0525 f_{ck}^{2/3}(1.6-d)(1.2+40\rho_l)+0.15\sigma_{cp})d-0.6pL \quad (4\text{-}4)$$

where ρ_l is the lesser of longitudinal tension reinforcement ratio and 0.02, D(0.005-0.02), σ_{cp} is equal to N/A_g where N is the axial force and A_g is the gross area of the cross section. If punching shear is to be checked in a slab or wall, then the term -0.6pL is replaced by the actual level of shear force acting on the loaded area, the term $0.15\sigma_{cp}$ is not included and the term $(1.2+40\rho_l)d$ is now multiplied by b_w, the length of the critical shear perimeter.

4.8.2.3 *Cracking in a flexural reinforced concrete member*
Cracking in concrete members in a seawater environment will accelerate the rate of chloride penetration and thus speed up the rate at which corrosion may first appear. Therefore, it is necessary to pay particular attention to prevent the development of cracks with widths of 0.3 mm of more.

$$g_3 = g_3(\eta, f_{tj}, \sigma_r) = 90(\eta f_{tj})^{1/2} - \sigma_r \quad (4\text{-}5)$$

where η is a coefficient which is equal to 1 for normal (round) bars and 1.6 for deformed bars, f_{tj} is the characteristic tensile strength of concrete (in MPa) and σ_r is the actual stress in the tensile reinforcement (also in MPa). Note that if $90(\eta f_{tj})^{1/2}$ is greater than $0.5f_e$, then σ_r should be compared against $0.5f_e$ where f_e is the stress in the reinforcement corresponding to the end of the elastic phase.

4.8.2.4 *Chloride penetration and corrosion in reinforced concrete elements*

$$g_5 = g_5(C_{cr}, C_o, x_c, D_c, t_l) = C_{cr} - C_o(1-\mathrm{erf}(x_c/2(D_c t_l)^{1/2})) \quad (4\text{-}6)$$

where C_{cr} is the critical chloride ion density (in $kg.m^{-3}$) when corrosion starts at the surface of the reinforcement, C_o is the measured chloride ion density at the surface of the concrete, x_c is the depth of the concrete cover, D_c is the chloride diffusion coefficient (in $m^2.s^{-1}$) and t_l is the lifetime of the structure (or the time at which an assessment is to be made) (Matsushima et al., 1998).

4.9 UNCERTAINTIES ATTRIBUTED TO THE LS EQUATIONS: MORE REFINED STRUCTURAL MODELS

In order to assess the uncertainties attached to the simplified LS equations presented above, the following series of analysis models have been studied. The three dynamic analysis methods (3-DOF, Shell FE and Continuum FE) represent a hierarchical progression towards greater realism. This sub-section concludes with an estimation of the precision which can be attached to the simplified, equivalent static SDoF beam type models.

4.9.1 *Simple 3-degree-of-freedom dynamic model*

In order to examine whether a full dynamic model is justified when analysing the front face of a vertical caisson breakwater subjected to a wave impact, a simplified 3-*degree of freedom* model may be used (Fig. 4-8). This model (Tan, 1998) builds upon the elastic translational and rotational models developed by Oumeraci & Kortenhaus (1994) and Pedersen (1997). Two of the degrees of freedom correspond to the rigid body horizontal translation and rotation; the third degree of freedom represents bending of the front wall.

The dynamic equation of motion for such a 3-DOF system is given by

$$\begin{bmatrix} m-m_w & 0 & 0 \\ 0 & \theta_t & 0 \\ 0 & 0 & m_w \end{bmatrix} \begin{Bmatrix} \ddot{x} \\ \ddot{\phi} \\ \ddot{x}_w \end{Bmatrix} + \begin{bmatrix} c_s & -c_s l_{h_4} & 0 \\ -c_s l_{h_4} & c_\phi + c_s (l_{h_4})^2 & 0 \\ 0 & 0 & m_w \end{bmatrix} \begin{Bmatrix} \dot{x} \\ \dot{\phi} \\ \dot{x}_w \end{Bmatrix} +$$

$$\begin{bmatrix} k_s + k_w & -k_s l_{h_4} & -k_w \\ -k_s l_{h_4} + k_w l_{h_2} & k_\phi + k_s (l_{h_4})^2 & -k_w l_{h_2} \\ -k_w & 0 & k_w \end{bmatrix} \begin{Bmatrix} x \\ \phi \\ x_w \end{Bmatrix} = \begin{Bmatrix} f_{h_1} + f_{h_3} \\ f_{h_1} l_{h_1} - f_{h_3} l_{h_3} + f_v l_v \\ f_{h_2} \end{Bmatrix} \quad (4\text{-}7)$$

where m is the total mass of the caisson, m_w is the mass of a full height rectangular panel on the front face, θ_t is the rotational inertia of the caisson, and x and x_w represent the horizontal displacement of the body of the caisson (minus the front wall) and displacement of the caisson including the front wall, respectively (over-dots and double over-dots signify first and second order differentiation with respect to time). ϕ indicates the rotation of the caisson and c_s the damping of the coupled foundation/fluid. L_{h4} is the level arm length between the point of horizontal reaction and the centre of rotation, c_ϕ is the rotational damping. k_x is the foundation stiffness, whereas k_w is the stiffness of the front wall and k_ϕ is the rotational stiffness.

Vertical equilibrium and motion are not addressed in this model (the rubble mound reaction equates to the vertical load resulting from the structure's self weight). The horizontal load is split into three parts; upper and lower forces which do not bear onto the front wall (f_{h1} and f_{h3}), and a mid force which acts on the wall (f_{h2}). l_{h1} is the vertical lever-arm distance from centroid of the caisson to f_{h1}, l_{h2} is the vertical lever-arm distance from centroid of the caisson to f_{h2} (shown as zero in Fig. 4-8) and l_{h3} is the vertical lever-arm distance from centroid of the caisson to f_{h3}. f_v is the vertical uplift force and finally l_v is the horizontal lever-arm distance from centroid of the caisson to f_v.

The model has been coded using a Newmark time-integration scheme as a 100-line MATLAB script (Tan, 1998). Figure 4-9 shows a simulation of the response of Genoa Voltri breakwater during a wave impact where peak pressures of 660 kPa are assumed. The natural frequency of the front wall in a typical Genoa Voltri caisson is calculated to be of the order of 140Hz, whereas the frequency of global rotation the structure is approximately 1Hz.

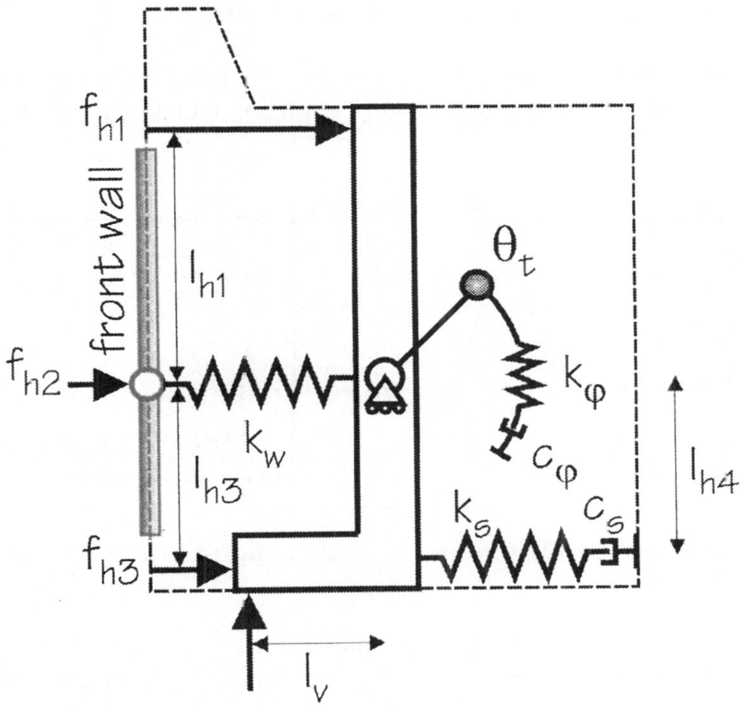

Figure 4-8. Diagram of 3-DOF dynamic lumped parameter model.

G_{rm}	200MPa
T_r	0.01s
T_d	0.0025s
d	0.7m
p	660kPa

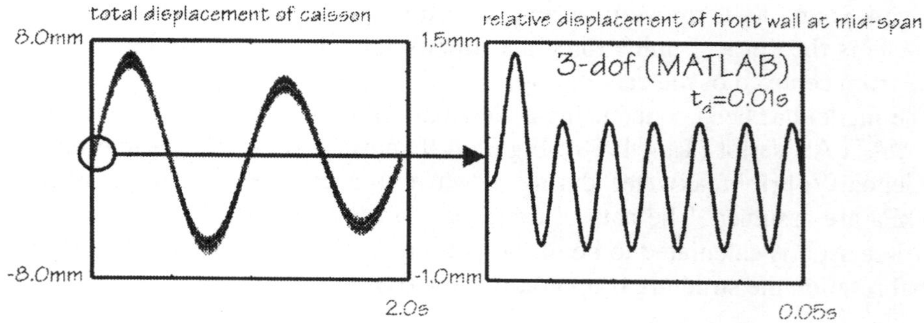

Figure 4-9. Horizontal motion of caisson under wave impact (Genoa Voltri).

Figure 4-9(a) shows the total displacement of the front wall under a triangular pressure pulse of duration, t_d, 0.01s and rise time, t_r, 0.0025s. Note that the peak displacement is of the order of 7mm.

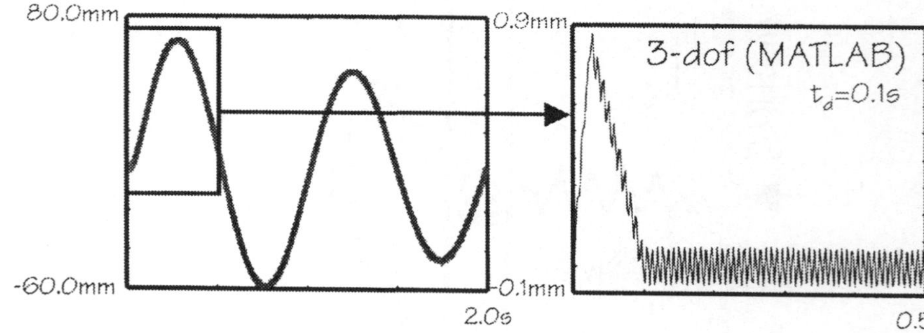

Figure 4-10. Horizontal motion of caisson under wave impact with longer rise time and impact duration (Genoa Voltri).

Figure 4-9(b) shows the relative mid-span displacement of the front wall under the same loading. The short duration impact causes a dynamic amplification (in this linear analysis) of approximately 1.63. This reduces to 1.33, if the duration remains constant, but the rise time increases to 0.005s.

Figure 4-10 shows a second simulation, with a tenfold increased impact duration (from 0.01 to 0.1s), but the same peak pressure. In this case, the global maximum caisson displacements are much larger (approximately 70mm), but the wall deflections are smaller (0.85mm compared to 1.4mm) as the external force no longer excites the wall bending mode so clearly. For this structure, wave impacts with a rise time greater than about 0.075s will lead to dynamic deformations essentially equal to the static deformation (0.85mm). Thus, an equivalent static analysis is reasonable when designing the front wall for moderate to longer duration wave impacts. Note that because of the low natural frequency associated with the rigid body motion of the caisson (compared to that of the wall), changes of even one or two orders of magnitude to the rubble mound stiffness have a negligible influence on the relative wall displacement. Thus, a minimal role is played by the foundation in influencing the wall's maximum bending moments, for the example considered here.

It is important to remember that even though the dynamic bending moments may be higher than the equivalent static moments, it does not imply that a section will fail if it has been designed to only just resist the static loads. In order to determine if dynamic rupture will occur, a non-linear analysis is required. The impact load will be on the structure for a very short time. Some of this load will be resisted by the inertial forces and there may be sufficient ductility in the section to partially yield without complete loss of load carrying capability.

4.9.2 *Layered shell non-linear FE models*

Whilst the 3-DoF model captures the basic dynamics of the front face of a caisson, the use of a layered shell, explicit FE analysis code provides a more advanced tool for relatively thin walled, curved structures. Within this framework, through thickness cracking may be simulated using an equivalent smeared approach operating at nine *Gauss* points in each layer of each element. Cracks may open (and close) normal to the shell layers. The rate at which softening (leading to a complete loss of tensile load capacity) occurs is controlled by the Specific Fracture Energy and the inelastic strain rate. Geometric non-linearity has also been included in the following example, to quantify the membrane stiffening effects under increased deformations. Dynamic equilibrium is expressed as

$$\sum_n [M]\{\ddot{d}\} + \sum_n [C]\{\dot{d}\} + \sum_n [K]\{d\} = \sum_n \{f\} \qquad (4\text{-}8)$$

where [M], [C] and [K] are the elemental mass, damping and stiffness matrices respectively, {f} are the external forces and the summation symbol implies addition of each elemental contribution to the global system of equations. The family of enhanced isoparametric shell elements used in the FE code were originally developed by Huang (1989). These elements exhibit superior characteristics (in the sense of a reduced tendency to *shear-lock* as the shell becomes thinner) over conventional shell elements. The reinforcement is treated as a stiffer layer within the shell. A total of ten layers through the thickness of the shell were adopted (6 layers to represent the concrete and 4 for the steel). A lumped mass matrix scheme is used whereby the element mass is distributed in proportion to the diagonal terms of the consistent mass matrix (Rock et al., 1976).

As in the 3-DoF studies, a rise time of 0.0025s and duration of 0.01s was assumed in these analyses. Note from the deformed plot the two-way bending action near the top and bottom of the wall panels. Figure 4-11 shows the horizontal displacement contours on the front face of a multi-celled caisson subjected to a triangular pressure pulse with a bi-linear vertical distribution. Figure 4-12 gives the corresponding displacement-time histories. The curve denoted *non-linear 0.7m* includes both material and geometric non-linearity effects. A second example analysis (using the same pressure pulse) illustrates the effect of using a reduced front wall thickness (*linear 0.5m* and *non-linear 0.5m*). Only one half of a single wall panel was considered; the panel being idealised as simply-supported on the bearing edges.

Hydrodynamic added mass and damping effects are neglected here, the latter is considered to have little influence on the peak displacement, which is realised very early in the analysis in this short duration impact.

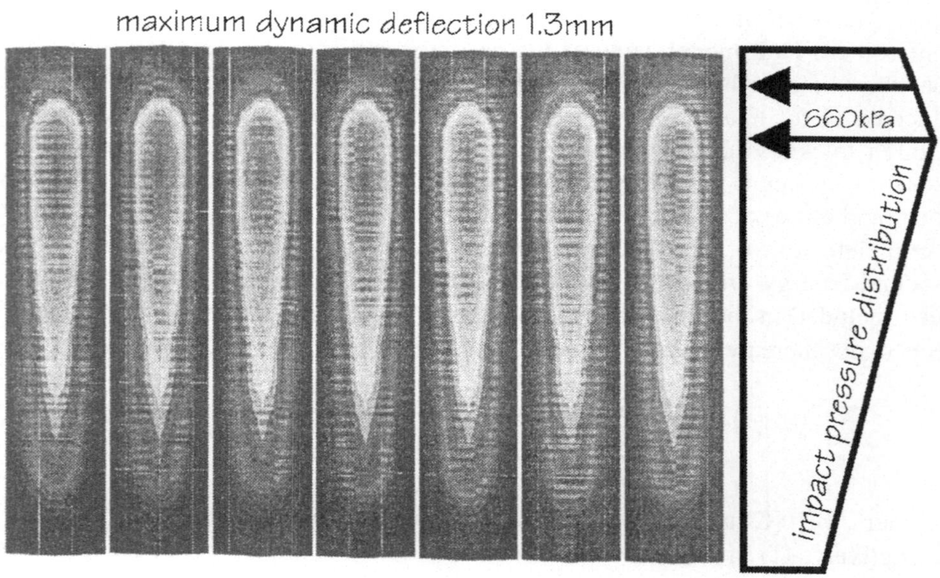

Figure 4-11. Contours of maximum horizontal displacement on front face of caisson under wave impact using FE shell analysis

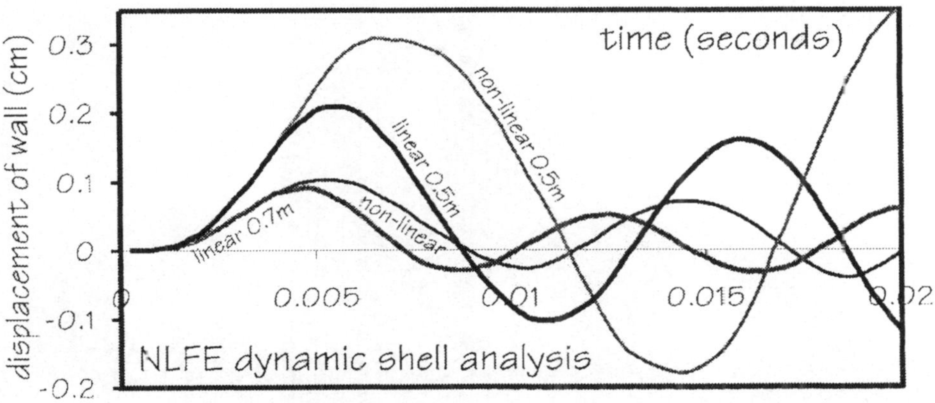

Figure 4-12. Horizontal motion of caisson under wave impact using FE shell analysis.

4.9.3 *Full 3-dimensional continuum FE models*

Fully 3-dimensional FE non-linear dynamic codes demand very significant computer resources, yet such techniques are needed in many structural analysis problems. For example, in the case of a caisson sitting on a rubble mount, over 20,000 20-noded brick elements may be needed to represent the structure in sufficient detail. If 1000 time-steps are to be followed and, on average, 10 non-linear iterations are required to reach dynamic equilibrium, then an analysis may take over 100 hours running on the latest generation Unix workstation. Even the condensed results from the analysis may consume well over 1GB of disc storage. It is strongly recommended that a linear analysis be undertaken prior to performing any non-linear analyses. Linear analyses will provide significant insight into the way the structure is transmitting the loads and the results should be used to verify if the mesh density, alignment and boundary conditions are appropriate[11].

At the University of Sheffield a new finite element code and caisson pre-processor have been specifically written for the PROVERBS research study. The new code, *yaFEc*, offers a number of unique features to provide a robust advanced simulation tool. The features include the use of a fast pre-conditioned gradient iterative solver within a Hilber-Hughes-Taylor time-stepping algorithm and the use of fully consistent tangent matrices and a Closest-Point return scheme (in a Newton-Raphson non-linear solution approach) for the hardening/softening plasticity model. The microplane constitutive formulation (Carol et al., 1992) is also included as an optional material model.

---

[11] See NAFEMS guidelines and Cook et al. (1989) for practical advice on performing FE analyses and Isenberg (1993), Comité Euro-international du Beton (1996) and Bazant (1989) for further guidance on modelling reinforced concrete.

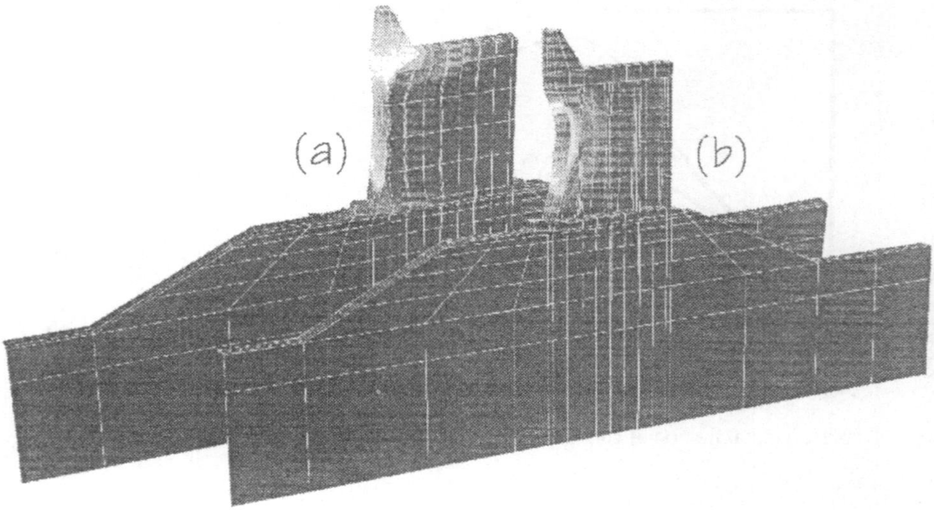

Figure 4-13. Deformed mesh showing contours of horizontal displacement using full 3-D FE dynamic analysis (Genoa Voltri Caisson, t=0.07s): (a) caisson with ballast fill in cells; (b) caisson without fill in cells.

Figure 4-13(a) shows the (exaggerated) deformed shape, with horizontal displacement contours superimposed (bending moments, or shear forces could be plotted in a similar manner) for the Genoa Voltri breakwater. In this example, the ballast fill inside the cells has been modelled by 3-d continuum elements. Figure 4-13(b) shows the comparable results for the situation where the ballast has not been included in the analysis. Both plots show the deformed structure at the same stage of loading under identical pressure pulses (p_1=660kPa, t_r=0.0025s and t_d=0.01s). Note the much higher bending deformation in the front wall in the second case. Such analyses illustrate the progressive transmission of the pressure pulse through the structure into the foundation. Figure 4-14 gives the corresponding horizontal displacement-time curves for the (ballast filled) front wall.

It should be remembered that this illustrative example neglects any restraint which may be offered by the adjacent cells and neighbouring caissons as the transverse cross-walls are considered as free to translate horizontally (but not laterally). Smaller displacements will result if the stiffening effect from the thick side walls and adjacent caissons, plus the spatially localised pressure distribution (rather than long crested assumption) were take into account.

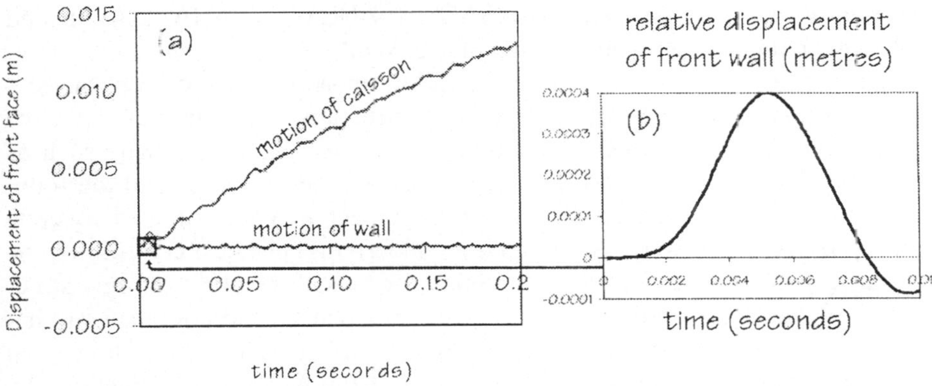

Figure 4-14. Transient wall displacement from 3-D FE analysis.

Care should be exercised before reading to much into the direct comparisons between the four structural models (simplified flexural LSE, 3-DoF dynamic model, non-linear FE shell analysis and 3d continuum dynamic FE analysis) because of the slightly different boundary conditions and material constants used in the runs. The bending moment predicted by the flexural LSE, over-estimates the static value given by the 3-DoF model by just 5%. This is because the bending moment factor for a multi-celled rectangular caisson has been rounded-up. The peak dynamic displacement in the wall predicted using the 3-DoF model is approximately 1.5 times the value given by the shell analysis. This is due to the full centre-to-centre span (of 4.18m) being using in the former analysis, whereas the clear span (or 4m) was used in the latter. It is also a result of a low elastic modulus (25GPa) being used in the 3-DoF but not FE shell analysis (where 30GPa was used). The shell analysis predicts a maximum wall displacement of twice that observed in the full 3-dimensional FE analysis. This last disagreement is largely due to the shell analysis assuming simply supported boundary conditions, whereas the continuum analysis considers the restraint generated by the cross walls. When one looks at the total horizontal displacements (3-Dof and 3-dimensional FE continuum analyses), the FE simulation predicts a peak translation of about 14mm, whereas the 3-DoF model gives just 7mm. This is due to the fact that the FE analysis adopted a lower berm stiffness than that used in the 3-DoF model, and the FE analysis reported here neglected the inertial mass contribution from the fluid (this was included, in a simplified manner, in the 3-Dof model). Note that in all three dynamic analyses, the times at which the wall reaches its maximum relative displacement are similar (approximately 0.005s, for an impact with a rise time of 0.0025s and duration 0.01s).

One may conclude that the three types of dynamic analyses show broad agreement. Today, sensitivity analyses using sophisticated 3-d linear (and to a lesser extent, non-linear) FE codes such as ABAQUS, ANSYS, DIANA, LUSAS and DYNA may be undertaken in any design office. Whilst an equivalent static analysis is appropriate for all but the shortest duration impacts (in the case of the rectangular, cellular breakwater described above), structural engineers now have the means to investigate the effect of including, or neglecting, phenomena such as material non-linearity in the soil, loss of contact between the base and the foundation during rocking and reduced steel reinforcement area and softened concrete (to simulate the corrosion). Although the underlying physics controlling the structural response is well understood, comparisons with results from real structures are still needed to provide greater assurance that all the dominant mechanisms have been addressed. Large scale laboratory investigations have shown that there remains more work to be done in the area of fluid-structure interaction. All analyses discussed in this chapter have assumed a pressure-time history for the wave impact which is independent of the motion of the structure. The following section discusses possible extensions to conventional structural analyses, whereby some of the hydrodynamic effects may be included.

4.9.3.1 *Dynamic fluid-soil-structure interaction*

Whilst the current state-of-the-art in CFD-FE modelling has not yet reached a level of maturity to include of a 3-dimensional fluid domain which is able to realistically capture the complex hydro-dynamics of wave breaking and slamming, it is relatively straightforward to introduce an inviscid compressible fluid to simulate the pressure transients in a coupled fluid-structure interaction analysis.

Full dynamic equilibrium for the coupled *fluid-structure* motion is given by the following expression

$$\begin{bmatrix} M & 0 \\ Q^T & \tilde{M} \end{bmatrix} \begin{Bmatrix} \ddot{d}_s \\ \ddot{p}_f \end{Bmatrix} + \begin{bmatrix} C & 0 \\ 0 & \tilde{C} \end{bmatrix} \begin{Bmatrix} \dot{d}_s \\ \dot{p}_f \end{Bmatrix} + \begin{bmatrix} K & -Q \\ 0 & \tilde{K} \end{bmatrix} \begin{Bmatrix} d_s \\ p_f \end{Bmatrix} = \begin{Bmatrix} F_s \\ 0 \end{Bmatrix} \tag{4-9}$$

All terms relating to the structure (identified by the subscript s) are identical to those given earlier for the shell analysis. The additional terms involving the sub-matrices $[C]$ refer to the fluid-structure coupling, or fluid domain alone $[\tilde{M}]$, $[\tilde{C}]$ and $[\tilde{K}]$. Further details may be found in Zienkiewicz and Taylor (Zienkiewicz & Taylor, 1991).

This coupled system leads to a non-symmetric form which may be solved for any set of input forces, $\{F_s\}$, by the generalised SST step by step algorithm. Looking to the future, it is clear that further research and development in numerical modelling studies will lead to ever greater realism in coupled CFD-FE simulations.

4.9.3.2 *Modelling the dynamic far-field*

Although the previous examples for a fairly typical caisson have confirmed basic engineering intuition (that role played by the foundation in influencing the bending response of the front wall is negligible) realistic models of the soil are needed if seismic analyses are to be performed. Recent innovative work by Wolf & Song (1996) has lead to a new method of treating the far-field in time-domain dynamic soil-structure-interaction studies. The following partitioned equation of motion includes additional terms which account for the behaviour of the soil beyond a boundary (identified with the subscripts b). This scheme requires the convolution of $\left[M^{\infty}\right]$ with previous velocities, however the procedure can readily be incorporated into a standard time-stepping scheme (such as Newmark, or HHT). Neglecting any explicit damping matrix (radiation damping is automatically satisfied and material damping can appear through a non-linear structural stiffness matrix) we have

$$\begin{bmatrix} [M_{ss}] & [M_{sb}] \\ [M_{bs}] & [M_{bb}] + \gamma\Delta t\left[M_1^{\infty}\right] \end{bmatrix} \begin{Bmatrix} \{\ddot{d}_s\}_n \\ \{\ddot{d}_b\}_n \end{Bmatrix} + \begin{bmatrix} [K_{ss}] & [K_{sb}] \\ [K_{bs}] & [K_{bb}] \end{bmatrix} \begin{Bmatrix} \{d_s\}_n \\ \{d_b\}_n \end{Bmatrix} = \begin{Bmatrix} \{f_s\}_n \\ -\{r\}_n + \{f_b\}_n \end{Bmatrix} \quad (4\text{-}10)$$

The method leads to very significant computational savings in large scale analyses, avoiding the use of huge extended meshes or inaccurate transmitting boundaries. Work is currently under way to efficiently incorporate this scheme within *yaFEc*. Note that the method may be applied to both the solid and the fluid phases.

4.9.3.3 *Quantifying the uncertainties*

The uncertainties associated with some of the material parameters are relatively well defined and their statistical characteristics have already be identified. The case for the structural models is not so simple. Whilst each of the approaches listed above demonstrate that the simplified beam analogy is useful for a preliminary design, the use of finite element analysis is strongly recommended when determining the direct stresses, bending moments and shear forces in a member. By way of example, the bending moments in the front face of the wall of a rectangular cellular caisson may differ by a factor of two, depending on the method of analysis (which largely controls the type of boundary conditions which can be treated). It is considered quite inappropriate to introduce *calibration* or *model* factors for, say, the beam model to cover all conceivable wall configurations (including curved front faces and perforated walls) and effective fixities. Conversely, because of its flexibility, the careful use of finite element models, by an experienced structural engineer, enables a detailed understanding of the structural action to be gained and also allows the engineer to study a range of design options. Adaptive meshing techniques which are linked to structural optimisation algorithms are already emerging in the field of FE technology. The future looks very exciting for coastal engineers involved in the structural design of caissons, as it is felt that they will soon have access to a new generation of advanced FE tools which include automatic adaptive meshing routines.

4.10 CONSTRUCTION ISSUES

A well detailed reinforced concrete caisson with a carefully designed concrete mix will provide many years of excellent service if appropriate supervision was provided during the construction phase. Looking to the future, there is now a move towards hybrid caissons with part-perforated walls and curved faces. There is also scope for caissons to significantly increase in size. Lengths of over 100m are perfectly plausible provided the global bending and torsional stiffnesses are sufficient to cope with the loads induced during towing. Once in place, the benefits of actively introducing shear transfer between adjacent caissons seem very clear. Use of an in-situ gravel filled pocket (plugged either side by concrete), sandwiched between the caissons appears to offer a good compromise between load sharing and providing a flexible coupling without requiring excessive precision in caisson alignment (see, for example, PIANC, 1997).

It is felt that the concerns of some engineers on the issue of material durability have sometimes been overstated (see Ozaki & Matsuya, 1986 for one example of the performance of sixty-year old marine concrete). In many cases the topic has been treated as a material problem alone with little understanding of the structural consequences. Unfortunately there remains considerable confusion in the minds of some engineers as to the real significance of cracking in a reinforced concrete structure. A cellular caisson with sufficient flexural and shear reinforcement (where attention paid to detailing for shrinkage, corners, laps and joints) offers a multitude of load paths to transmit the forces. This is true of many existing caissons, as wall sections tend to have been over-sized as a result of over-conservative, simplified analyses. Local spalling and even significant corrosion in certain areas can often have little real effect on the overall stability of the structure. The consequence of material degradation is intimately tied to the precise location in the structural frame. One concept which is not always appreciated is that by increasing the cover to the reinforcing steel in a flexural member, one is not automatically improving the durability of the section as the likelihood of cracking on the tensile face is increased. The interested reader is referred to Maage et al.,1996, Price et al., 1989, Tumidajski & Chan, 1996, Mackechnie & Alexander, 1997, and Taywood Engineering Research Labs, 1988 (and the references cited within) for further information on the use of reinforced concrete in a marine environment.

A number of techniques have recently emerged whereby the likelihood of corrosion in a reinforced concrete structure may be markedly reduced. Methods such as cathodic protection, de-salination and re-alkalisation as well as the used of coatings and migratory corrosion inhibitors are finding greater use. Another area which will see increased activity, is the use of non-ferrous reinforcement in marine structures.

As the availability of ever cheaper microprocessors continues, engineers will have greater opportunities to reliably instrument, log and control the performance of their structure. It is argued here that all engineers should insist on some form of permanent monitoring instrumentation, as it will lead to greater understanding in the way the structures are loaded and the manner in which they respond throughout their service life. The benefits of a well devised monitoring programme are not only those of providing meaningful data to help define a maintenance strategy but also their value in helping engineers move forward to develop more economical, safer structures.

REFERENCES TO VOLUME IIc

Chapter 1: CROUCH, R.S. (1999): Introduction.

Chapter 2: MARTINEZ, A.; KOVARIK, J.-B.; BERDIN, D. (1999): Structural design of vertical breakwaters - limitations of current practice and existing design codes. 37 pp.

Chapter 3: VROUWENVELDER, A.W.C.M.; BIELECKI, M. (1999): Caisson reliability during transport and placing. 36 pp.

Chapter 4: CROUCH, R.S. (1999a): In-service behaviour of cellular reinforced concrete caissons under severe wave impact. 39 pp.

Chapter 5: CROUCH, R.S. (1999b): Some observations on the durability and repair of concrete structures in a marine environment. 30 pp.

REFERENCES

Bazant, Z.P. (chairman) 1989. Fracture mechanics of concrete: concepts, models and determiniation of material properties. Report by ACI Committee, 446 pp.

Carol, I.; Prat, P.C.; Bazant, Z.P. 1992. New explicit microplane model for concrete: theoretical aspects and numerical implementation. *International Journal of Solids and Structures*, vol. 29, no. 2, pp. 1173-1191.

Cook, R.D.; Malkus; D.S.; Plesha, M.E. 1989. Concepts and applications of finite element analysis. John Wiley & Sons, 3rd edition, 630 pp.

Comité Euro-international du Beton 1996. RC elements under cyclic load, CEB 210, 190 pp.

Crouch, R.S.; Wolf, J.P. 1994. Unified 3-d critical state bounding surface plasticity model incorporating a continuous plastic response under cyclic paths: part I & II. *International Journal for Numerical and Analytical Methods in Geomechanics*, vol. 18, pp. 735-784.

Huang, H.-C. 1989. Static and dynamic analyses of plates and shells: theory, software and applications. London, U.K.: Springer-Verlag, 194 pp.

Isenberg, J. (ed.) 1993. Finite element analysis of reinforced concrete structures II, ASCE, 717 pp.

Maage, M.; Helland, S.; Poulsen, E.; Vennesland, O.; Carlsen, J.E. 1996. Service lift prediction of existing concrete structures exposed to marine environment. *ACI Materials Journal*, ACI, Farmington, Michigan, USA, vol. 93, pp. 602-608.

Mackechnie, J.R.; Alexander, M.G. 1997. Exposure of concrete in different marine environments. *Journal Mat Civil Engineer*, ASCE, vol. 9, no. 1, pp. 41-44.

Matsushima, M.; Seki, H.; Matsui, K.A. 1998. Reliability approach to landing pier optimum repair level. *ACI Materials Journal*, ACI, Farmington, Michigan, USA, vol. 95, no. 3, pp. 218-225.

O'Brien, E.J.; Dixon, A.S. 1995. Reinforced and pre-stressed concrete design: the complete process. Harlow, U.K.: Longman Scientific and Technical, pp. 259-261.

Oumeraci, H.; Kortenhaus, A. 1994. Analysis of dynamic response of caisson breakwaters. *Coastal Engineering, Special Issue on 'Vertical Breakwaters'*, Eds.: Oumeraci, H. et al., Amsterdam, The Netherlands: Elsevier Science Publishers B.V., vol. 22, nos. 1/2, pp. 159-183.

Ozaki, S.; Matsuya, M. 1986. Sixty-year old concrete in a marine environment. *Transactions of the Japan Concrete Institute*, Tokyo, Japan, vol. 8, pp. 169-174.

Pedersen, J. 1997. Dynamic response of caisson breakwaters subjected to impulsive wave loading - design diagrams for static load factors. *Proceedings 1st Overall Project Workshop, MAST III, PROVERBS-Project: Probabilistic Design Tools for Vertical Breakwaters*, Las Palmas, Gran Canaria, Annex 2C, 22 pp.

PIANC PTC II - WG28 1997. Recommendations for the construction of breakwaters with vertical and inclined concrete walls. *Permanent International Association of Navigation Congresses (PIANC). Working Group 28, Report of Sub-Group C*, Diaz Rato, J.L. et al., Final Report, 58 pp.

Price, W.I.J.; Hamby, E.C.; Tricklebank, A.H. 1989. Review of fatigue in concrete marines structures. *Offshore Technology Report: OTR / Department of Energy*, Her Majesty's Stationery Office (HMSO), OTH 87 235, London, U.K.

Rock, T.; Hinton, E.; Zienkiewicz, O.C. 1976. A note on mass lumping and related processes in the finite element method. *Journal of Earthquake Engineering and Structural Dynamics*, vol. 4, pp. 245-249.

Tan, T.-W. 1998. Simplified dynamic model of RC caisson response under wave attack. *MSc Dissertation, University of Sheffield*, Sheffield, U.K., 103 pp.

Taywood Engieering Research Labs 1988. Effectiveness of concrete to protect steel reinforcement from corrosion in marine structures. *Offshore Technology Report: OTR / Department of Energy*, Her Majesty's Stationery OfFranzius-Institut für Wasserbau und Küsteningenieurwesen der Universität Hannoverce (HMSO), OTH 87 247, London, U.K.

Tumidajski, P.J.; Chan, G.W. 1996. Boltzmann-Matano analysis of chloride diffusion into blended cement concrete. *Journal Mat Civil Engineer*, ASCE, vol. 8, no. 4, pp. 195-200.

Van der Meer, J.W.; Benassai, E. 1984. Wave forces and impacts on a circular and square caisson. *Proceedings International Conference Coastal Engineering (ICCE)*, ASCE, Houston, Texas, USA, no. 19, Volume 3, pp. 2920-2932.

Wolf, J.P. 1994. Foundation vibration analysis using simple physical models. Englewood Cliffs/N.J., USA: Prentice-Hall, 423 pp.

Wolf, J.P.; Song, C. 1996. Finite element modelling of unbounded media. Chichester, U.K.: John Wiley, 331 pp.

Zienkiewicz, O.C.; Taylor, R.L. 1991. The finite element method. Volume 2: Solid and fluid mechanics dynamics and non-linearity, London, U.K.: McGraw Hill, 4th edition, 807 pp.

CHAPTER 5

Probabilistic design tools and applications

Han Vrijling and Hessel Voortman

5.1 INTRODUCTION

In this chapter the probabilistic design tools will be presented. This encompasses the methods to combine the inherent uncertainty of the natural boundary conditions, the uncertainty due to lack of information of the natural environment, the quality of the structure and the engineering models into the measure of a failure probability that expresses the reliability of a structural system. Also a decision has to be taken whether the structural reliability is sufficient in view of the economic and societal functions of the structure. As an aid to this decision a safety philosophy has been formulated.

As the application of the probabilistic design method requires considerable effort and resources, a simpler approach using partial safety factors is derived within the probabilistic framework.

This chapter aims at designers of vertical breakwaters with an interest in the probabilistic design method and the background of a partial safety factor code. It may also serve as an introduction for researchers who want to familiarise themselves with the theoretical backgrounds. These readers however, are referred to Volume II for more detailed information.

This chapter treats the description of the failure modes of a vertical breakwater including the uncertainties in section 5.2.2 and 5.3.2 and the methods to calculate the probabilities of failure of each failure mode in section 5.2.3. In section 5.2.4 the methods to gain insight in the performance of the structure as a whole are treated. The ways to combine the failure probabilities of the various modes into the reliability of the structure are given in section 5.2.5.

The framework to decide on the optimal failure probability is dealt with in section 5.2.6 together with the partial safety factor system. In section 5.4 the theory is applied to a number of case studies. Finally, some perspectives are given in section 5.5.

5.2 GENERAL INTRODUCTION OF PROBABILISTIC METHODS

5.2.1 *Introduction*

The most important and most clear difference of probabilistic design compared to conventional (deterministic) design is that in probabilistic design one takes explicitly account of the uncertainties involved in the behaviour of the structure under consideration. Over the years considerable progress has been made in the development of probabilistic methods. This section attempts to introduce shortly the probabilistic working method, independent of the application to vertical breakwaters. For further reading, several textbooks have been written. As a start for further reading the following references can be used:

- Thoft-Christensen & Baker, 1982;
- Madsen, et al. 1986;
- Ditlevsen & Madsen, 1996.

Mentioning these three references is in no way meant to imply any judgement of the value of other references on the same subject.

5.2.2 *Limit state equations and uncertainties*

5.2.2.1 *The concept of limit states*

The first step in a reliability analysis of any structure is defining its functions. When the functions of the structure are defined, the ways in which malfunctioning of the structure can occur are defined. These ways of malfunctioning are called failure modes. The failure modes are described in such a way that they are fit for mathematical treatment. A function that describes functioning or failure of a structure or one of its components is called a reliability function or limit state equation. A general limit state equation is denoted *g* and can be written as:

$$M = g(\mathbf{X}) = R(\mathbf{X}) - S(\mathbf{X}) \tag{5-1}$$

in which $\mathbf{X}$ is a vector of random variables describing the geometry of the structure, the loads that are applied, the strength of materials etc; M is a random variable, usually referred to as the safety margin; R($\mathbf{X}$) is the strength (Resistance) of the structure as a function of $\mathbf{X}$; S($\mathbf{X}$) is the load (Solicitation) on the structure as a function of $\mathbf{X}$.

The limit state equations are defined in such a way that negative values of realisations of M indicate failure and positive values indicate safe states.

For a general structure several limit states can be defined. Take for example the simply supported concrete beam with a point load in the middle in Figure 5-1.

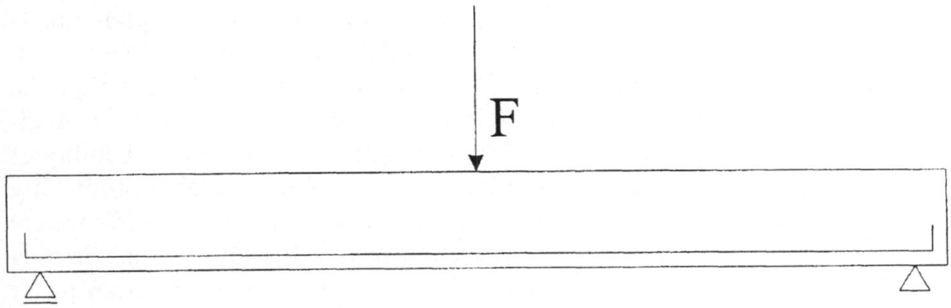

Figure 5-1. Simply supported concrete beam.

In the situation where the position of the load is fixed, the following limit states can be defined for this simple structure:

- Exceedance of the ultimate bending moment in the middle of the beam;
- Exceedance of the shearing strength near one of the supports;
- Exceedance of the admissible deformation in the middle of the beam;
- Cracking of the concrete on the lower side of the beam, which may lead to corrosion of the reinforcement bars in an aggressive environment;
- Chloride ingress through the uncracked concrete, followed by corrosion of the reinforcement;
- Collapse during construction due to lower strength during construction or extreme loading.

Observation of these six failure modes shows that the consequences are not of equal magnitude in all cases. The first two failure modes lead to immediate collapse of the beam, while the other three threaten its functioning (inadmissible deformation) or may introduce failure after a period of time in which repair can still be made (cracking and chloride ingress).

Therefore, in general four kinds of limit states are defined (see e.g. Eurocode 1, Basis of Design, 1994):

- Ultimate Limit States (ULS), describing immediate collapse of the structure;
- Serviceability Limit States (SLS), describing loss of function of the structure without collapse (a beam with a too large deformation might not be able to support a load which is sensitive to this deformation, while the beam is still able to withstand the load);
- Construction Limit States (CLS). During construction load and strength may deviate considerably of the design values, leading to failure during construction;
- Accidental Limit States (ALS), describing failure under accident conditions (collisions, explosions).

The acceptable probability of failure for a certain limit state depends on its character. Usually the highest safety requirements are set for Ultimate Limit States. Accepted failure probabilities for Serviceability Limit States might be considerably higher, especially if the effects of failure are easily reversed. Accidental Limit States can be treated like Ultimate Limit States, or the probability of occurrence of the accident can be taken into account. The acceptable probability of failure also depends on the time in which it is possible that a certain failure mode occurs. For failure during the construction phase, this time is considerably shorter than for the other types of failure. Therefore, for the construction phase characteristic values with a smaller return period are defined.

5.2.2.2 *Uncertainties related to the limit state formulation*

Basically, the limit state equation is a deterministic model indicating functioning or failure of the structure. Uncertainties are generally related to the input of the limit state equation. The following types of uncertainty are discerned (see also Vrijling & van Gelder, 1998):

1. Inherent uncertainty;
2. Model uncertainty;
3. Statistical uncertainty.

Ad 1: The uncertainty that is part of the described physical process is called inherent uncertainty. This uncertainty exists even if unlimited data is available. For instance: even if the wave height at a certain location is measured during an infinite period of time, the wave heights will still be uncertain in the future. A probability distribution can be used to describe the inherent uncertainty.

Ad 2: Model uncertainty can be distinguished into two subtypes. The first type of model uncertainty is related to the limit state equation itself. The model describing the physical process is a schematisation of the true process. Due to the schematisation, parts of the process are left out under the assumption that they are not important to the final result. This leaving out of parts of the process introduces a scatter (uncertainty) when comparing the model to measurements. Using a more sophisticated model, which describes more accurately the physical process, can reduce this type of uncertainty. The second kind of model uncertainty is related to the distribution function of the input variables. Also parametric distribution functions are schematisations of some real (unknown) distribution. Model uncertainty related to the input variables means that the chosen model might not be the true or best model. This type of uncertainty is reduced if more measurements are available, since then the correct distribution type becomes more clear.

Ad 3: When fitting a parametric distribution to limited data, the parameters of the distribution are also of random nature. The uncertainty in the parameters is generally referred to as statistical uncertainty. This type of uncertainty reduces when the number of data points increases.

5.2.3 *Reliability analysis on level II and III*

5.2.3.1 *Introduction*

The reliability of a structure or component is defined as the probability that the structure or component is able to fulfil its function. Reversibly the probability of failure is defined as the probability that the structure does not function. The properties of the structure (load, strength, geometry) are modelled by a vector of random variables, called the basic variables. The space of the basic variables is divided in a safe set and a failure set by the limit state equation(s) (Fig. 5-2).

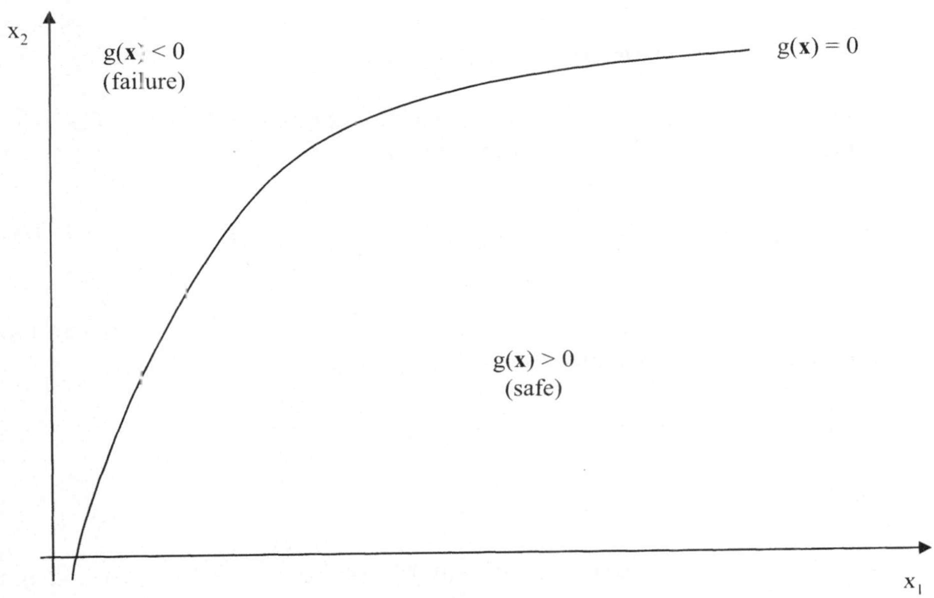

Figure 5-2. Limit state equation, safe set and failure set in the space of basic variables.

The probability of failure equals the probability that a combination of values of the basic variables lies in the failure domain. In formula:

$$P_f = P(\mathbf{X} \in F) \tag{5-2}$$

in which **X** is the vector of basic variables; and F is the failure domain.

Evaluation of this probability comes down to the determination of the volume of the joint probability density function of the basic variables in the failure domain. In formula:

$$P_f = \int_F f_X(\mathbf{x})d\mathbf{x} = \int_{g(\mathbf{x})\leq 0} f_X(\mathbf{x})d\mathbf{x} \tag{5-3}$$

in which $f_X(\mathbf{x})$ is the joint probability density function of the basic variables; and $g(\mathbf{x})$ is the limit state equation.

In general it is not possible to solve the integral analytically. Several numerical methods have been developed in the past. Section 5.2.3.2 introduces the direct integration methods. Section 5.2.3.3 introduces approximating methods. More details are given in (Thoft-Christensen & Baker, 1982; Madsen et al. 1986; Ditlevsen & Madsen, 1996).

5.2.3.2 *Direct integration methods (Level III)*

Riemann integration

Standard numerical integration methods can be applied to Equation (5-3). In that case the probability of failure is estimated by:

$$P_f \approx \sum_{i_1=0}^{m_1}\sum_{i_2=0}^{m_2}\ldots\sum_{i_n=0}^{m_n} 1(g(\mathbf{x}))f_X(x_{01}+i_1\Delta x_1, x_{02}+i_2\Delta x_2,\ldots,x_{0n}+i_n\Delta x_n)\Delta x_1\Delta x_2\ldots\Delta x_n \tag{5-4}$$

in which m_i is the number of steps for variable number i; n is the number of basic variables; $1(g(\mathbf{x}))$ is the indicator function defined as

$$1(g(\mathbf{x}))=1, \quad \text{if } g(\mathbf{x})\leq 0; \tag{5-5}$$

$$1(g(\mathbf{x}))=0, \quad \text{if } g(\mathbf{x})>0 \tag{5-6}$$

The calculation time depends on the number of basic variables (n) and the number of calculation steps to be taken (m). The total number of iterations can be written as:

$$N = \prod_{j=1}^{n} m_j \tag{5-7}$$

This indicates that the calculation time increases rapidly with an increasing number of basic variables. Furthermore the calculation time as well as the accuracy of the method depend strongly on the number of calculation steps per variable. Importance sampling methods have been proposed to increase the calculation speed as well as the accuracy of the calculation method. These methods are not elaborated upon here. Reference is made to (Ouypornprasert, 1987; CUR, 1997).

Monte Carlo simulation

A different method which uses the joint distribution of the basic variables is Monte Carlo simulation. In this method a large sample of values of the basic variables is generated and the number of failures is counted. The number of failures equals:

$$N_f = \sum_{j=1}^{N} 1(g(\mathbf{x}_j)) \tag{5-8}$$

in which N is the total number of simulations. In Equation (5-8) the same indicator function is used as in Equation (5-4).

The probability of failure can be estimated by:

$$P_f \approx \frac{N_f}{N} \tag{5-9}$$

The coefficient of variation of the failure probability can be estimated by:

$$V_{P_f} \approx \frac{1}{\sqrt{P_f N}} \tag{5-10}$$

in which P_f denotes the estimated failure probability.

The accuracy of the method depends on the number of simulations (CUR, 1997). The relative error made in the simulation can be written as:

$$\varepsilon = \frac{\frac{N_f}{N} - P_f}{P_f} \tag{5-11}$$

The expected value of the error is zero. The standard deviation is given as:

$$\sigma_\varepsilon = \sqrt{\frac{1 - P_f}{N P_f}} \tag{5-12}$$

For a large number of simulations, the error is Normal distributed. Therefore the probability that the relative error is smaller than a certain value E can be written as:

$$P(\varepsilon < E) = \Phi\left(\frac{E}{\sigma_\varepsilon}\right) \tag{5-13}$$

$$N > \frac{k^2}{E^2}\left(\frac{1}{P_f} - 1\right) \tag{5-14}$$

The probability of the relative error E being smaller than $k\sigma_\varepsilon$ now equals $\Phi(k)$. For desired values of k and E the required number of simulations is given by:

Requiring a relative error of $E = 0.1$ lying within the 95 % confidence interval ($k = 1.96$) results in:

$$N > 400\left(\frac{1}{P_f} - 1\right) \tag{5-15}$$

Equations (5-14) and (5-15) show that the required number of simulations and thus the calculation time depend on the probability of failure to be calculated. Most structures in coastal engineering possess a relatively high probability of failure (i.e. a relatively low reliability) compared to structural elements/systems, resulting in reasonable calculation times for Monte Carlo simulation. The calculation time is independent of the number of basic variables and therefore Monte Carlo simulation should be favoured over the Riemann method in case of a large number of basic variables (typically more than five). Furthermore, the Monte Carlo method is very robust, meaning that it is able to handle discontinuous failure spaces and reliability calculations in which more than one design point are involved (see below).

The problem of long calculation times can be partly overcome by applying importance sampling. This is not elaborated upon here. Reference is made to (Bucher, 1987; Ditlevsen & Madsen, 1996; CUR, 1997).

5.2.3.3 *Approximating methods (Level II)*

First Order Reliability Method (FORM)

In the FORM-procedure the value of the volume integral (Equation (5-3)) is estimated by an approximating procedure. The following procedure is followed to estimate the probability of failure:

- A transformation $\mathbf{X} = T(\mathbf{U})$ is carried out, mapping all the random variables in the space of standard normal-distributed variables (*U*-space);
- The reliability function is also transformed to the *U*-space and is replaced by its first-order Taylor approximation in a certain point.

Hasofer & Lind (1974) have shown that the calculated probability of failure is invariant for the formulation of the limit state equation, if the limit state is linearised in the point with the highest value of the joint probability density of the basic variables (design point, see Fig. 5-3).

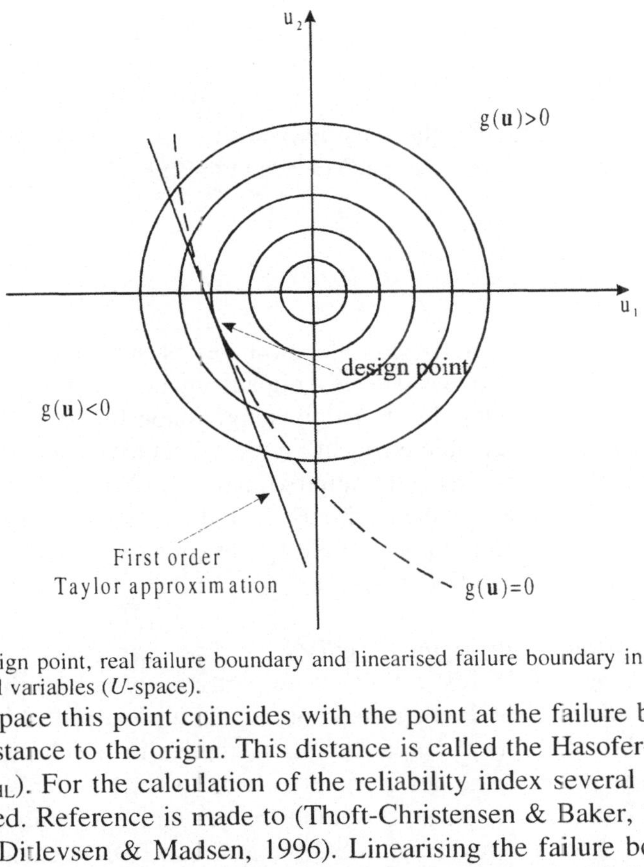

Figure 5-3. Design point, real failure boundary and linearised failure boundary in the space of the standard-normal variables (*U*-space).

In the U-space this point coincides with the point at the failure boundary with minimum distance to the origin. This distance is called the Hasofer-Lind reliability index (β_{HL}). For the calculation of the reliability index several methods have been proposed. Reference is made to (Thoft-Christensen & Baker, 1982; Madsen et al. 1986; Ditlevsen & Madsen, 1996). Linearising the failure boundary in U-space leads to the following approximation of the probability of failure

$$P_f = P(M < 0) \approx P\left(\beta_{HL} - \boldsymbol{\alpha}^T \mathbf{U} < 0\right) = \Phi(-\beta_{HL}) \qquad (5\text{-}16)$$

This solution is exact only if the reliability function is linear in the basic variables and if the basic variables have Normal distributions[12]. As noted above, the elements in the $\boldsymbol{\alpha}$-vector are a measure of the importance of the random variables. $\alpha_i^2 \cdot 100\%$ gives the relative importance in % of the random variable x_i.

Generally, the limit state equation is dependent on a number of deterministic parameters, which can be collected in the vector **p**. The elements in **p** can be statistical parameters like expected values and standard deviations and it can be e.g. geometrical quantities with negligible uncertainty. The limit state equation in U-space is written as:

$$g(\mathbf{U},\mathbf{p})=0 \tag{5-17}$$

For the reliability index β_{HL} the sensitivity with respect to the parameter p_j can easily be obtained in the form, see e.g. Madsen et al (1986):

$$\frac{d\beta_{HL}}{dp_j} \tag{5-18}$$

FORM calculations generally provide estimates of the failure probability in relatively short calculation times. This is a big advantage over level III methods in general. However, if the reliability function is highly non-linear, FORM-estimates of P_f may possess a considerable error. In Figure 5-3 a curved limit state equation is shown together with its first order approximation. In this case the estimate of the probability of failure obtained with FORM underestimates the real failure probability. In case of discontinuous failure spaces, FORM procedures may fail to give a correct failure probability at all.

Second Order Reliability Method (SORM)

The disadvantages of FORM estimates are partly overcome by the Second Order Reliability Method (SORM). Instead of calculating the probability of failure directly from the reliability index, a formula is applied in which the curvature of the limit state equation at the design point is used to get a better estimate of the probability of failure. This working method implies that a reliability index and a design point are known. Therefore, in SORM at first a FORM calculation is performed. For more details reference is made to (Breitung, 1984). Regarding the description of discontinuous failure spaces, the same disadvantages as for FORM apply.

[12] Probability distributions other than Normal will lead to non-linearity of the safety margin after transformation to *U*-space.

5.2.4 *Fault tree analysis*

5.2.4.1 *General system analysis by fault tree*

The methods for reliability calculations given in the previous section are all based on the analysis of one limit state only. However, even for a very simple structure generally several failure modes are relevant. A set of limit states can be presented as a series system, a parallel system or a combination thereof. The probability of failure of the system is determined by the properties of the system as well as by the individual failure probabilities of the components (limit states).

If the order of occurrence of the failure modes is not of consequence for the failure of the structure, the system of failure modes can be described by a fault tree. An example is given in Figure 5-4. The system representation is also given.

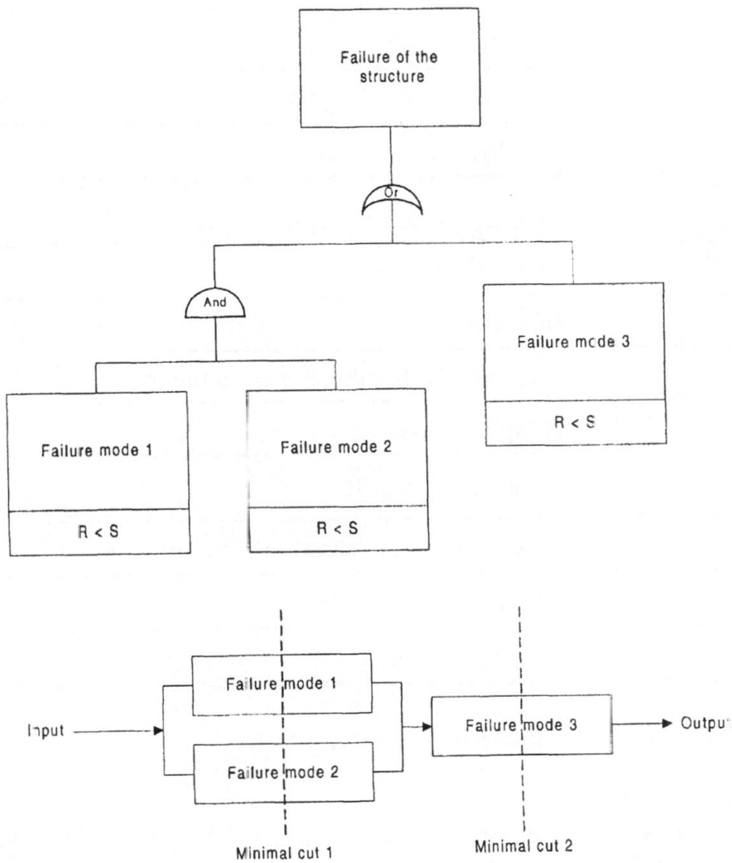

Figure 5-4. Fault tree and system representation of a system of three failure modes.

The two presentations of the system are completely equivalent. An and-gate coincides with a parallel system of failure modes and an or-gate coincides with a series system of failure modes.

An overview of the notation used in the fault trees is given in Table 5-1 and Table 5-2.

Table 5-1. Overview of gate symbols in fault trees (Andrews & Moss, 1993).

Symbol	Description
	"And" gate
	"Or" gate
	Voting gate
	Inhibit gate
	Exclusive "or" gate
	Priority "and" gate

Table 5-2. Overview of event symbols in fault trees (Andrews & Moss, 1993).

Symbol	Description
	Base event
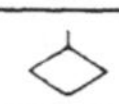	Event not further developed in the tree
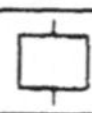	Compound event I
	Compound event II
	Conditional event (used in combination with inhibit gate)
	Normal event (house event)
	Reference symbol

5.2.5 *Calculation of system probability of failure*

5.2.5.1 *Introduction*

In the previous section the fault tree presentation of a system of failure modes is introduced. The fault tree indicates whether failure modes are part of a parallel system or part of a series system of failure modes. An analytical expression for the probability of system failure is possible only if the failure modes are uncorrelated or fully correlated. The formulae are given in Table 5-3.

Table 5-3. Overview of system probability of failure for combinations of correlation and system type.

System type	Upper bound	Independent components	Lower bound
Series	$P_f = \sum_{i=1}^{n} P_{f_i}$	$P_f = \prod_{i=1}^{n} P_{f_i}$	$P_f = \max(P_{f_i})$
Parallel	$P_f = \min(P_{f_i})$	$P_f = 1 - \prod_{i=1}^{n}(1 - P_{f_i})$	$P_f = 0$

The formulae given in Table 5-3 are also bounds on the real system probability of failure. For a series system the upper bound is given by the uncorrelated case and the lower bound by the case with full correlation. For a parallel system the upper bound is given by the case with full correlation and the lower bound by the uncorrelated case.

For arbitrary correlation between limit state equations, an analytical solution is no longer possible. In that case one has to use numerical methods to obtain the system probability of failure. Section 5.2.5.2 introduces direct integration methods. Section 5.2.5.3 introduces a few approximating methods for system failure.

5.2.5.2 *Direct integration methods for systems*

Riemann integration

The Riemann process introduced in section 5.2.3.2 can also be applied to evaluate the probability of system failure. Similar to the case with on limit state equation the probability of failure is estimated by:

$$P_f \approx \sum_{i_1=0}^{m_1}\sum_{i_2=0}^{m_2}\cdots\sum_{i_n=0}^{m_n} 1(\mathbf{g}(\mathbf{x})) f_{\mathbf{X}}(x_{01}+i_1\Delta x_1, x_{02}+i_2\Delta x_2, \ldots, x_{0n}+i_n\Delta x_n)\Delta x_1\Delta x_2\ldots\Delta x_n \quad (5\text{-}19)$$

In which $\mathbf{g}(\mathbf{x})$ is a vector of limit state equations; m_i is the number of steps for variable number i; and n is the number of basic variables.

Since now the system probability of failure has to be calculated, the indicator function is defined in a different way. For a series system the indicator function is given as:

$$1(\mathbf{g}(\mathbf{x})) = \begin{cases} 1 \text{ if } \min\limits_{i=1,n}(g_i(\mathbf{x})) \le 0 \\ 0 \text{ if } \min\limits_{i=1,n}(g_i(\mathbf{x})) > 0 \end{cases} \quad (5\text{-}20)$$

For a parallel system the indicator function is:

$$1(\mathbf{g}(\mathbf{x})) = \begin{cases} 1 \text{ if } \max\limits_{i=1,n}(g_i(\mathbf{x})) \le 0 \\ 0 \text{ if } \max\limits_{i=1,n}(g_i(\mathbf{x})) > 0 \end{cases} \tag{5-21}$$

In general the evaluation of more limit state equations in a calculation step will not take much extra time. Therefore, the performance of the Riemann integration for more limit states is comparable to the performance for one limit state (section 5.2.3.2).

<u>*Monte Carlo simulation*</u>

Like Riemann integration, Monte Carlo simulation is also applicable for systems. The number of failures is determined by applying Equation (5-8). Also in this case the indicator function is replaced by one of the system indicator functions given in the previous section. The evaluation of extra limit states is in general not very time consuming. Therefore, the performance of the Monte Carlo method is not heavily influenced by the application to more limit states. Also for systems the Monte Carlo method proves to be a very robust, but not very fast method.

5.2.5.3 *Approximating methods for systems*

<u>*Fundamental bounds on the system probability of failure*</u>

When the failure probabilities per limit state are known, it is always possible to provide a lower and upper bound for the failure probability (see Table 5-3). In several cases these bounds prove to give a reasonable range in which the real probability of system failure is to be found. In some cases however, the fundamental bounds provide a too wide range. In that case one has to use more advanced methods.

<u>*Ditlevsen bounds for the system probability of failure*</u>

An alternative calculation method for the bounds of the probability of failure of a series system is developed by (Ditlevsen, 1979). The bounds according to Ditlevsen are given by:

$$P(g_1(\mathbf{x})<0) + \sum_{i=2}^{n} \max\left[\left(P(g_i(\mathbf{x})<0) - \sum_{j=1}^{i-1} P(g_i(\mathbf{x})<0 \cap g_j(\mathbf{x})<0)\right), 0\right] \le P_f$$

$$P_f \le \sum_{i=1}^{n} P(g_i(\mathbf{x})<0) - \max_{j<i}\left(P(g_i(\mathbf{x})<0 \cap g_j(\mathbf{x})<0)\right) \tag{5-22}$$

These bounds are more narrow than the fundamental bounds, but in case of a large number of limit states they may still be too wide for practical application.

First-order method for systems
An approximating procedure which is able to provide a more accurate estimate of the system probability of failure of a system is proposed by (Hohenbichler & Rackwitz, 1983). In this method the correlated limit state equations are transformed to a set of uncorrelated limit states. The system probability of failure can then be calculated using fundamental rules for the probability of system failure and first or second order estimates of the probability of failure per limit state. Generally, an approximation is obtained for $\Phi_n(\boldsymbol{\beta}, \boldsymbol{\rho})$ in which $\boldsymbol{\beta}$ denotes the vector of reliability indices for every limit state and $\boldsymbol{\rho}$ denotes the correlation between the failure modes. The failure probability of a parallel system is then given by:

$$P_{f;parallel} \approx \Phi_n(-\beta, \rho) \tag{5-23}$$

And for a series system by:

$$P_{f;series} \approx 1 - \Phi_n(\beta, \rho) \tag{5-24}$$

This method in general has shorter calculation times than the level III methods. However, due to the approximations estimates of the failure probability might show considerable errors (Schuëller & Stix, 1987). As with all approximating methods one should be aware of this disadvantage.

5.2.6 *Choice of safety level*
To construct a breakwater that is always performing its function and is perfectly safe from collapse is at least an uneconomical pursuit and most likely an impossible task. Although expertly designed and well constructed, there will always be a small possibility that the structure fails under severe circumstances (Ultimate Limit State). The acceptable probability of failure is a question of socio-economic reasoning.

In a design procedure one has to determine the preferred level of safety (i.e. the acceptable failure probability). For most civil engineering structures the acceptable failure probability will be based on considerations of the probability of loss of life due to failure of the structure.

In general two points of view for the acceptable safety level can be defined (Vrijling et al., 1995):

The individual accepted risk. The probability accepted by and individual person to die in case of failure of the structure; In Western countries this probability is of the order 10^{-4} per year or smaller

The societal accepted risk. Two approaches are presented, depending on the relative importance of the total number of lives lost in case of failure on the one hand and the total economic damage on the other. If the number of potential casualties is large the likelihood of failure should be limited accordingly. The accepted probability of occurrence of a certain number of casualties in case of failure of a structure is then restricted proportional to the inverse of the square of this number (Vrijling et al., 1995). If the economic damage is large, an economic optimisation equating the marginal investment in the structure with the marginal reduction in risk should be carried out to find the optimal dimensions of the structure.

The two boundary conditions based on the loss of human lives form the upper limits for the acceptable probability of failure of any structure. In case of a breakwater without amenities the probability of loss of life in case of failure is very small. In that case the acceptable probability of failure can be determined by economical optimisation, weighing the expected value of the capitalised damage in the life of the structure (risk) against the investment in the breakwater. The next section provides more background on this concept.

If, for a specific breakwater, failure would include a number of casualties, the economic optimisation should be performed under the constraint of the maximum allowable probability of failure as defined by the two criteria related to loss of life.

The explicit assessment of the acceptable probability of failure as sketched above is only warranted in case of large projects with sufficient means. For smaller projects a second approach is generally advised. This second approach to the acceptable safety level is based on the evaluation of the safety of existing structures supplemented by considerations of the extent of the losses involved in case of failure. Consequently the assumption is made that the new structure should meet the safety requirements that seem to be reasonable in practice. This approach is found in many codes where a classification of the losses in case of failure leads to an acceptable probability of failure. Most structural codes provide safety classes for structures (NKB 1978, Eurocode 1). The structure to be designed might fit in one of these safety classes providing an acceptable probability of failure. It should be noted however that for most structural systems loss of life is involved contrary to breakwaters. Therefore the following classification and table with acceptable probabilities of failure was developed especially for vertical breakwaters:

- *Low safety class*, where failure implies no risk to human injury and very small environmental and economic consequences;

- *Normal safety class*, where failure implies no risk to human injury and some environmental and economic consequences;
- *High safety class*, where failure implies risk to human injury and significant environmental pollution or high economic or political consequences;
- *Very high safety class*, where failure implies risk to human injury and extensive environmental pollution or very high economic or political consequences.

Table 5-4. Overview of safety classes.

Limit state type	**Safety class**			
	Low	**Normal**	**High**	**Very high**
SLS	0.4	0.2	0.1	0.05
ULS	0.2	0.1	0.05	0.01

5.2.7 *Reliability based design procedures*

5.2.7.1 *General formulation of reliability based optimal design*

Generally, in a design process one pursues the cheapest design that fulfils the demands defined for the structure. The demands can be expressed in two fundamentally different ways:

- The total expected lifetime costs of the structure consisting of the investment and the expected value of the damage costs are minimised as a function of the design variables;
- If a partial safety factor system is available, one can optimise the design by minimising the construction costs as a function of the design variables under the constraint that the design equations related to the limit state equations for all the failure modes are positive.

The minimisation of the lifetime costs can be formalised as follows (Enevoldsen & Sørensen, 1993):

$$\begin{aligned}
&\min_{\mathbf{z}} C_T(\mathbf{z}) = C_I(\mathbf{z}) + C_{F;ULS} P_{F;ULS}(\mathbf{z}) + C_{F;SLS} P_{F;SLS}(\mathbf{z}) \\
&\text{s.t} \quad z_i^L \le z_i \le z_i^U \qquad i = 1,\dots,m \\
&\qquad P_{F;ULS}(\mathbf{z}) \le P_{F;ULS}^U \\
&\qquad P_{F;SLS}(\mathbf{z}) \le P_{F;SLS}^U
\end{aligned} \tag{5-25}$$

In which:

$\mathbf{z} = (z_1, z_2, \dots, z_m)$: The vector of design variables;

$C_T(\mathbf{z})$: The total lifetime costs of the structure;

$C_I(\mathbf{z})$: The investment in the structure as a function of the design variables $\mathbf{z}$;
$C_{F;ULS}$: The damage in monetary terms in case of ULS failure;
$C_{F;SLS}$: The damage in monetary terms in case of SLS failure;
$P_{F;ULS}(\mathbf{z})$: The probability of ULS failure as a function of the design variables;
$P_{F;SLS}(\mathbf{z})$: The probability of SLS failure as a function of the design variables;
z_i^L, z_i^U: The lower and upper bound of design variable i;
$P_{F;ULS}^U, P_{F;SLS}^U$: The upper bound of the failure probability for ULS failure and SLS failure respectively.

Generally the design variables will be subjected to constraints. For instance, all geometrical quantities should be greater than zero. Furthermore, the failure probabilities can be subject to constraints, especially for structures where human lives are involved. In that case the maximum failure probabilities are enforced by regulations. In cases that loss of human lives is not involved in case of failure of the structure, formally the constraint on the failure probabilities can be set to 1 and the acceptable failure probability as well as the optimal design are completely decided by the lifetime costs only. If relevant, maintenance costs and inspection costs can be added to the total expected lifetime costs.

Obtaining accurate assessments of the damage in case of failure is not always practically possible. In that case, the optimal design can be found by minimising a cost function which only comprises the investment and imposing a constraint on the failure probability which expresses a qualitative idea of the economic optimal failure probability.

If the design is performed using a code based partial safety factor system, the following optimisation problem is applicable:

$$\begin{aligned} &\min_{\mathbf{z}} C_T(\mathbf{z}) = C_I(\mathbf{z}) \\ &\text{s.t} \quad z_i^L \le z_i \le z_i^U \\ &\qquad G_i(\mathbf{z}, \mathbf{x}^C, \gamma) > 0 \end{aligned} \tag{5-26}$$

In which:

$C_I(\mathbf{z})$: The investment in the structure as a function of the design variables $\mathbf{z}$;
$G_i(\mathbf{z},\mathbf{x}^C,\gamma)$: The limit state function for failure mode *i* as a function of the design variables $\mathbf{z}$, the characteristic values of the random variables as defined in the partial safety factor system $\mathbf{x}^c$ and the vector of partial safety factors γ.

Generally, partial safety factors are available for several target probabilities of failure or safety classes (see below). Since the choice of the safety factors involves implicitly the choice of a target probability of failure and expected costs of failure, the same optimal design should be obtained from (23) and (24).

5.2.7.2 *Cost optimisation*

If loss of life in case of failure of the structure is not an issue for the structure under consideration, no constraint is set on the failure probability and the acceptable probability of failure equals the economic optimal probability of failure. A procedure for probabilistic optimisation of vertical breakwaters has been developed (Sørensen et al, 1994; Voortman et al, 1998, Volume IId, section 4.1).

The optimisation for a vertical breakwater can be written as:

$$\min_{\mathbf{z}} C(\mathbf{z}) = C_{I;0} + C_I(\mathbf{z}) + \sum_{n=1}^{N}\left(\frac{365 C_{F;SLS} P_{F;SLS}(\mathbf{z}) + C_{F;ULS} P_{F;ULS}(\mathbf{z})}{(1+r'-g)^n} + \frac{C_{ma\,int}}{(1+r')^n}\right) \qquad (5\text{-}27)$$

$$\text{s.t} \quad 0 \le z_i$$

In which:

$\mathbf{z}$:	The vector of design variables;
$C_{I;0}$:	Initial costs, not depending on the design variables;
$C_I(\mathbf{z})$:	Construction costs as a function of the design variables;
$C_{F;SLS}$:	Costs per day in case of serviceability failure;
$P_{F;SLS}(\mathbf{z})$:	The probability of serviceability failure per day;
$C_{F;ULS}$:	Costs per event in case of ultimate limit state failure;
$P_{F;ULS}(\mathbf{z})$:	The probability of ultimate limit state failure per year;
C_{maint}:	Maintenance costs for the breakwater per year;
r':	The net interest rate per year;
g:	The yearly rate of economical growth, expressing growth and development of the harbour;
N:	The lifetime of the structure in years.

Inspection of Equation (5-27) shows that the total lifetime costs consist of investment costs and the expected value of the damage costs. In principle, for every year of the structure's lifetime, the expected damage has to be taken into account. Not all the costs are made at the same time. Therefore, the influence of interest, inflation and economical growth has to be taken into account in order to make a fair comparison of the different costs (van Dantzig, 1956).

The expected value of the damage costs is a function of the failure probability. The failure probability is a function of the design variables. Therefore, minimisation of Equation (5-27) results in the optimal geometry and at the same time the optimal failure probability. Ready at hand minimisation algorithms can be applied to find the optimal set of design variables.

When implementing the cost function in any programming language, the failure probability as a function of the design variables has to be included. For this part of the cost function one of the probabilistic procedures introduced in section 5.2.3 or section 5.2.5 can be used. Due to the specific character of the optimisation process, the choice of the probabilistic procedure is not an arbitrary one. One should be aware of the following points:

- The minimisation process comprises a large number of evaluations of the cost function, each evaluation involving a probability calculation. Therefore, time-consuming methods should be avoided;
- The values of the cost function for any given point should be stable. Especially the Monte Carlo procedure provides probability estimates that contain an error, which is inherent to the procedure. This (small) error generally presents no problem, but in this case it causes variations of the cost function, which disturb the optimisation process (see Fig. 5-5).

The points of attention mentioned above lead to the conclusion that level II methods are suitable for application in an optimisation process. Level III methods will generally lead to too much computational effort or will disturb the optimisation process. They can be applied to perform a final check on the calculated failure probability of the optimal design.

The procedure described above has been applied to a fictitious design case of a vertical breakwater in a water depth of 25 m. Three design variables are considered: the height and width of the caisson and the height of the rubble berm.

As a first step a deterministic optimisation for chosen wave heights has been performed. This step is meant to show the connection between the deterministic optimisation for a given safety level and the full probabilistic approach. Because of this, the choice of the input values (comparable to the characteristic values in Equation 5-26) is not corresponding to the choice made for the partial safety factor system (see below). Furthermore, all safety factors have been set to 1 and the berm height is fixed at a value of 6 m. For this situation it is possible to find a minimum caisson width as a function of the crest height for every single failure mode. Once the crest height and the caisson width are known, the construction costs of the caisson breakwater can be calculated (see Fig. 5-7).

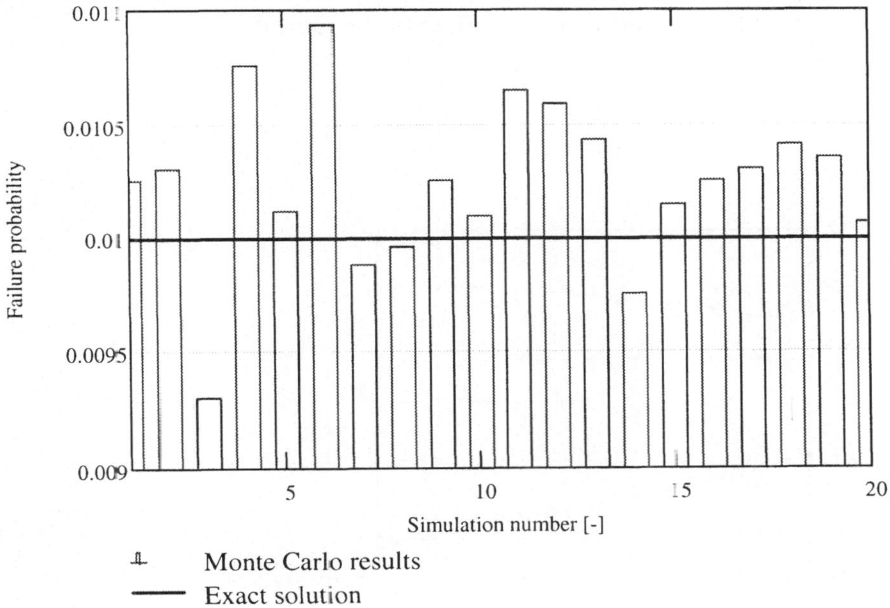

Figure 5-5. Result of 20 calculations of failure probability by Monte Carlo.

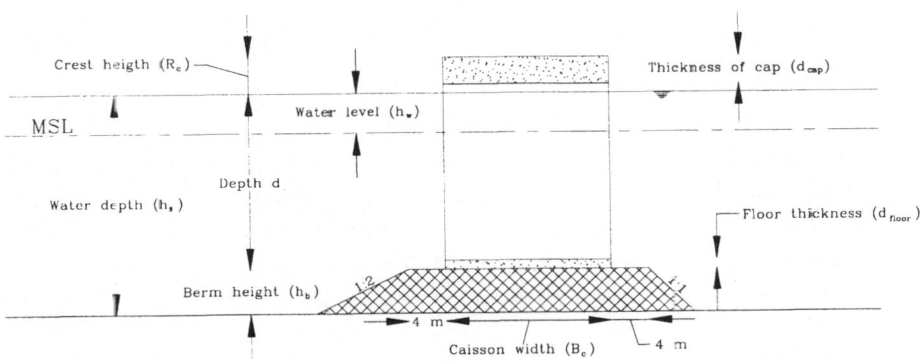

Figure 5-6. Overview of conceptual breakwater design for economic optimisation (Voortman et al., 1998).

Generally, bearing capacity failure of the subsoil shows the largest minimum caisson width. Furthermore, the results show that in general a lower crest height leads to a more narrow caisson and thus to lower construction costs. However, the minimum crest height required is determined by wave transmission. In the deterministic approach the minimum crest height related to wave transmission imposes a constraint on the crest height. Thus, the optimal geometry is decided by wave transmission and by bearing capacity failure of the subsoil.

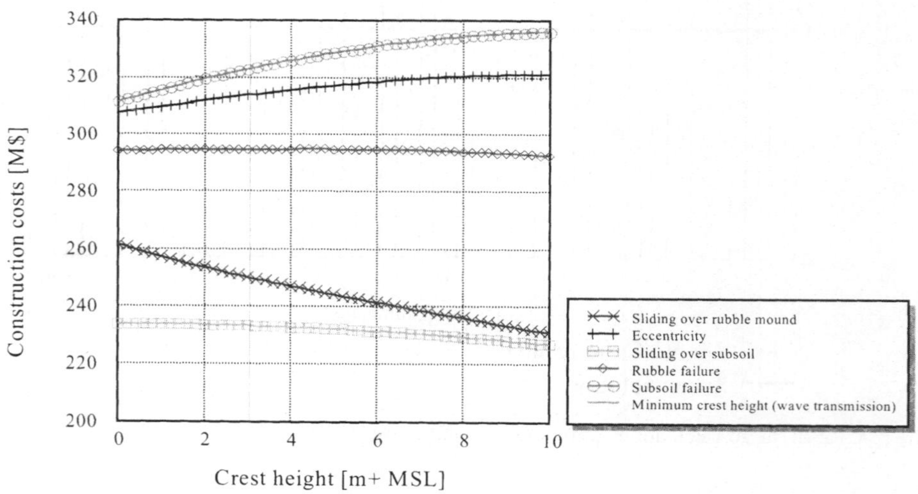

Figure 5-7. Construction costs of the breakwater as a function of the crest height (Berm height 6 m) (Voortman et al., 1998).

While at first sight it seems reasonable to have an equal failure probability of failure for all the failure modes in the system, probabilistic optimisation shows that, like in the deterministic approach, bearing capacity failure of the subsoil largely determines the probability of ultimate limit state failure (see Fig. 5-8).

Inspection of the lifetime costs as a function of crest height and caisson width indicates that also in the probabilistic approach the crest height is limited by wave transmission (see Fig. 5-9).

The optimal probability of system failure is quite low in comparison to existing structures ($8 \cdot 10^{-4}$). This could be caused by the choice of the cost figures or by a limited spreading of the random variables.

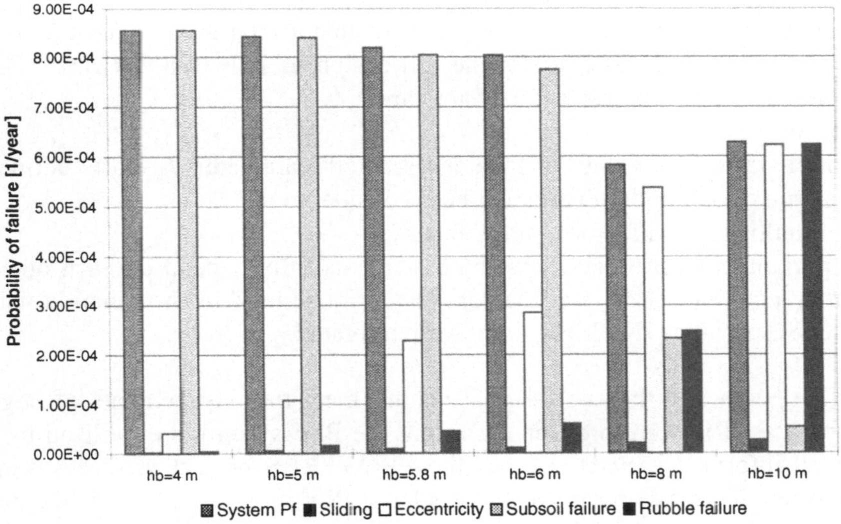

Figure 5-8. Overview of ULS failures probabilities for several berm heights (Voortman et al., 1998).

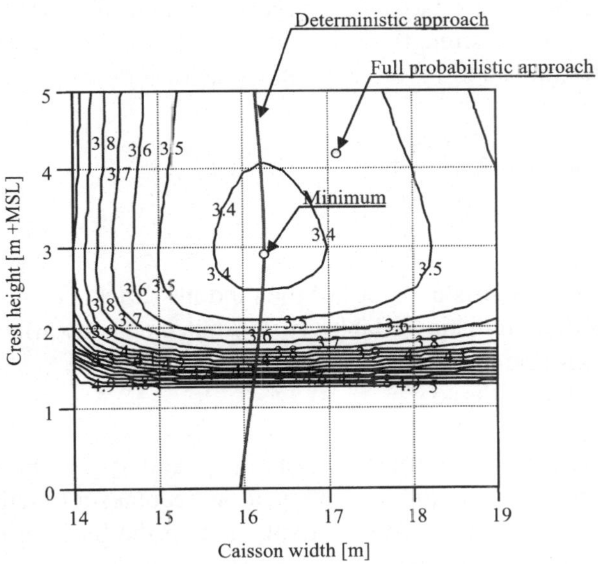

Figure 5-9. Contour plot of total lifetime costs in 10^8 US $ (Random wave height only) and optimal geometries for different levels of modelling. Taken from Voortman et al., (1998), (see also Volume IId, section 4.1).

5.2.7.3 *Partial Safety Factor System*

A partial safety factor system (PSF) may conveniently be used in the first design step of a large breakwater, or for the full design of a small breakwater. The key elements of a partial safety factor system are:

- A set of limit states;
- A set of characteristic values of load and strength parameters (usually defined as a quantile of the random variable under consideration;
- A set of multiplication factors (safety factors).

It is important to realise that these three elements fully depend on each other, meaning that a certain set of safety factors may only be used in combination with the limit states for which the PSF system was calibrated.

The PSF system and the code format should meet the requirements of engineering practice. It is therefore important that the PSF system is establised by a team of well-informed experts. Within the PROVERBS project, a PSF system was established. This was meant to show what a PSF system may look like and how it is derived. There are indications that in the current system, the uncertainty in soil properties and soil strength models are underestimated, leading to too low partial safety factors for these aspects.

Generally, partial safety factors are used to create a distance between the load on the structure and the strength of the structure. Put in the general form of equation (5-1) the design equation for each limit state *i* becomes:

$$G_i\left(\mathbf{x}^c, \mathbf{z}, \gamma_R, \gamma_S\right) = \frac{R\left(\mathbf{x}^c, \mathbf{z}\right)}{\gamma_R} - \gamma_S \cdot S\left(\mathbf{x}^c, \mathbf{z}\right) \qquad (5\text{-}28)$$

Where:

$\mathbf{x}^c$: vector of characteristic values of the random variables,
$\mathbf{z}$: vector of deterministic variables (for instance design variables);
γ_R: Safety factor on the strength;
γ_S: Safety factor on the load.

Depending on the code format, the factors γ_R and γ_S may be split into several partial factors. In that case, the design equation becomes a function of a vector of partial safety factors. This concept is explained on the basis of the calibration of the PSF system for vertical breakwaters, which is outlined in the remainder of this section.

The partial safety factors for vertical breakwaters have been calibrated on a probabilistic basis. The calibration is performed for a given class of structures, materials and/or loads in such a way that for all structure types considered the reliability level obtained using the calibrated partial safety factors for design is as close as possible to a specified target reliability level. Procedures to perform this type of calibration of partial safety factors are described in for example Thoft-Christensen & Baker, 1982, Madsen et al. (1986) and Ditlevsen & Madsen (1996).

A code calibration procedure usually includes the following basic steps:

1) definition of scope of the code; here : vertical wall breakwaters
2) definition of the code objective; here : to minimise the difference between the target reliability level and the reliability level obtained when designing different typical structures using the calibrated partial safety factors
3) selection of code format, see below
4) selection of target reliability levels, see section 5.2.6
5) calculation of calibrated partial safety factors, see below
6) verification of the system of partial safety factors, see below

The partial safety factors can be calibrated as follows. For each failure mode one or more limit state functions are established:

$$g_i(\mathbf{x}, \mathbf{z}) = 0 \qquad (5\text{-}29)$$

where $\mathbf{x}^c = (x_1^c, ... x_n^c)$ are realisations of n stochastic variables $\mathbf{X} = (X_1, ..., X_n)$ and $\mathbf{z} = (z_1, ... z_N)$ are the deterministic design variables.

Corresponding to (5-29) a design equation is established from which $\mathbf{z} = (z_1, ... z_N)$ are determined:

$$G_i(\mathbf{x}^c, \mathbf{z}, \gamma) = 0 \qquad (5\text{-}30)$$

where $\mathbf{x}^c = (x_1^c, ... x_n^c)$ are characteristic values of $\mathbf{X} = (X_1, ..., X_n)$ and $\gamma = (\gamma_1, ..., \gamma_M)$ are partial safety factors. Usually design values for loads are obtained by multiplying the characteristic with the partial safety factors and design values for resistances are obtained by dividing the characteristic values with the partial safety factors. On the basis of the limit state functions in (1) element reliability indices β_i or a system reliability index β_s can be determined for the structure considered, see e.g. Madsen et al. (1986).

The partial safety factors γ are calibrated such that the reliability indices corresponding to L example structures are as close as possible to a target reliability index β^2. This is formulated by the following optimisation problem

$$\min \quad W(\gamma) = \sum_{i=1}^{L} w_i \left(\beta_i(\gamma) - \beta^t\right)^2 \tag{5-31}$$

where w_i i = 1, 2, ..., L are weighting factors indicating the relative frequency of appearance of the different design situations. Instead of using the reliability indices in Equation 5-31 to measure the deviation from the target, the probabilities of failure can be used. $\beta_i(\gamma)$ is the system reliability index for example structure i with a design **z** obtained from the design equations using the partial safety factors γ and the characteristic values. Usually the partial safety factors are constrained to be larger than or equal to 1.

Within PROVERBS; code calibration has been performed primarily for geotechnical limit states only. Therefore, only a partial safety factor for the wave force γ_H is calibrated, which in this case replaces the general safety factor γ_S. The partial safety factor for the wave force γ_H in case of vertical breakwaters depends upon:

a) the uncertainty related to the wave height and other load parameters
b) the expected lifetime of the structure
c) the uncertainty with respect to the model used to find the wave force

Ad a) uncertainty related to load parameters: the uncertainties related to these parameters in the limit state functions are taken into account in a deterministic design through one sub partial safety factor. The factor is calibrated in such a way that a large factor is used in the case of large uncertainties and a small partial safety factor is used when the uncertainties are relatively small.

Ad b) expected lifetime: the expected lifetime T_L for vertical breakwaters can be quite different. Therefore three different expected lifetimes are considered: T_L =20 years, T_L = 50 years and T_L = 100 years.

Ad c) uncertainty about the wave force model: the wave forces can be predicted with the equations presented in this report. This is no more than a rough estimate for a specific site and specific breakwater design. The corresponding model uncertainty is relatively large, especially where it concerns the force due to wave impacts. Sometimes laboratory tests are performed in order to estimate the wave forces more accurately. In that case the uncertainty related to the wave forces is reduced and it is therefore reasonable to decrease the sub partial safety factor.

Again, due to the fact that primarily geotechnical failure modes are considered, only a partial safety factor is calibrated for the soil strength. Therefore, only a partial safety factor for the foundation strength γ_m is calibrated, which in this case replaces the general safety factor γ_R. The partial safety factor for the soil strength γ_m depends upon:

d) the uncertainty related to the soil parameters and other strength parameters
e) the uncertainty with respect to the model used to find the strength
f) the amount of quality control during construction
g) the safety class as discussed in section 5.2.6[13]
h) the type of limit state[13]

Ad d) uncertainty about soil parameters: this uncertainty depends on whether detailed field and laboratory tests have been performed and several other factors as discussed in section 3.8.2.

Ad e) uncertainty about strength model: the strength model may consist of the relatively simple limit state functions for geotechnical failure presented in section 3.7.2. More sophisticated equations are presented in Volume IIb, chapter 6. And the uncertainty can be reduced to the minimum if a combination of these models, numerical models and centrifuge testing is applied.

Ad f) quality control: finally, the uncertainty can also depend on the amount of control at the construction site.

Ad h) type of limit state: two types of limit states may be considered, namely: ULS (Ultimate Limit State; e.g. foundation failure, failure of significant part of caisson concrete structure) and SLS (Serviceability Limit State, e.g. overtopping, settlement of foundation soil). Acceptable probabilities of failure could be as indicated in Table 5-4.

The characteristic values are suggested to be the mean value for self weight and permanent actions, for wave heights the expected largest significant wave height in the design lifetime T_r, 5% fractiles for structural strength parameters and the mean values for geotechnical strength parameters. An example of the partial safety factor system is shown in Table 5-5.

Table 5-5. Partial safety factors.

p.s.f.	**Parameter**	**Tentative values - γ**
Loads		
γ_{G_1}	Self weight	1.0
γ_{G_2}	Permanent actions, e.g. ballast	1.1
γ_H	Wave force	See below
Strength		
γ_ϕ	Effective friction angle	See below

[13] The partial factors for the safety class and the type of limit state do not necessarily apply to the strength side, but may also be defined as applicable to the loading side. Another alternative is to apply a separate (third) factor, giving the effect of safety class and limit state type.

p.s.f.	**Parameter**	**Tentative values - γ**
γ_{C_u}	Undrained shear strength	See below
γ_c	Concrete strength	1.6
γ_r	Reinforcement	1.3

The partial safety factor for the wave force is determined from :

$$\gamma_H = \gamma_{H_0} \gamma_T \gamma_{H_2} \tag{5-32}$$

where:

γ_{H0} takes into account the uncertainty related to the wave height and other hydraulic parameters

γ_T takes into account the influence of the expected lifetime. For example $\gamma_T = 1$ for $T = 50$ years, $\gamma_T > 1$ for $T = 100$ years and $\gamma_T < 1$ for $T = 20$ years

γ_{H2} takes into account the effect of model tests used to estimate the wave force. $\gamma_{H_2} = 1$ if no model tests are performed

The partial safety factor for the geotechnical parameters is determined from :

$$\gamma_m = \gamma_0 \gamma_1 \gamma_2 \gamma_3 \tag{5-33}$$

where:

γ_0 takes into account the uncertainty related to the soil strength parameters

γ_1 takes into account the effect of the safety class and the type of limit state[14]

γ_2 takes into account the effect of model uncertainties related to the strength models. $\gamma_2 = 1$ for sliding over the base and for the preliminary design models described in Volume IIb, chapter 6. $\gamma_2 = 1.1$ for the limit state equations g_b through g_e in section 3.7.2. In cases where liquefaction or degradation are expected the value of γ_2 should be taken higher[15]

γ_3 takes into account the amount of control. $\gamma_3 = 1$ if normal control is used

[14] May also be applied on the strength side or as a separate factor.

[15] A safety factor for these conditions has not been calibrated.

The factors in equations (5-32) and (5-33) are calibrated using representative values for the soil strength parameters of fairly solid soil[16] and with wave climates from Bilbao, Sines, Tripoli and Fallonica (see Volume IId, section 4.2) and the limit state functions presented in Volume IIb, chapter 6. The full stochastic model used in the calibration is shown in Table 5-5.

The calibrated partial safety factors corresponding to normal safety class in ULS and high safety class in SLS:

Wave force:	$\gamma_{H0}=1.1$
Geotechnical failure modes with drained subsoil:	$\gamma_0=1.2$
Geotechnical failure modes with undrained subsoil:	$\gamma_0=1.3$

The following factors defined above are derived (no factors are derived for γ_3 taking into account the amount of control). The factors for γ_l can be used to obtain partial safety factors for all safety classes in ULS and SLS.

Table 5-6. Safety factor γ_T, which takes into account the influence of the expected lifetime T.

T	**20 years**	**50 years**	**100 years**
γ_T	0.98	1.0	1.05

Table 5-7. Safety factor γ_{H_2} which takes into account the effect of the wave force model

Model	**1**	**2**
γ_{H2}	1.0	0.85

Table 5-8. Safety factor γ_l , which takes into account the effect of the safety class and type of limit state.

	Safety class				
ULS		**Low**	**Normal**	**High**	**Very high**
SLS	**Low**	**Normal**	**High**	**Very high**	
P_f	0.4	0.2	0.1	0.05	0.01
γ_l	0.75	0.9	1.0	1.1	1.25

The safety classes correspond to the ones given in Table 5-5. The verification of the partial safety factors is described in Volume IId, section 4.2.

[16] Meant is a soil where no complications like liquefaction or degradation are expected.

5.3 PROBABILISTIC METHODS APPLIED TO VERTICAL BREAKWATERS IN GENERAL

5.3.1 *Fault tree for a vertical breakwater*

As stated in section 5.2.2 the first step in developing the fault tree and the limit states for a structure is defining the functions that are to be performed. In this project the main function of a breakwater has been defined as:

"Providing sufficiently tranquil water for ship manoeuvring and berthing"

There are several ways in which a breakwater might fail to fulfil this main function. This could be caused by wave energy entering the harbour through the entrance by refraction and diffraction, wave energy passing over the breakwater due to severe overtopping or collapse of a part of the structure, leading to a breach. Wave energy entering through the entrance is a matter of design of the layout of the entrance and is not part of PROVERBS. It is however part of the system of failure modes and is therefore included in the fault tree. In this section only the top part of the fault tree is presented. The complete fault tree is given in section 4.2 of Volume IId.

5.3.2 *Specific limit states for vertical breakwaters*

5.3.2.1 *Introduction*

For a vertical breakwater several limit states can be discerned. Within PROVERBS an extensive set of failure modes has been created. This set consists of Ultimate as well as Serviceability Limit States. Accidental Limit States have been omitted.

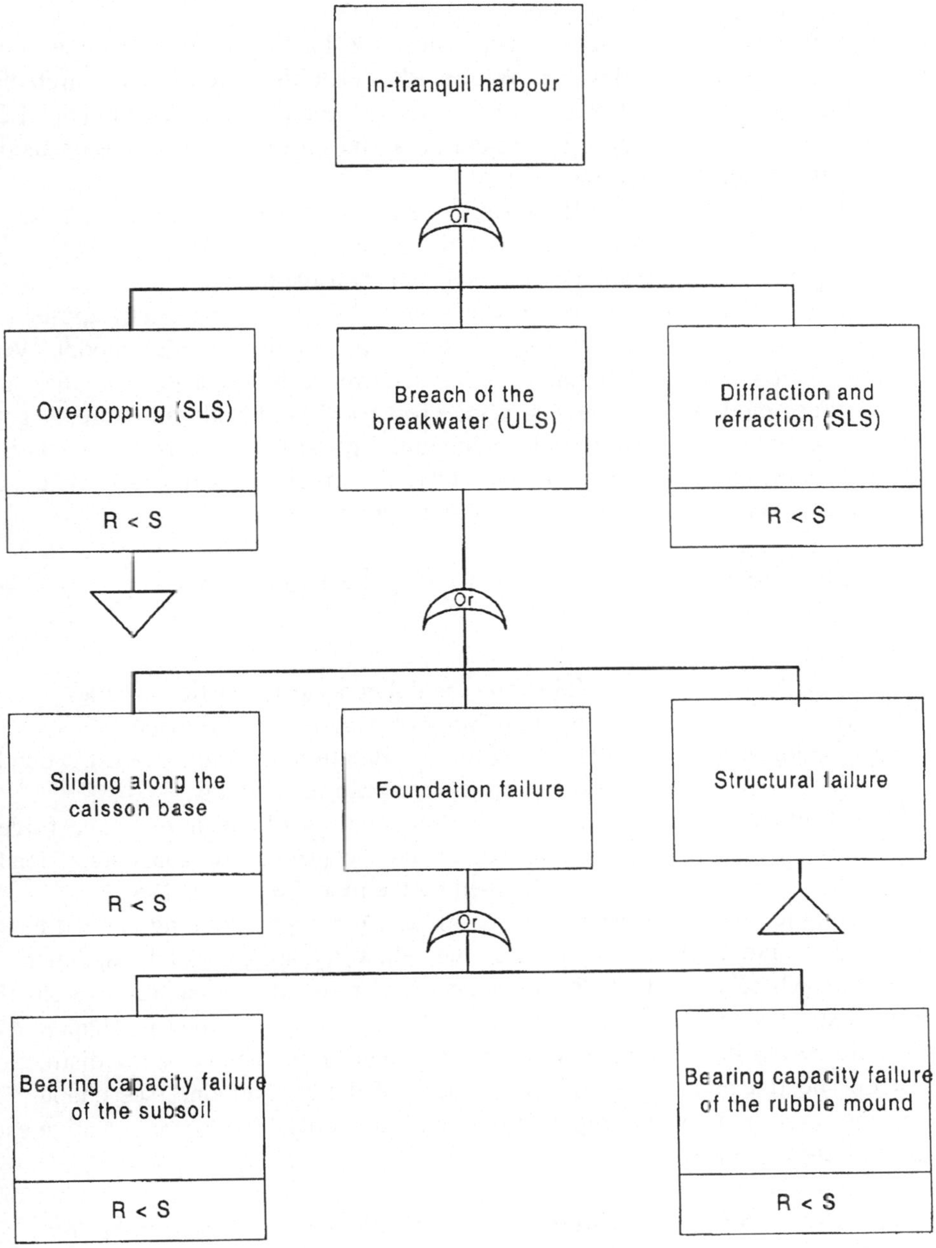

Figure 5-10. Fault tree of a vertical breakwater.

5.3.2.2 *Loading of the breakwater*

For all limit states except wave transmission, the loading is given by the wave forces exerted at the breakwater. Within PROVERBS, extensive research has been directed to this aspect of breakwater design (details are given in chapter 2). In the description given here, the emphasis is laid on the implementation of the load models in a probabilistic framework.

The models used in PROVERBS can be categorised in three types:

1. Load models describing quasi-static (pulsating) wave forces;
2. Load models describing dynamic (impact) wave forces;
3. Decision models, indicating what type of load model should be applied.

On a deterministic level, the parameter map and the breaker model (Volume IIa, sections 2.2 and 2.3) indicate the occurrence of impacts. In reality, the input to these models (water levels, wave properties) is of random nature. A general vertical breakwater will therefore encounter quasi-static loads as well as impact loads during its lifetime. Theoretically, the distribution function of all the wave forces exerted at the breakwater is therefore written as:

$$P(F<f)=P_{impact}\cdot P(F<f\mid impact)+\left(1-P_{impact}\right)\cdot P(F<f\mid no\ impact) \qquad (5\text{-}34)$$

In which:

F:	Wave force modelled as a stochastic variable
P_{impact}:	The probability of occurrence of impacts;
$P(F<f\mid impact)$:	The distribution function of impact loads, conditional on the occurrence of impacts;
$P(F<f\mid no\ impact)$:	The distribution function of pulsating wave forces, conditional on the occurrence of quasi-static loads, e.g. obtained by the model of Goda (1985).

The complex nature of especially the models for impact loading has led to the choice of Monte Carlo simulation for the reliability analysis of breakwaters in conditions where impact loads might occur. For organisational reasons, in the probabilistic framework an early version of the model described in chapter 2 of this Volume has been adopted. What has not been implemented is the distinction of steep and flat bed slopes and the position of the breaker line (see chapter 2). Omitting these parts of the impact model will generally lead to conservative estimates of the wave forces.

5.3.2.3 *Serviceability limit states related to performance of the breakwater*

Generally, breakwaters are built to provide protection against waves. Beside collapse of the breakwater itself, the breakwater can fail to provide protection because of refraction and diffraction effects or due to wave transmission. Refraction and diffraction should be limited by the design of the layout of the breakwater and the harbour entrance. Wave transmission has particularly influence on the cross section of the breakwater.

Wave transmission over vertical breakwaters can be described by means of the model of Goda (1969), modified for different caisson shapes by Heijn (1997).

The limit state for wave transmission has been defined as:

$$g(\mathbf{x}) = H_{s;acc} - K_t H_s \qquad (5\text{-}35)$$

In which:

$H_{s;acc}$:	The acceptable significant wave height inside the harbour basin;
K_t:	Transmission coefficient calculated by the Goda-Heijn model (Goda, 1969, Heijn, 1997);
H_s:	Significant wave height on the sea side of the breakwater.

5.3.2.4 *Foundation limit states*

The geotechnical failure modes have been defined and formulated in chapter 3. In short the failure modes are:

- Sliding along the base of the caisson;
- Bearing capacity failure of the rubble mound;
- Bearing capacity failure of the subsoil.

The limit state equations describing the different forms of failure are available on two levels of sophistication (see chapter 3). In this section the set of limit state equations for preliminary design are adopted as the standard form of modelling the foundation. However, a comparison with other levels of modelling will be shown.

The general form of the foundation limit states is similar to the general form given in section 5.2.2. It is typical for foundation limit states that loading and strength can not always be completely separated. For instance, the weight of the caisson acts as a load but at the same time increases the effective stress in the foundation, thus increasing the shear strength. Next to the stress level, the strength of the foundation is primarily decided by the properties of the soil, such as the friction angle and the cohesion. An overview of the input for the soil models as well as the uncertainties related to the soil properties is given in chapter 3.

5.3.2.5 *Structural limit states*

Like the modelling of the foundation, also the structural limit states of vertical breakwaters can be defined on different levels of sophistication (see chapter 4). In principal, only the beam models given in chapter 4 are suitable for application in a probabilistic calculation. A detailed description of these models is given in chapter 4. The structural strength of the breakwater is primarily decided by the material properties, as well as the sizing of the elements. It is important to note that processes like chloride ingress or cracking of the concrete may lead to corrosion of the reinforcement, which lowers in the strength in time.

5.4 CASE STUDIES

5.4.1 *General*

Within the PROVERBS project, all the elements for a full reliability analysis of vertical breakwaters have been defined. The case studies serve as a demonstration of the application of the developed tools. All case studies comprise a limited setup of the analysis. The case studies generally only deal with:

- ➢ Landward forces, quasi-static as well as impact;
- ➢ Failure due to one of the geotechnical failure modes defined in chapter 3.

Although in several cases, the above mentioned elements of the analysis are found to be decisive for the sizing of the breakwater, it is not guaranteed that this will always be the case. The analyses presented below may relatively easy be extended by including the physical modelling presented in chapters 2 through 4 in the overall concept presented in this chapter. The formulation of more limit states is the primary way to do so.

5.4.2 *Genoa Voltri (Italy)*

5.4.2.1 *The case*

Voltri breakwater represents a typical Italian and European design. It is an existing breakwater that is supposed to be rather safe; the supposition is based on experience on similar and smaller breakwaters existing in Genoa since more than 40 years that suffered no damage; many other breakwaters, similar in shape and design criteria, are present in several harbours in Italy. The assessed failure probability (related to a lifetime of 50 years) can then be confronted with evidence in prototype conditions.

Information required for the analysis regards structure geometry, wave climate, foundation characteristics and construction method. Being a real breakwater an interpretation of available documents was necessary. As long time passed since design and construction, some information might have been lost or misinterpreted. Results are the best we could obtain within the project deadline, but no responsibility is assumed on them.

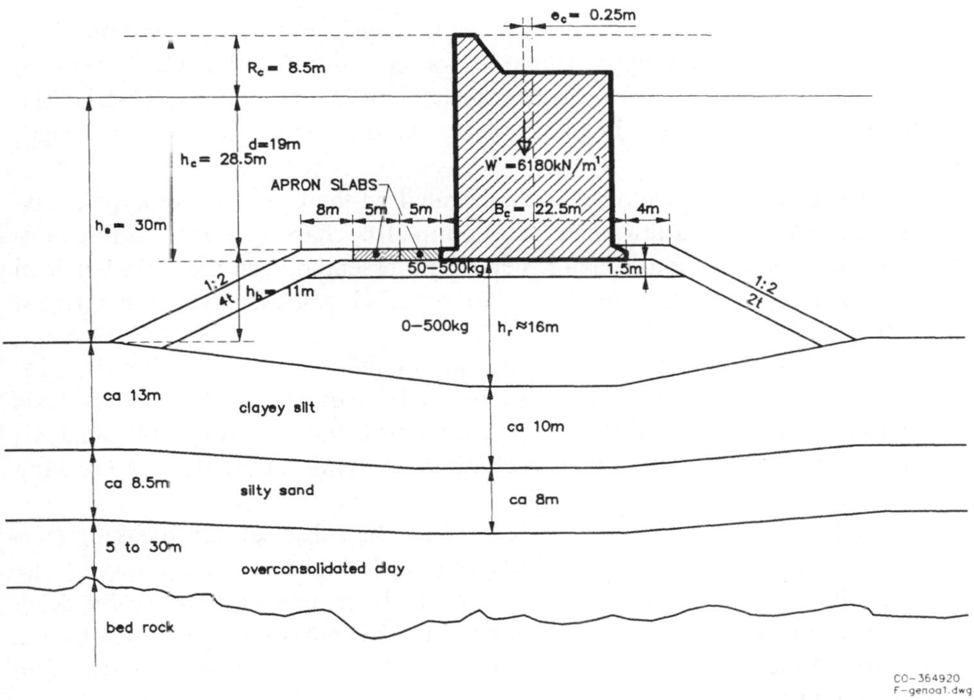

Figure 5-11. Breakwater section and soil layers in Genoa Voltri.

5.4.2.2 *Wave forces*

Breaking wave conditions were in principle excluded at design time by assuring a sufficient water depth in front of the wall; in order to verify the effect of this assumption breaking conditions are regarded as possible and the related hazard is analysed.

In most part of the analysis non breaking conditions are considered and the two set of formulae given by Goda (1985) are used in order to transform waves from off-shore to the structure site and to evaluate the maximum equivalent static force.

As the wave increases, breaking conditions may be reached. In breaking conditions the equivalent static force is computed using the new set of formulae developed within PROVERBS. The parameter map (section 2.2.2) is used in order to understand which is the typical wave action that characterises the site. For the case of Voltri the parameter map shows that breaking waves are possible only for extreme wave conditions. Breaking wave height is evaluated according to (section 2.2.3) and from this the percentage of waves breaking at the structure. Due to foreshore water depth the percentage of broken waves is negligible.

The force time pattern is evaluated on the basis of few parameters (section 2.5.1), whose probability distribution was assessed by Van Gelder (1998) with the help of a data base of measures obtained by several physical models. The longitudinal extension of the breaker is considered to be the full caisson length (section 2.5.3).

The dynamic model, developed and calibrated on the basis of prototype measurements carried out on a few caissons (Volume IIb, chapter 3), is used to evaluate the sliding force response factor, depending essentially on the ratio between force duration and relevant natural oscillation period; static equivalent horizontal and vertical force are then obtained.

Even for the same incident wave, breaking or non breaking conditions correspond to quite different static equivalent applied forces; the level II risk analysis procedure is unable to cope with this discontinuous behaviour. A level III analysis (Monte Carlo simulation) was then performed applying the model for breaking waves.

A level II analysis was anyway performed for the subset of non breaking conditions: the discontinuity was artificially removed extending the validity of the Goda formulae. The design point was found to be actually below the breaking limit. The complete validity of the non breaking hypothesis is confirmed by the comparison of the assessed failure probability with the Monte Carlo simulation in dynamic conditions.

5.4.2.3 *Failure functions*

Among the failure modes developed under PROVERBS (Volume IIb chapter 6), those related to caisson, rubble-mound and cohesive soil interaction were implemented; the relative failure probability is computed through a level II analysis.

The following failure modes are analysed by failure functions representing static force balance:

- caisson sliding along the base;
- bearing capacity failure in rubble mound, i.e. sliding within rubble mound (this mode dominates systematically the traditional overturning failure mode);
- bearing capacity failure in subsoil, i.e. sliding through the high rubble foundation and the silt-sand subsoil.

The analysis can not be considered exhaustive, since several failure modes were not implemented (settlement, scour, deterioration and corrosion of reinforcement due to chloride ingress through the concrete or through cracks, structural failure due to wave loading) and some hydraulic responses were not analysed (overtopping, wave transmission, wave reflection).

Some failure modes are limit sub-cases of others: Rubsan2 is a sub-case of Rubsand4, and Slidclay3 of Rubclay8. When model errors are not considered in system hazard assessment, we can examine all of them at the same time and include them in the list of series element of the system. When model uncertainties are described only the best model is used for each failure mode (Rubsand4 and Rubclay8).

The system reliability is analysed and the lower bound and upper bound of the system failure probability are assessed with Ditlevsen method.

Bishop method is used as alternative to Rubsand4-Ruclay8 series to represent foundation stability, accounting for friction and cohesion in the rubble mound and for the likely configuration of the rubble-mound subsoil interface.

In dynamic conditions Monte Carlo simulations were carried out only for the sliding failure, the typical failure mode induced by breaking waves, Oumeraci (1994).

5.4.2.4 *Variable statistics*

In principle all the parameters are stochastic variables in a hazard analysis, but some show very small variation and can be considered as deterministic with almost no effect on the final result. These are almost all the geometric parameters and unit weights and a few others.

The shape of the statistical distribution of variable parameters is frequently known a-priori based on wider experience and on the nature of the variable. For instance the mean water level is the sum of a random meteorological effect and several sinusoidal astronomical components; the result is well described by a Gaussian variable.

The offshore significant wave height follows an extreme distribution which was better identified on the basis of the available dataset. Despite the importance of the harbour and probably due to the pluri-secular qualitative experience instrumental wave records are not available for Genoa. Wave measurements were sporadic and not public. The assumed statistics is based on KNMI visual observation and MetOffice hindcasting and is confronted with the qualitative experience, as described in Martinelli (1998). The offshore significant wave height was adapted to a GEV distribution (and resulted almost equivalent to a Gumbel distribution).

In all the cases where the variables were physically limited at one side, it was chosen to assume the LogNormal distribution associating null probability below the physical limit. For most cases the variance is small and the assumed distribution becomes almost equal to the Normal distribution.

When there was no clear information on the variance but only on reasonable extreme values, the simple rule was adopted that two standard deviations account for the difference between the maximum (minimum) and the mean value.

The assumed statistics for parameters relative to the hydraulic aspects is:

Table 5-9. Statistics for hydraulic boundary conditions for Genoa Voltri breakwater.

GENOA VOLTRI (50 years)	**Distribution**	**Mean value**	**Standard dev.**	**Notes**
Off-shore significant wave height H_{os}	Gumbel	6.2 m	12%	Based on experienced storms
Deep water wave steepness s_{op}	LogNormal	0.035	10%	Based on steepness of most severe storms
Water level on M.S.L.	Normal	0.1 m	0.2 m	HHW=0.5 m

The statistics of geotechnical parameters assumed for Voltri is:

Table 5-10. Statistics of geotechnical parameters for Genoa Voltri breakwater.

GENOA VOLTRI	**Distribution**	**Mean value**	**Stand. dev.**	**Notes**
Sliding coefficient μ	LogNormal	0.64	10%	Takayama & Ikeda (1992); $\mu - 2\sigma = 0.5$
Rubble mound friction angle	Sum of two LogNormals	$E(\phi_{1r}+\psi_1)$	$\sigma(\phi_{1r}+\psi_1)$	$\mu - 2\sigma = 38°$
Rubble mound residual friction angle, ϕ_{1r}	LogNormal	28°	2°	
Rubble mound dilation angle, ψ_1	LogNormal	16°	2°	
Rubble mound cohesion, for use in Cohsand2*	LogNormal	20 kPa	10%	Japanese praxis, Tanimoto & Takahashi (1994)
Cayey silt cohesion	LogNormal	110 kPa	30%	$\mu - 2\sigma = 50$ kPa
Silty sand friction angle, ϕ_2	Sum of two LogNormals	$E(\phi_{2p}+\psi_2)$	$\sigma(\phi_{2r}+\psi_2)$	$\mu - 2\sigma = 27°$
Silty sand residual friction angle, ϕ_{2r}	LogNormal	25°	2°	
Silty sand dilation angle, ψ_2	LogNormal	8°	2°	
Silty sand undrained cohesion	LogNormal	190 kPa	30%	Estimated from vertical pressure

The statistics of the main geometrical stochastic parameters influencing the resistance of the foundation is:

Table 5-11. Statistics for geometric parameters governing the resistance of the structure.

GENOA VOLTRI	**Distribution**	**Mean value**	**Standard dev.**	**Notes**
Differences in superstructure thickness	Shifted LogNornal	2.2 m	0.25 m	Minimum value 1.0 m
Depth reached by rubble mound below original bottom, used with Bishop failure function	LogNormal	6.0 m	1.0 m	Evaluation of settlement performed before construction

5.4.2.5 *Model uncertainties*

In the attempt of finding all the sources of uncertainty, models and formulae used in the computations were examined and the effect of assumptions and approximations estimated.

Some models are physically based; variables involved are well known and no relevant error is associated to them. This is for instance the case of the formula that evaluates the caisson weight starting from the dimensions and the specific weights.

Some models (e.g. the Goda set of formulae for the evaluation of the maximum force and moments) are based on empirical relations and show a certain scatter (see Volume IId, section 4.1).

Other models (e.g. the geotechnical models) are physically based but rely on assumptions or simplifications; in this case an idea of the induced bias and uncertainty was obtained by applying different models.

In any case a calibration factor is applied to the result of the formula providing the true value. This factor is a random variable whose statistics represents bias and uncertainty of the formula.

Uncertainties in Goda wave force formulae

Goda formulae as most design formulae are biased in order to provide a safe relation, rather than the mean value; this must be accounted for in a reliability analysis.

Van der Meer et al. (1994) gave an estimate of the bias and standard deviation of the Goda formulae on the basis of model tests. Vrijling (1996) pointed out how the derived standard deviation included also statistical uncertainty of the maximum wave height in the experiment realisation. Vrijling arrived to the conclusions presented in the following table.

Table 5-12. Calibration factors for wave loading derived for the Goda model.

GODA Model	**Distribution**	**Mean value**	**Standard dev.**	**Notes**
Calibration factor for horizontal forces	LogNormal	0.90	0.20	IId, section 4.1
Calibration factor for uplift forces	LogNormal	0.77	0.20	IId, section 4.1
Calibration factor for horizontal moments	LogNormal	0.72	0.37	IId, section 4.1
Calibration factor for uplift moments	LogNormal	0.72	0.34	IId, section 4.1
Calibration factor for seepage horizontal force	LogNormal	0.65	0.30	On the basis of the uncertainties given above

An analysis of the effect of apron slab provides results consistent with calibration factors reported in the table. Depending on gaps between the apron slabs, under-pressure can be smaller than predicted by Goda formula and, due to uncertainty of the spatial distribution, its shape is not always triangular. The uncertainty of moments is slightly greater than the uncertainty of forces.

The great uncertainty is mainly related to breaking waves conditions. Several model tests show that if the formula is restricted to non breaking waves the standard deviation is much lower (6-8% for horizontal force).

The statistics of the calibration factor for seepage forces, not defined by Goda's formulae nor checked by van der Meer or Vrijling, was derived from the linear pressure distribution suggested by Goda. The effect of a filtration length longer than the caisson base is greater on seepage forces than on uplift, the bias and uncertainty of this calibration factor are therefore assumed proportionally greater than for uplift.

For the breaker impact model we assume the statistics presented by van Gelder (1998):

Table 5-13. Statistics for stochastic parameters in the impact model.

	Distribution	**Mean value**	**Standard dev.**	**Notes**
Coefficient 'k' for impulse of breaking wave	LogNormal	0.086	97%	Van Gelder (1998)
Coefficient 'c' for total duration of breaking wave force	LogNormal	2.17	50%	Van Gelder (1998)

Uncertainties in geotechnical failure functions

For all geotechnical models the calibration factor Z is assumed as homogeneous to a failure load (or safety) factor: the failure function has always the form R/S-Z. Definitions of resistance R and load S and statistics for Z are provided below.

The effect of approximations and hypotheses were roughly evaluated assuming that the foundation geometry is correctly schematised by the failure functions.

The most realistic failure functions based on the upper limit theory are credited of a 2% average error towards the unsafe side. A similar error is credited to Bishop method, Lancellotta (1995).

The effect of friction among the unstable body and the adjacent stable ones is evaluated assuming horizontal pressure equal to 60% of the vertical pressure. Under non breaking waves it is assumed that instability is extended to a 100 m long reach of the breakwater, whereas under breaking waves it is assumed that instability regards a single caisson.

Table 5-14. Calibration factors for various failure modes in the subsoil incl. uncertainties.

Failure function	**Definition of the calibration factor**	**Expected value**	**Uncertainty**	**Comments**
Slidtak1	$(F_G-F_U)\tan\phi / F_H$	1.00	1%	Horizontality err.
RubSand2	$\{W_1+(F_G-F_U)\omega_{1V}/$ $\{(F_H+F_{HU})\omega_{1H}\}$	1.10 1.05	10% 15%	Non breaking Breaking
CohSand2	$\{W_{int}+W_1+(F_G-F_U)\omega_{1V}/$ $\{(F_H+F_{HU})\omega_{1H}\}$	1.10 1.05	10% 15%	Non breaking Breaking
SlidClay3	$W_1/(F_H+F_{HU})$	0.96 0.85	4% 10%	Non breaking Breaking
Rubsand4	$\{\Sigma W_i+(F_G-F_U)\omega_{1V}/$ $\{(F_H+F_{HU})\omega_{1H}\}$	0.97 0.87	3% 8%	Non breaking Breaking
RubClay8	$\{\Sigma W_1+(F_G-F_U)\omega_{4V}/$ $\{(F_H+F_{HU})\omega_{4H}\}$	0.96 0.85	4% 10%	Non breaking Breaking
Bishop	M_R/M_S	0.97 0.87	3% 10%	Non breaking Breaking

5.4.2.6 *System failure probability*

The considered failure modes are caisson sliding, rubble mound and subsoil failure. Failure probabilities and design point co-ordinates are given in the following table.

Table 5-15. Genoa Voltri hazard analysis for separate modes.

		Design point of main variables				
Failure function	**Failure prob. in 50 years**	**H_{so}**	**Sliding coeffi-cient μ**	**Rubble mound friction angle ϕ_1**	**Rubble mound cohesion**	**Cohesion subsoil c_u'**
Slidtak1	**2.3%**	**7.6 m**	**0.60**			
Rubsand2	8.0%	7.0 m		43.0°		
CohSand2	2.9%	7.5 m		43.0°	19.8 kPa	
Rubsand4	**11.5%**	**6.8 m**		**43.0°**		
Slidclay3	8.2%	6.8 m		43.6°		84.1 kPa
RubClay8	**8.1%**	**6.8 m**		**43.5°**		**83.6 kPa**
Bishop	1.5%	7.8 m		42.8°	19.8 kPa	110.0 kPa

Since model errors are considered, only the best equation (equation providing the least uncertainty) is used for each independent failure mode in system failure analysis. We judged that Slidtak1 is representative of caisson sliding over the base, Rubsand4 of rubble mound bearing capacity failure and Rubclay8 of subsoil bearing capacity failure. The overall failure probability results to be 15% (the lower and upper bounds are 14.7% and 15.2% respectively). The value should be taken as typical for structures of this kind recently designed in Europe.

Rubsand4, i.e. bearing capacity failure in rubble mound causing sliding along a curved surface, results to be the most important failure mode. This does not agree with the experienced failures (see for instance Oumeraci, 1994) since failure in the rubble mound is not commonly reported as a cause of breakwater failure (not as much as sliding at the caisson base). This could be due in some proportion to the specific design of Voltri rubble mound, that is evidently less wide at the harbour side than at the offshore side, or possibly to an erroneous evaluation of calibration factor statistics, for instance to an underestimation of systematic effect of lateral friction, or else this failure mode that can represent an important horizontal displacement of caissons can be confused in prototype with caisson sliding on its base.

The design wave conditions correspond actually to non breaking waves.

5.4.2.7 *Sensitivity analysis*

A sensitivity analysis of failure probability showed that future wave load is the greatest cause of uncertainty; second is the complex of model uncertainties and third foundation characteristics.

The effect of the berm width at the harbour side was analysed showing that a wider berm would result in a significantly safer breakwater.

The effect of a real or apparent (representing the curvature of the failure boundary) cohesion in the rubble mound shows that even a smaller value than suggested by Japanese guidelines (12 compared to 20 kPa) combined with a cautious tangent friction angle of 35° result in a significantly stronger rubble mound than for the equivalent secant friction angle 37° (8.2 % against 12.8 % failure probability).

Similarly the use of Bishop failure function combined with a realistic curved contact surface between rubble mound and subsoil shows that failure probability estimated by Rubclay8 for a conventional geometry (plane horizontal base) is significantly overestimated. As a consequence the system failure probability should be controlled by rubble mound failure more than shown by previous figures.

5.4.2.8 *Effect of breaking*

A Monte Carlo simulation was performed including the effect of possible breaking waves for sliding failure mode. In extreme conditions breakers do actually occur but failure probability does not increase since caisson would slide yet for lower non breaking waves. A similar result occurs obviously also for the critical rubble mound failure mode, showing that globally breaking even if possible does not influence the hazard of the analysed breakwater.

5.4.2.9 *Conclusions*

- The structure failure probability was found to be around 10-15%.
- The critical failure mode is bearing capacity failure in the rubble mound.
- A wider berm at the harbour side would significantly reduce the hazard.
- The greater hazard originates from the intrinsic uncertainty of future waves.
- Uncertainty in Goda formulae is relevant, but relevance would decrease if the better behaviour of the formulae for non breaking waves is considered.
- Different model uncertainties, both for wave action and for foundation behaviour, should be applied for non breaking and breaking waves.

5.4.3 *Easchel breakwater*

5.4.3.1 *Introduction*

The Easchel breakwater is a fictitious breakwater, placed on a thin bedding layer. The sea bottom consists of well-described Eastern Scheldt sand. The objective of this case study is twofold:

1. Investigate the influence of a few geometric parameters of the breakwater on the probability of caisson instability (sliding along the base, bearing capacity failure of the rubble mound, bearing capacity failure of the subsoil);

2. Investigate the influence of model variations on the probability of caisson instability.

Ad 1: The width of the caisson and the height of the top slab have been varied. The results lead to a modification of the original design;

Ad 2: For the modified design, the following variations of models have been studied:

- Two different models for wave breaking (Goda and linear wave theory);
- Two alternatives for the loading of the breakwater (including or neglecting of impacts, see chapter 2);
- Two alternatives for the probability calculations (First Order Reliability Method or Monte Carlo, this chapter);
- Foundation modelling on three levels of sophistication (feasibility level design, preliminary design and finite element modelling, see also chapter 3).

In this case study all failure probabilities are expressed per year.

5.4.3.2 *Breakwater geometry and boundary conditions*

The Easchel breakwater is built on a mildly sloping seabed. The slope equals 0.4% for a long distance offshore. The depth at the toe is 12.6 m with respect to mean sea level. An overview of the breakwater cross section is given in Figure 5-12.

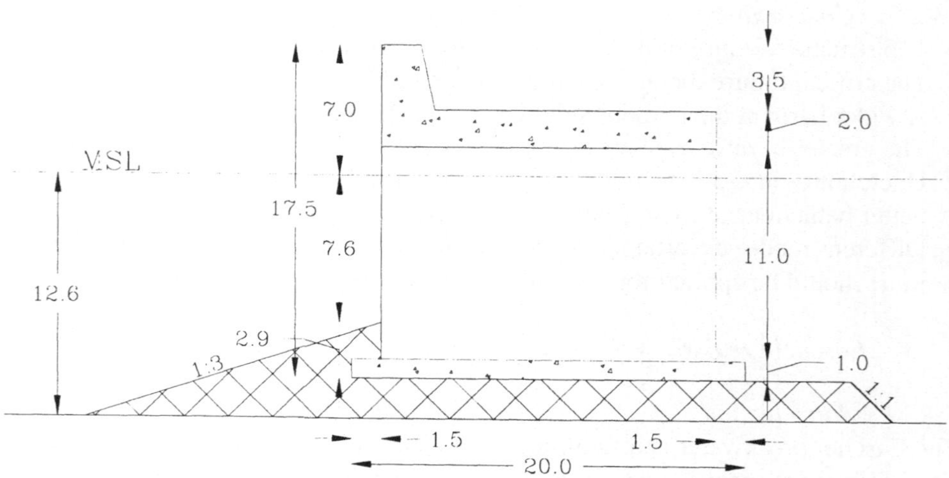

Figure 5-12. Overview of breakwater cross section.

The subsoil consists of Eastern Scheldt sand, which was thoroughly investigated in the design phase of the Eastern Scheldt Storm Surge Barrier in the Netherlands. For the properties of the rubble, expert estimates have been used.

Table 5-16. Overview of distributions of rubble properties.

Variable	Distribution type	Mean	Standard deviation
Friction angle	Normal	43	2.5
Dilatancy angle	Lognormal	10	1.4
Density	Deterministic	21	-
Cohesion	Deterministic	0	-

Table 5-17. Overview of distributions of subsoil properties.

Variable	Distribution type	Mean	Standard deviation
Friction angle	Normal	37	2
Dilatancy angle	Lognormal	8	1.4
Density	Deterministic	21	-
Cohesion	Deterministic	0	-

The hydraulic boundary conditions are given in Table 5-18.

Table 5-18. Overview of hydraulic boundary conditions on deep water.

Variable	Distribution type	Mean	Standard deviation
Water level (h_w)	Normal	0.2	0.2
Significant wave height (H_{s0}) [m]	Gumbel	5.05	0.63
Wave steepness (s_{0p}) [-]	Lognormal	0.027	0.0068

The breakwater is placed in an area with a negligible tidal difference. A small variation of the water level has been assumed.

Generally, the wave height and the wave period are highly correlated. In a probabilistic calculation this correlation has to be taken into account. A description of the correlation by means of a physical relationship is generally to be preferred. Analysis of buoy measurements shows that the equivalent deep water wave steepness, defined as:

$$s_{0p} = \frac{H_{s0}}{\frac{g}{2\pi} T_p^2} \tag{5-36}$$

is virtually statistically independent of the wave height. Therefore, the wave steepness has been used as input in the probabilistic calculation and the wave period is derived from the wave height and the wave steepness.

5.4.3.3 *Inshore wave climate*

For the reliability analysis, the wave conditions just in front of the breakwater are relevant. Several models for wave transformation in shallow water have been proposed (Goda, 1985; Vrijling & Bruinsma, 1980; Battjes & Janssen, 1978). In this study the model of Goda and the model of Vrijling & Bruinsma (denoted as the Eastern Scheldt model, see section 2.2.2) have been used. A sample of 10.000 wave heights is shown in Figure 5-13.

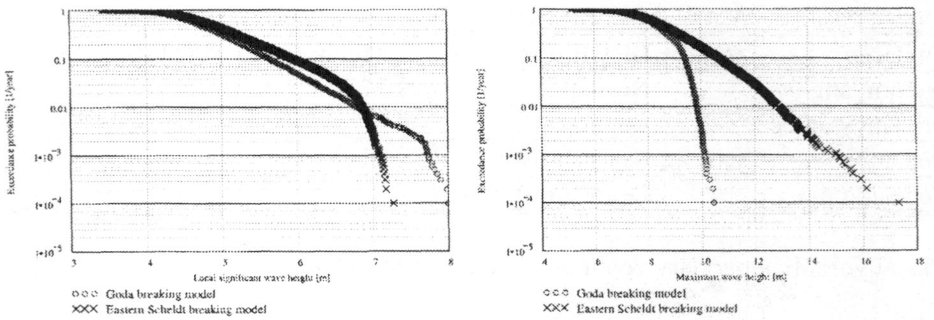

Figure 5-13. Empirical distributions of significant wave height and maximum wave height in front of the breakwater.

It appears that the Goda model results in slightly higher significant wave heights in front of the structure. However, the maximum wave heights, which are relevant for the force calculations, are higher for the Eastern Scheldt model. Analysis showed that the differences regarding the maximum wave height may be caused by one parameter which is chosen as a fixed value in the Goda model and which is treated as a random variable in the Eastern Scheldt model (Volume IId, section 5.2).

5.4.3.4 *Loading of the structure*

The loading of the structure has been determined in two different ways:

- According to the classical Goda model, including the model uncertainty derived by Bruining (1994);
- According to the model developed in the PROVERBS project, as described in chapter 2.

The combination with the wave transformation model leads to a total of four alternative force distributions. The resulting horizontal forces are shown in Figure 5-14.

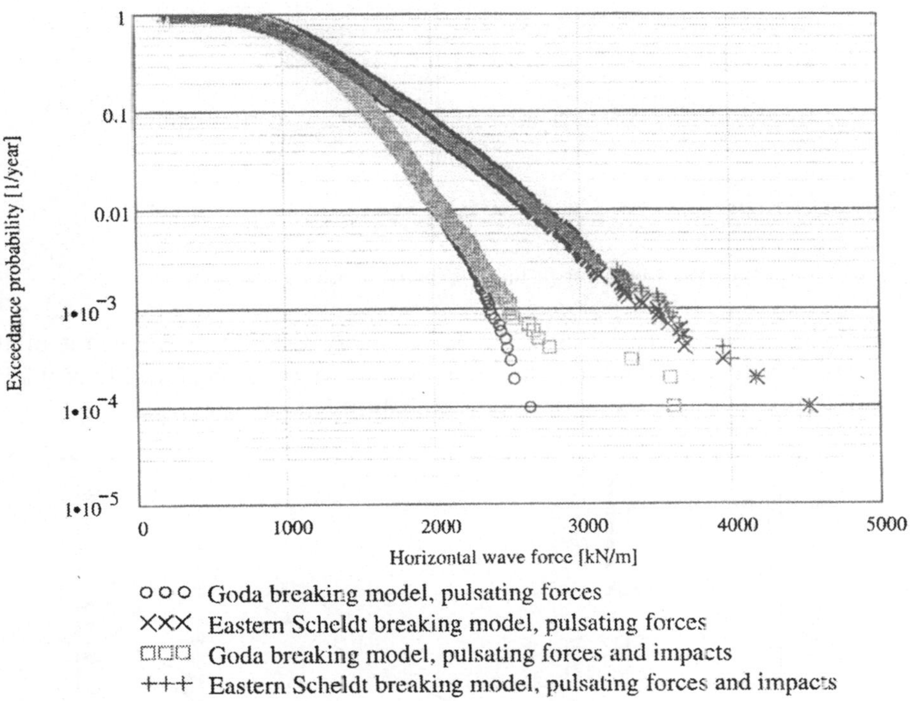

Figure 5-14. Empirical distributions of horizontal wave force for four model combinations.

For pulsating forces only, the influence of the higher maximum wave heights resulting from the Eastern Scheldt model is clearly visible. In case of the Goda wave breaking model, application of the impact model leads to an increase of the forces in the upper region of the distribution. In case of the Eastern Scheldt model the influence on the horizontal force is small. The influence is mainly found in the region around the 10^{-3} quantile. Since this is in the order of magnitude of the failure probability of the structure, application of the impact model does influence the failure probability in case of the Eastern Scheldt breaker model (see below).

5.4.3.5 *Influence of the breakwater geometry on the probability of caisson instability*

The influence of the caisson width and the thickness of the top slab have been investigated for the model combination Goda wave breaking / pulsating forces. The influence of the caisson width is shown in Figure 5-15.

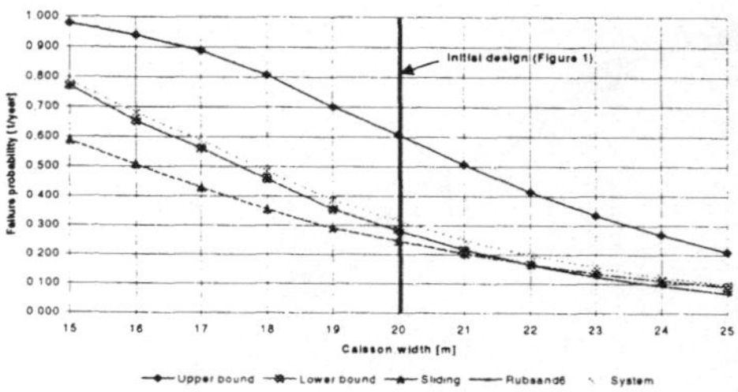

Figure 5-15. Influence of caisson width on the probability of caisson instability.

The top slab thickness is varied in such a way that the crest height of the breakwater remains constant, i.e. the crown wall is stepwise replaced by a thicker top slab. Figure 5-16 shows the resulting probability of caisson instability.

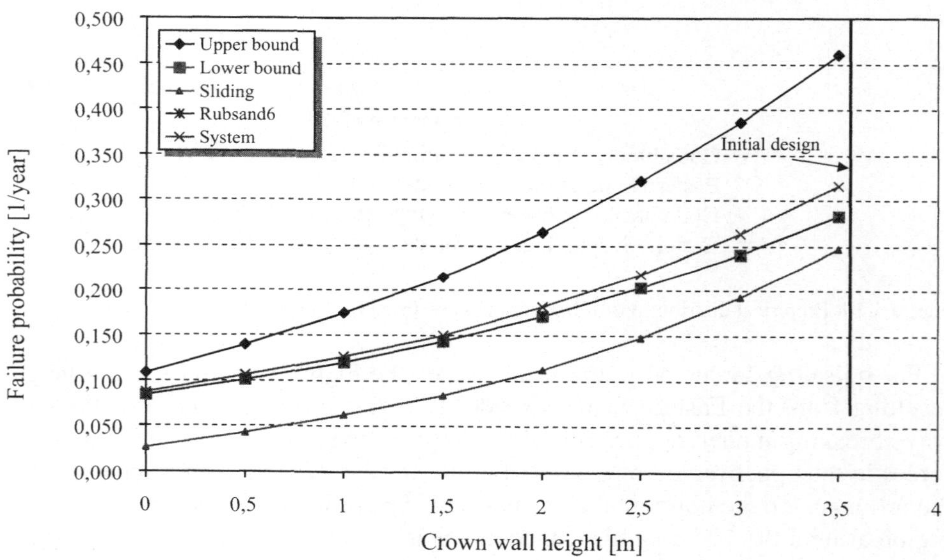

Figure 5-16. Influence of the top slab thickness on the probability of caisson instability.

The results of the sensitivity analysis lead to a modification of the design. See Figure 5-17.

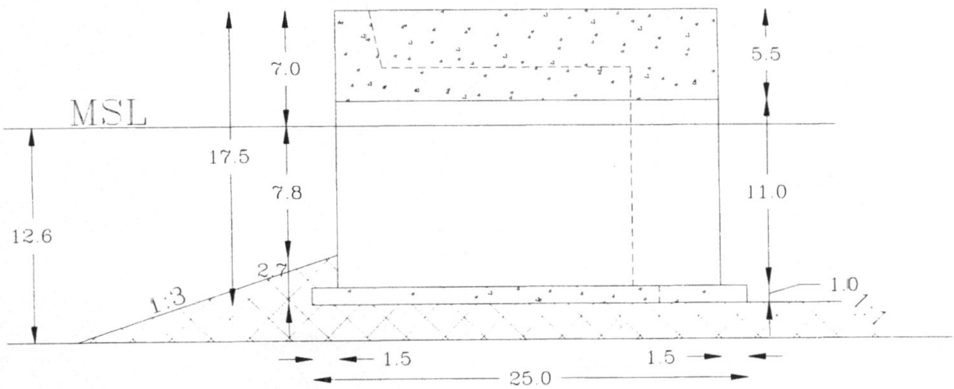

Figure 5-17. Cross section of the modified design of the Easchel breakwater (dotted line: original design).

5.4.3.6 *Comparison of model combinations for pulsating wave forces*

For pulsating loads, eight model combinations have been used. An overview is given in Table 5-19.

Table 5-19. Overview of tested model combinations for quasi-static loads.

No.	Foundation model	Wave breaking model	Integration method
1	Preliminary	Goda	FORM
2			Monte Carlo
3		Eastern Scheldt	FORM
4			Monte Carlo
5	Feasibility	Goda	FORM
6			Monte Carlo
7		Eastern Scheldt	FORM
8			Monte Carlo

An overview of the results is given in Figure 5-18.

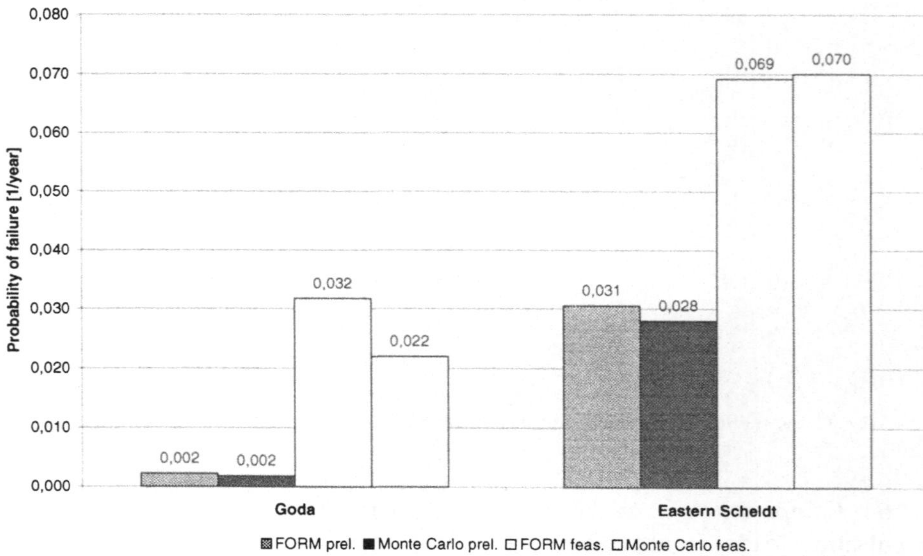

Figure 5-18. Overview of calculated failure probabilities for different combinations of calculation method and foundation models (pulsating forces only).

As expected, the feasibility design level models for the foundation tend to be more conservative than the preliminary design level models. This results in higher failure probabilities for this type of foundation models. Furthermore, the higher maximum wave heights for the Eastern Scheldt model lead to higher failure probabilities. Generally, the calculated failure probability is in the same order of magnitude for FORM and Monte Carlo.

Using the design points from the FORM calculations in a finite element model (PLAXIS) supports the conclusion that the feasibility models are the most conservative. The PLAXIS results show that the feasibility models underestimate the wave forces at collapse by approximately 30 % (compared to PLAXIS) and the preliminary design models by approximately 10 % (Volume IId, section 5.2).

5.4.3.7 *The influence of impact loading*

The probability of failure under impact loading has been calculated using Monte Carlo analysis. As expected, application of impact loading leads to an increase of the failure probability.

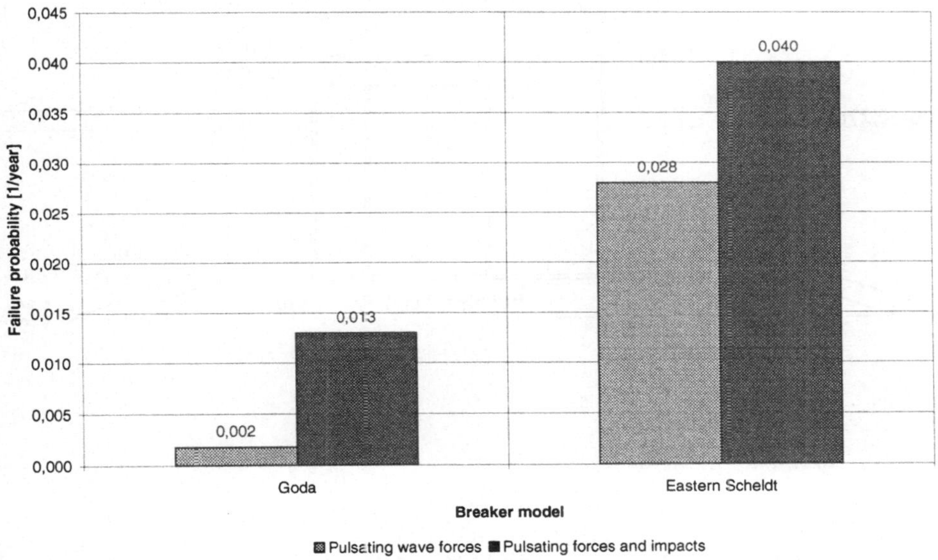

Figure 5-19. Calculated failure probabilities for pulsating loads and for mixed loading (calculation method: Monte Carlo, Foundation model: preliminary design level).

5.4.4 *Reliability analysis of geotechnical failure modes for the Mutsu-Ogawara West breakwater*

5.4.4.1 *Introduction*

A reliability analysis is performed with a breakwater with the same geometry as the Mutsu-Ogawara West breakwater in Japan. The geometry is shown in Figure 5-20. The wave conditions and the subsoil strength parameters are not known such that a detailed stochastic model for these can be formulated. Therefore a representative wave climate corresponding to the known design wave height is assumed. Further, weak and strong subsoil models are formulated which represent typical strength parameters for sand (drained) and clay (undrained) subsoils. The design lifetime is 50 years. In this case study all failure probabilities are expressed per lifetime (see Figure 6.5, page 119 in Christiani, 1997).

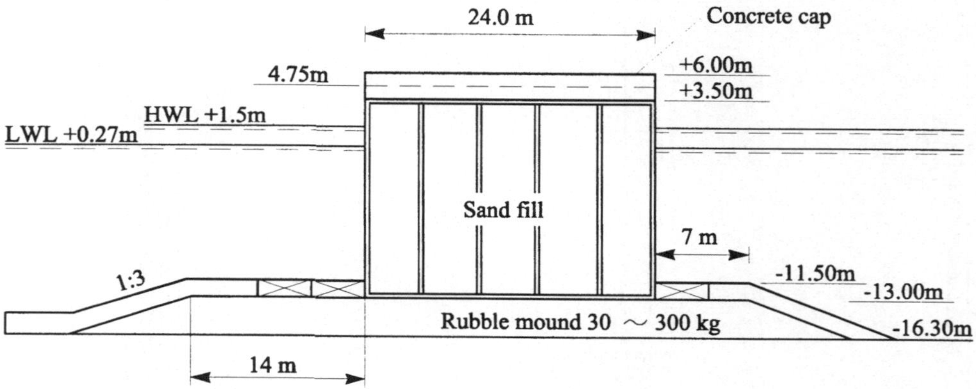

Figure 5-20. Mutsu-Ogawara breakwater.

5.4.4.2 *Stochastic models*

Wave height

The maximum significant wave height H_s^T in the design lifetime T is assumed to be modelled on the basis of a limited number N of wave height observations. An extreme Weibull distribution is used:

$$F_{H_S^T}(h) = \left[1 - \exp\left(-\left(\frac{h - H'}{u}\right)^{\alpha}\right)\right]^{\lambda T} \tag{5-37}$$

where λ is the number of observations per year. α, u and H' are parameters to be fitted to the observed data. In order to model the statistical uncertainty u is modelled as a Normal distributed stochastic variable with coefficient of variation

$$V_u = \frac{1}{\sqrt{N}} \sqrt{\frac{\Gamma(1 + 2/\alpha)}{\Gamma^2(1 + 1/\alpha)} - 1} \tag{5-38}$$

The design significant wave height is 8.0 m. It is assumed that N = 20 data has been used, $\lambda = 1$, $\alpha = 0.90$, u = 1.71 m and H' = 0 m.

The model uncertainty related to the quality of the measured wave data is modelled by a stochastic variable F_{Hs} which is assumed to be normally distributed with a coefficient of variation equal to 0.05 corresponding to good wave data.

Wave forces

Both pulsating and impact wave loading are considered:

- Pulsating wave forces are estimated using the Goda formula, see section 2.4.1. The model uncertainties related to horizontal and uplift wave forces are assumed to be fully correlated. Also model uncertainties related to horizontal and uplift wave moments are assumed to be fully correlated.
- Impact loading consisting of a horizontal and a vertical (uplift) part is estimated by the model described in section 2.5.1. The model uncertainties related to the impact loads are described in detail in section 2.5.4 and modelled by the stochastic variables no. 18, 19, 20 and 21, see Table 5-20.

Geotechnical parameters

The geotechnical parameters for the rubble mound material and the subsoil are not known for the Mutsu-Ogawara breakwater. Therefore representative strength parameters are used corresponding to weak and strong subsoils. Both sand (drained) and clay (undrained) subsoil models are considered. Further different values of the coefficient of variation of the strength parameters are investigated, see Table 5-20. Since the undrained shear strength for clay is modelled by a stochastic field the correlation structure has to be specified. The expected value function $E[c_u(x, z)]$ and the covariance function $C[c_u(x_1, z_1), c_u(x_2, z_2)]$ are assumed to be:

$$E[c_u(x,z)] = c_{u0} + c_{u1}z \tag{5-39}$$

$$C[c_u(x_1,z_1), c_u(x_2,z_2)] = \sigma_{c_u}^2 \exp(-|\alpha(z_1 - z_2)|)\exp(-(\beta(x_1 - x_2))^2) \tag{5-40}$$

where (x_1, z_1) and (x_2, z_2) are two points in the soil. z is the vertical coordinate and x is the horizontal coordinate. $\sigma_{cu} = 30$ kPa, $\alpha = 0.33\ \text{m}^{-1}$ and $\beta = 0.033\ \text{m}^{-1}$ are used. For weak clay: $C_{u0} = 79$ kPa and $c_{u1} = 1$ kPa/m. For strong clay: $c_{u0} = 173$ kPa and $c_{u1} = 0$ kPa/m. The statistical parameters for weak and strong sand subsoil can be seen in Table 5-20.

The complete list of stochastic variables is shown in Table 5-20.

Table 5-20. Statistical model for Mutsu-Ogawara breakwater. W: Weibull, N: Normal, LN: Log-Normal.

i	X_i	Description	Mean	Standard Deviation	Distribution
1	H_s	Significant wave height $[m]$	see above	See above	W
2	u	Weibull parameter $[m]$	2	See above	N
3	F_{hs}	Model uncertainty on wave height	1	0.05	N
4	s_M	Wave steepness factor	1	0.25	N
5	ζ	Tidal elevation, maximum $\zeta_0 = 0.8$ m			Cosine
6	U_{F_H}	Model uncertainty horizontal force	0.90	0.2	N
7	U_{F_U}	Model uncertainty uplift	0.77	0.2	N
8	U_{M_H}	Model uncertainty horizontal moment	0.81	0.40	N
9	U_{M_U}	Model uncertainty uplift moment	0.72	0.37	N
10	ρ_c	Average density of caisson $[t/m^3]$	2.23	0.11	N
11	φ_1	Effective friction angle - rubble mound	46°	4.6°	LN
12	ψ_1	Angle of dilation – rubble mound	16.7°	1.67°	LN
13	φ_2	Effective friction angle - sand subsoil	39.0°/42.8° weak/strong	3.9° / 4.3° weak/strong	LN
14	ψ_2	Angle of dilation – sand subsoil	10.2°/15.3°	1.0° / 1.5°	LN
15	U	Clay strength	1	0	N
16	f	Friction coefficient	0.636	0.0954	LN
17	c_u	Undrained shear strength for impact load	400 kPa	80 kPa	LN
18	k	Factor for impact load	0.086	0.084	LN
19	c	Factor for impact load	2.17	1.08	LN
20	R	Model uncertainty factor for impact rise time	1	0.3	LN
21	U_I	Model uncertainty factor for impact forces	1	0.5	LN

5.4.4.3 *Reliability analysis*

The failure modes described in section 5.3.2 are used and the probability of failure within the design lifetime T is estimated. Eleven limit state functions are formulated for the following failure modes:

- Sliding :
  1. sliding along the caisson base
- Bearing capacity failure of the rubble mound:
  2. rupture in rubble along bottom of caisson
  3. rupture in rubble mound - straight rupture line
  4. rupture in rubble mound – curved rupture line
- Bearing capacity failure of the subsoil (sand)
  5. rupture in subsoil along bottom of rubble mound
  6. rupture in rubble mound and sand subsoil – mode 1
  7. rupture in rubble mound and sand subsoil – mode 2
- Bearing capacity failure of the subsoil (clay)
  8. rupture in subsoil along bottom of rubble mound
  9. rupture in rubble mound and clay subsoil – mode 1
  10. rupture in rubble mound and clay subsoil – mode 2

In the case of impact loads and sand subsoil, the subsoil is assumed to behave as undrained and the following failure mode is investigated:

- Bearing capacity failure of the subsoil (undrained, impact loads only):
  11. rupture in rubble mound and sand subsoil – mode 2

Tables 5-21 and 5-22 show the results of a reliability analysis of the breakwater. Three different coefficients of variation for the soil strength parameters are investigated. The reliability analysis is performed using Monte Carlo simulation with 10.000 samples. The system probability of failure is taken as the maximum probability of failure for the individual failure modes. It is seen that if impact loading is taking into account then the probability of failure is very high, especially for weak clay subsoil. Further it is also seen that for clay subsoil it is very important if the strength is weak or strong. The importance of the coefficient of variation of the soil strength is only important for strong clay subsoil. Finally, it is seen that reasonable low and acceptable probabilities of failure are obtained for strong subsoils if model tests to determine the wave forces more accurate are performed.

Table 5-23 shows the probability of failure for the different failure modes in the case of no impact loading taking into account and strong subsoil with coefficients of variation equal to 0.15 for clay and 0.08 for sand. It is seen that if no model tests are performed the most important failure modes are no. 7 (sand) and 10 (clay). If model tests are performed also failure mode 1 (sliding) can be important. In Volume IId, section 5.3, a more detailed description of the reliability analyses is given including a description of the most important stochastic variables.

Table 5-21. Probability of failure in a lifetime of 50 years for sand subsoil.

Coefficient of variation	**No impact load**				**Impact load**			
	No model tests		**Model tests**		**No model tests**		**Model tests**	
	Weak	**Strong**	**Weak**	**strong**	**Weak**	**Strong**	**Weak**	**Strong**
0.05	0.065	0.037	0.0023	0.0023	0.29	0.27	0.25	0.25
0.08	0.065	0.037	0.0023	0.0023	0.29	0.27	0.25	0.25
0.10	0.065	0.037	0.0023	0.0023	0.29	0.27	0.25	0.25
Probability of impact					0.26			

Table 5-22. Probability of failure in a lifetime of 50 years for clay subsoil.

Coefficient of variation	**No impact load**				**Impact load**			
	No model tests		**Model tests**		**No model tests**		**Model tests**	
	Weak	**Strong**	**Weak**	**Strong**	**Weak**	**Strong**	**Weak**	**Strong**
0.10	0.71	0.027	Weak	0.0023	0.76	0.27	0.82	0.25
0.15	0.71	0.036	0.77	0.0037	0.76	0.27	0.80	0.25
0.20	0.71	0.051	0.74	0.0180	0.76	0.28	0.78	0.26
Probability of impact					0.26			

Table 5-23. Probability of failure for the case of no impact loading and strong subsoil with coefficients of variation equal to 0.15 for clay and 0.08 for sand.

Failure mode	**no model tests**	**model tests**
1 rubble mound	$1.76\ 10^{-2}$	$2.31\ 10^{-3}$
2	$4.02\ 10^{-4}$	$2.32\ 10^{-7}$
3	$4.79\ 10^{-3}$	≈ 0
4	$4.80\ 10^{-7}$	≈ 0
5 sand subsoil	$2.00\ 10^{-4}$	≈ 0
6	$2.31\ 10^{-2}$	$1.00\ 10^{-4}$
7	$3.73\ 10^{-2}$	$3.00\ 10^{-4}$
System / sand subsoil	$\mathbf{3.73\ 10^{-2}}$	$\mathbf{2.31\ 10^{-3}}$
8 clay subsoil	$1.12\ 10^{-2}$	≈ 0
9	$1.63\ 10^{-2}$	$2.00\ 10^{-4}$
10	$3.62\ 10^{-2}$	$3.70\ 10^{-3}$
System / clay subsoil	$\mathbf{3.62\ 10^{-2}}$	$\mathbf{3.70\ 10^{-3}}$

5.5 PERSPECTIVES

5.5.1 *Durability*

An important consideration in the design of concrete structures in seawater is that of durability. Chloride ingress, thought to be aided by loading induced cracking of the concrete threatens the steel reinforcement. Over time, the cross-section of the reinforcement will be reduced by corrosion, spalling will take place and finally the strength of the concrete structure will be impaired.

Models to describe chloride ingress as a diffusion process are available. At this moment one has to rely however on measurements at the completed structure to estimate the necessary values of the diffusion coefficient. It is not yet possible to estimate these values just on the basis of the concrete specification.

Also models to estimate the rate of the corrosion and the related reduction of the size of the reinforcement still have to be developed. In practice this is solved by designing the reinforcement on the basis of SLS conditions and limited crack-width (typically 0.3 mm) with a safety factor of approximately 1.5. The result is that the ULS load can be easily withstood by the concrete structure and that the amount of reinforcement is decided by the limitation of the crack-width in the SLS condition.

To avoid uneconomic over-design, better models have to be developed that relate chloride ingress, which is thought to be accelerated by cracking, to the final failure of the structure with reduced reinforcement.

5.5.2 *Impacts*

The maximum force and the duration of impacts in prototype are still a matter of discussion. This will be solved by further research.

In the mean time a designer should try to avoid creating impact conditions in front of the vertical breakwater. A study of the parameter map (chapter 2) shows that impact can mostly be avoided by choosing the appropriate geometry for the caisson and the mound on which it is founded.

5.5.3 *Construction*

When, during the design of a breakwater one is presented with the choice between a rubble mound type and a vertical caisson breakwater a thorough consideration of the construction method and sequence should be made. It is clear that the transport and the placement of huge caisson structures without proper consideration is more risky than the handling of classical armour elements on a rubble mound breakwater. Also unexpected settlements during construction are more easily accommodated by rubble mound structures than caisson structures.

5.5.4 *Reflection*

Without further measures, the reflection of the incoming waves by a vertical breakwater is nearly 100%, resulting in a confused sea in front of the structure that may cause hindrance to shipping. Perforating the front wall can to some extent reduce this effect. The effectiveness of the reduction for random seas with a peak frequency changing over time should be further studied.

5.5.5 *Shear keys*

A philosophy for the design of shear keys has to be developed. The central question is to what extent a caisson may call on his neighbours during ULS conditions in short crested seas. It should be noted that a large dependency on shear key action could lead to repeated movement of a single caisson during a subsequent wave crest and through

ACKNOWLEDGEMENTS

The following people are gratefully acknowledged for their contribution to this chapter:

H.F. Burcharth	Aalborg University, Denmark
J. Dalsgaard Sørensen	Aalborg University, Denmark
P.H.A.J.M. van Gelder	Delft University of Technology, The Netherlands
A. Lamberti	University of Bologna, Italy
L. Martinelli	University of Bologna, Italy

REFERENCES TO VOLUME IId

Chapter 1: VRIJLING, J.K. (1999): Introduction.

Chapter 2: VRIJLING, J.K. (1999): Fault tree analysis of a vertical breakwater. 9 pp.

Chapter 3: SØRENSEN, J.D.; BURCHARTH, H.F. (1999): Limit state equations including uncertainties. 26 pp., 1 Annex.

Chapter 4.1: VOORTMAN, H.G.; KUIJPER, H.K.T.; VRIJLING, J.K. (1999): Economic optimal design of vertical breakwaters. 17 pp.

Chapter 4.2: SØRENSEN, J.D.; BURCHARTH, H.F. (1999): Partial safety factor system. 24 pp.

Chapter 5.1: LAMBERTI, A.; MARTINELLI, L.; DE GROOT, M.B.; GOLÜCKE, K.; VAN HOVEN, A.; ZWANENBURG, C. (1999): Hazard analysis of Genoa Voltri breakwater. 40 pp.

Chapter 5.2: VOORTMAN, H.G.; VRIJLING, J.K. (1999): Reliability analysis of the Easchel breakwater. 29 pp.

Chapter 5.3: SØRENSEN, J.D.; BURCHARTH, H.F. (1999): Other representative structures: Mutsu-Ogawara, Niigata East and West. 19 pp.

REFERENCES

Andrews, J.D.; Moss, T.R. 1993. Reliability and risk assessment. Harlow, U.K.: Longman Scientific and Technical. 368 pp.

Battjes, J.A.; Janssen, J.P.F.M. 1978. Energy loss and set-up due to breaking of random waves. *Proceedings International Conference Coastal Engineering (ICCE)*, Hamburg, Germany, no. 16, Volume 1, pp. 569-587.

Breitung, K. 1984. Asymptotic approximations for multinormal integrals. *Journal of Engineering Mechanics*, ASCE, Engineering Mechanics Division, vol. 110, pp. 357-366.

Bruining, J.W. 1994. Wave forces on vertical breakwaters. Reliability of design formula. *Delft Hydraulics Report*. Emmeloord, The Netherlands, 65 pp.

Bucher, G.C. 1987. Adaptive sampling, an iterative fast Monte Carlo procedure. *Internal Report, University of Innsbruck*, Innsbruck, Austria.

Christiani, E. 1997. Application of reliability in breakwater design. *Ph.D. thesis, Hydraulics & Coastal Engineering Laboratory, Department of Civil Engineering, Aalborg University*, Aalborg, Denmark, Series Paper 14, 179 pp.; 1 Appendix.

CUR 1997. Risk analysis in civil engineering. *Centre for Civil Engineering Research, Codes and Specification*, CUR Research Committee E10 "Risk Analysis", no. 190, Vol. 1: Probabilistic Design in Theory, Gouda, The Netherlands. In Dutch.

Ditlevsen, O. 1979. Narrow reliability bounds for structural systems. *Journal of Structural Mechanics*, vol. 7, pp. 453-472.

Ditlevsen, O.; Madsen, H.O. 1996. Structural reliability methods. Chichester et al.: John Wiley & Sons, 372 pp.

Enevoldsen, I.; Sørensen, J.D. 1993. Reliability-based optimisation of series systems and parallel systems. *Proceedings IFIP WG 7.5 Conference on Reliability and Optimisation of Structural Systems*, Thoft-Christensen, P.; Ishikawa, H., Takamatsu, Kagawa, Japan, no. 5, pp. 3[illegible]-46.

Eurocode 1 1994. Basis of design and actions on structures. Part 1: Basis of Design.

Goda, Y. 1969. Re-analysis of laboratory data on wave transmission over breakwaters. *Report. Port and Harbour Research Institute (PHRI)*, Tokyo, Japan, vol. 8, no. 3.

Goda, Y. 1985. Random seas and design of maritime structures. Tokyo: University of Tokyo Press, 323 pp.

Hasofer, A.M.; Lind, N.C. 1974. An exact and invariant first order reliability format. *Journal of the Engineering Mechanics Division*, ASCE, vol. 100, pp. 111-121.

Heijn, K.M. 1997. Wave transmission at vertical breakwaters. *Master Thesis, Delft University of Technology*, Delft, The Netherlands.

Hohenbichler; Rackwitz, R. 1983. First-order concepts in system reliability. *Structural Safety*, vol. 1, pp. 177-188.

Lancellotta, R. 1995. Geotechnical engineering. Rotterdam, The Netherlands: Balkema, rev. and updated edition, 436 pp.

Madsen, H.O.; Krenk, S.; Lind, N.C. 1986. Methods of structural safety. Englewood Cliffs, N.J., USA: Prentice Hall Inc., 403 pp.

Martinelli, L. 1998. Risk analysis of Genoa Voltri breakwater: the case of Genoa Voltri. *Ph.D. Thesis, Università di Bologna*, 15 December 1998, Bologna, Italy.

NKB 1978. Recommendations for loading and safety regulations for structural design. *NKB report*, no. 36.

Oumeraci, H. 1994. Review and analysis of vertical breakwater failures - lessons learned. *Coastal Engineering, Special Issue on 'Vertical Breakwaters'*, Eds.: Oumeraci, H. et al., Amsterdam, The Netherlands: Elsevier Science Publishers B.V., vol. 22, nos. 1/2, pp. 3-29.

Ouypornprasert, W. 1987. Adaptive numerical integration in reliability analysis. *Internal Report, University of Innsbruck*, Innsbruck, Austria.

Schueller, G.I.; Stix, R. 1987. A critical appraisal of methods to determine failure probabilities. *Structural Safety*, Amsterdam, The Netherlands: Elsevier Science Publishers B.V., vol. 4, pp. 293-309.

Sørensen, J.D.; Burcharth, H.F.; Christiani, E. 1994. Reliability analysis and optimal design of monolithic vertical wall breakwaters. *Proceedings IFIP WG 7.5 Conference on Reliability and Optimisation of Structural Systems*, Rackwitz, R.; Augusti, G.; Borri, A. (eds.), Assisi, Italy, no. 6, Part 2: Technical Contributions, Paper 27, pp. 257-264.

Takayama, T.; Ikeda, N. 1992. Estimation of sliding failure probality of present breakwaters for probabilistic design. *Report. Port and Harbour Research Institute (PHRI)*, Tokyo, Japan, vol. 31, no. 5, pp. 3-32.

Tanimoto, K.; Takahashi, S. 1994. Design and construction of caisson breakwaters - the Japanese experience. *Coastal Engineering, Special Issue on 'Vertical Breakwaters'*, Eds.: Oumeraci, H. et al., Amsterdam, The Netherlands: Elsevier Science Publishers B.V., vol. 22, nos. 1/2, pp. 57-77.

Thoft-Christensen, P.; Baker M.J. 1982. Structural reliability theory and its applications. Berlin/Heidelberg/New York: Springer, 267 pp.

Van Dantzig, D. 1956. Economic decision problem for flood prevention. *Econometrica*, vol. 24, pp. 276-287.

Van der Meer, J.W.; d'Angremond, K.; Juhl, J. 1994. Probabilistic calculations of wave forces on vertical structures. *Proceedings International Conference Coastal Engineering (ICCE)*, ASCE, Kobe, Japan, vol. 23.

Van Gelder, P.H.A.J.M. 1998. Analysis of wave impact data on vertical breakwaters. *Proceedings Task 4 Workshop, MAST III, PROVERBS-Project: Probabilistic Design Tools for Vertical Breakwaters*, Delft, The Netherlands, Version 27/28 August 1998, Annex IV, 4 pp.; 8 figs.

Voortman, H.G.; Kuijper, H.K.T.; Vrijling, J.K. 1998. Economic optimal design of vertical breakwaters. *Proceedings International Conference Coastal Engineering (ICCE)*, ASCE, Copenhagen, Denmark, no. 26, 14 pp.

Vrijling, J.K.; Bruinsma, J. 1980. Hydraulic boundary conditions. In: *Hydraulic aspects of coastal structures: Developments in hydraulic engineering related to the design of the Oosterschelde storm surge barrier in the Netherlands*, Delft, The Neth, pp. 109-133.

Vrijling, J.K.; Van Hengel, W.; Houben, R.J. 1995. A framework for risk evaluation. *Journal of Hazardous Materials*, Amsterdam, The Netherlands: Elsevier Science B.V., vol. 43, pp. 245-261.

Vrijling, J.K. 1996. Evaluation of uncertainties and statistical descriptions. *Proceedings Task 4 Workshop Hannover, MAST III, PROVERBS-Project: Probabilistic Design Tools for Vertical Breakwaters*, Annex 3, Hannover, Germany, 15 pp.; 1 Annex.

Vrijling, J.K.; Van Gelder, P.H.A.J.M. 1998. The effect of inherent uncertainty in time and space on the reliability of flood protection. Lydersen, S.; Hansen, G.K.; Sandtorv, H. (eds.): Safety and Reliability, Proceedings of the ESREL'98 Conference, Volume 1, Trondheim, Norway, A.A.Balkema, Rotterdam, pp. 451-458.

CHAPTER 6

Conclusions of the overall project

Hocine Oumeraci

The integrated character of the research strategy which has been particularly pursued in PROVERBS, as well as the multi-disciplinary research efforts to link together the hydrodynamic, geotechnical, structural and probabilistic design aspects represent a substantial departure from the existing, essentially empirical approach, and thus constitutes a very important step towards a largely rational discipline. In this respect, the results will not only enable to lay down the scientific basis for authoritative design guidelines for coastal structures, but also constitutes a scientific platform for continuous improvements, even after completion of the MAST Programme.

The process oriented research adopted in PROVERBS also provides a physically sound departure and scientific basis for the development of innovative alternative constructions which can fulfil technical, economical and ecological criteria.

During the course of the project a number of candidate issues for future research have been identified with regards to the hydraulic, geotechnical, structural, probabilistic and safety aspects.

6.1 HYDRAULIC ASPECTS

Beside the development of numerical models and the generation of a large data base for hydraulic data, substantial improvements and major advances have particularly been achieved with respect to breaking criteria and wave load classification (parameter map), wave impact loading and model scaling, seaward impact and pulsating wave loads, hydraulic performance and loading of perforated structures, wave loads of crown walls and innovative high mound composite breakwaters, the damping effect of armour layers on impact and pulsating wave loads and on 3D-effects on wave loading of caisson breakwaters. Furthermore, the uncertainties associated with the prediction models and the impact parameters have also been quantified. Despite the amount of new knowledge generated, much research work remains to be done with respect to the following items.

(i) The parameter map for wave loading should be supplemented by models providing the probability of occurrence of each of the four loading cases as a function of the distributions of the input hydraulic parameters and the most relevant geometric parameters. For this purpose, more refined probabilistic models for wave transformation up to the structure have to be developed.

(ii) For cyclic wave loading which is strongly needed for the treatment of fatigue, progressive failure and durability, a concept of "design storm" similar to that used in offshore engineering, but more refined, should be worked out.

(iii) For impact loading, further research should be directed towards improving the developed force-time formula and the occurrence frequency of impact with due consideration of the statistical distribution of the input hydraulic parameters as mentioned under item (i) above. Furthermore, a representative parametric time history of impact pressure distribution should be developed on the basis of the statistical force distribution and of the force history obtained from the force-time formula. It is also particularly important to obtain the local information on impact pressure required for structural dynamic analysis.

(iv) For the scaling of impact loading, the empirical relationships obtained in PROVERBS for the evaluation of force correction factors when using FROUDE scaling should be refined on the basis of theoretical analysis, so that more simple and more generic formulae can be obtained for the correction factors as a function of the compressibility of the air-water-mixture. Practical guidance to assess the air-content as a function of the local conditions and the resulting compressibility is also needed.

(v) The results obtained from 2D-wave flume experiments and related numerical modelling of the impact loading towards the seaside induced by wave overtopping must be extended to account for 3D-effects. Moreover, the form of the overtopping plume including air entrapment for given waves and structure, the frequency of overtopping events and the associated proportion likely to generate significant seaward impulse must also be addressed. The research should also include the distribution of pressure impulse both over the rear face and beneath the bottom slab (uplift) of the caisson breakwater.

(vi) A further refinement and extension of the achieved results on 3D-effects on wave impact loading is required by addressing more systematically the effect on the onset of breaking, the occurrence frequency, the severity and the spatial distribution of impact for a parameterised geometry of the structure and the foreshore, wave obliquity and short-crestedness. Support by theoretical analysis is also required to develop more generic and semi-empirical formulae to predict these effects.

(vii) For perforated caisson breakwaters, it would be useful to bring the various approaches applied in PROVERBS for the development of load prediction formulae together into an unique procedure which will lead to an "universal" set of simple load formulae applicable to one chamber caissons as well as to multiple chamber systems. The same should apply for the prediction of the hydraulic performance (reflection, overtopping, transmission). Indirect methods based on correction factors, although very practical, have their inherent limitations since a plain wall and a perforated caisson work differently.

(viii) For crown walls, a more generic set of simple load formulae (for horizontal and vertical forces) based on large-scale model tests as well as on theoretical analysis supported by numerical modelling and validated by prototype data should be developed on the basis of the empirical formulae derived in PROVERBS.

(ix) Since the "High Mound Composite Breakwater (HMCB)" concept is basically different from the much smaller crown wall and the much larger composite caisson breakwater, further research on this new type of structure is particularly needed on the refinement of the breaking criterion and the classification of wave loads which should also apply for various structure configurations under random wave attack. The spatial and temporal pressure distribution on the superstructure (uplift and horizontal pressure) should also be made predictable.

(x) Wave loading has yet been addressed without due consideration of the rocking motion of the structure. Future research should include this aspect, particularly for severe impact conditions where significant motions of the structure are expected (load-structure-interaction model).

(xi) Wave overtopping has yet been investigated in terms of average and individual overtopping rates. Because of the problem associated with the use of these quantities for the assessment of the structural and stability implications of overtopping, future research should rather be directed towards the full description of the detailed flow field of the overtopping water mass.

6.2 GEOTECHNICAL ASPECTS

Beside the generation of a unique soil data base and the development of numerical models, new knowledge and considerable improvements have been achieved with respect to the mechanisms responsible for geotechnical failures, their prediction and the associated uncertainties. This is particularly true for the dynamic response to impact and cyclic loading, the pore pressure development in the rubble foundation and subsoil, the effect of the pore water flow generated in the foundation, the characteristics of the rupture surfaces, the relative effects of the various parameters on the stability of the foundation and alternative solutions to improve this stability. Despite these great advances and these systematic studies which are unique in the coastal engineering field there are still a number of items waiting for further research:

(i) The very good experience made in the development of simple engineering "dynamic" models for sliding and rotational motions, including incremental permanent displacements, should be extended to the bearing capacity failure mode responsible for seaward tilting. More focus herein should be put on stepwise failure and the cumulative mechanisms leading to ultimate collapse of the structure. Because of their relative importance for the outputs of the "dynamic" models the formulae to predict the soil stiffness and the added mass parameters may also be improved.

(ii) Especially for extreme wave impact loading situations where the structure is subject to severe rocking motions the wave-structure-foundation interaction must be fully accounted for in future modelling, because the loading is substantially affected by the motions of the structure (Fig. 6-1). In fact, ignoring this interaction would certainly lead to conservative results with respect to the wave loads (e.g. up to 30% for uplift force) that are transmitted to the foundation. Furthermore, the effects of the induced flow at the structure foundation interface on the progressive degradation of the foundation must also be investigated.

(iii) Models for instantaneous pore pressures directly induced by wave fluctuations in sandy and silty subsoil which are based on the assumption of incompressible pore water generally yields too optimistic results. Therefore, more information on gas content in prototype soils will be needed and incorporated into future models. Moreover, a reliable procedure to introduce the pore pressure along the potential failure surface is still required.

(iv) Although rupture surfaces are generally three-dimensional in nature the models commonly adopted are based on the conservative assumption of 2D-rupture surfaces. Detailed investigations are therefore required to refine and validate the approximate formula developed in PROVERBS to predict the 3D-effects. This is important because these effects might yield an increase in soil strength up to 40%.

(v) For the case of a breakwater on an undrained subsoil subject to a wave train of extreme high waves the soil strength should necessarily be described by a combination of the average component τ_a and the cyclic component τ_{cycl} of the shear resistance. However, a procedure on how this combination should be applied in stability analysis is still lacking (Volume IIb, chapter 5).

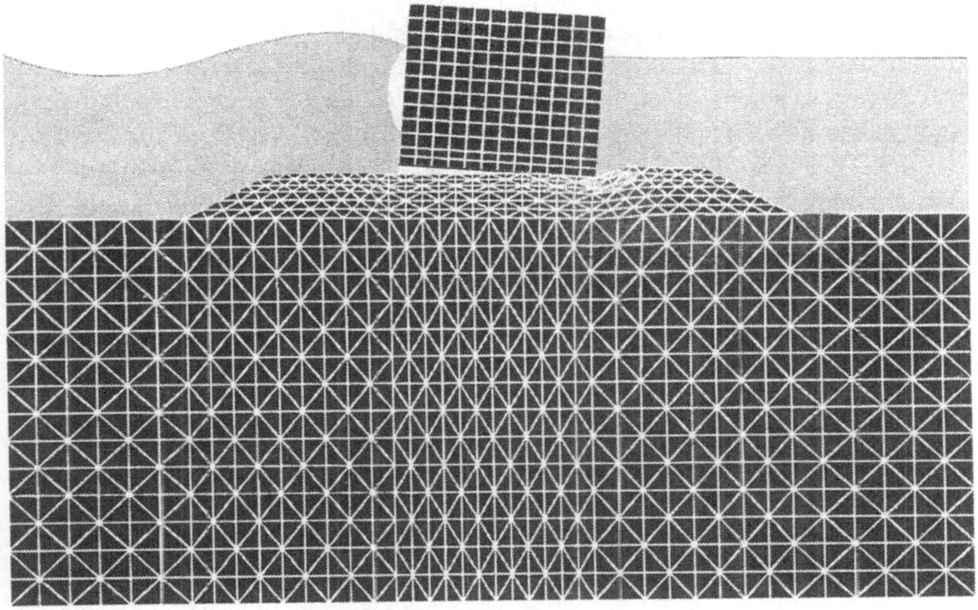

Figure 6-1. Wave impact loading of structure and foundation.

(vi) The forces and mechanisms responsible for seaward tilting of vertical breakwaters have been identified. Further detailed investigations to better understand and predict these forces and mechanisms are necessary.

(vii) One of the most important alternative solutions suggested to improve caisson breakwater stability consists in providing a skirt foundation. Detailed investigations on the failure modes associated with this type of foundation would be required before adopting this technique which has commonly been used for offshore platforms.

6.3 STRUCTURAL ASPECTS

The studies on structural aspects of caisson breakwaters represent a first and unique contribution of its kind in the sense that no systematic investigations were performed before on this topic. Considerable advances have particularly been achieved on finite element structural modelling related to in-service failure modes, on the treatment of pre-service failure modes, degradation as initiated by chloride ingress and cracking. Moreover, criteria and requirements for a future unified code for structural design have been developed on the basis of a comprehensive analysis of the existing codes. Nevertheless, further research is still needed on the following items:

(i) The suggested framework and approach for a unified code of practice should be verified by practitioners and committees which are in charge of code writing.

(ii) A further extension of the structural analysis for pre-service conditions is needed to include further failure modes which have not been addressed in PROVERBS. Moreover, the use of models for fluid-structure-interaction to describe the structural behaviour of floating and towed caissons would also result in considerable improvements.

(iii) Although PROVERBS has systematically improved the physical understanding of the most relevant processes which govern the structural response of concrete caisson breakwaters, more verification with prototype data are still needed to better ensure that all relevant failure mechanisms have been addressed.

(iv) Most of the approaches used to predict wave impact loading, like in PROVERBS, are based on the assumption that the impact load history is independent of the motions of the hit structure members. The results of large-scale model tests have shown that more research remains to be performed on the interaction of waves and structure members. For this purpose much effort is needed to couple the computational fluid dynamic models and the finite element structural models.

(v) Since adaptive meshing techniques linked to structural optimisation algorithms are emerging in finite element technology, careful use of these new advanced FE models by experienced structural engineers would be very useful to improve the quantification of the uncertainties associated with structural models.

(vi) Because of the limited area subject to wave loads in real sea state along the breakwater longer caissons (over 100 m) might represent a proper alternative to substantially reduce the required caisson width. However, much work is needed to ensure that global bending and torsional stiffness are sufficient to cope with the expected loads during towing and for in-service conditions (differential settlements). A further alternative to increase load sharing is the use of in-situ gravel filled pockets between two adjacent caissons. This needs to be investigated because it also permits to avoid excessive requirements related to caisson alignments.

(vii) Since most of the structural failures during in-service conditions are progressive, the development of well-deviced programmes for permanent monitoring instrumentation based on the results of PROVERBS would be very useful in the sense that it will help not only to define a repair and maintenance strategy but also to develop optimal structures.

(viii) Chloride ingress represents the most relevant process for durability of the reinforced concrete caissons in sea water. To use the prediction models suggested in PROVERBS the knowledge of the diffusion coefficient is required. However, no methods are yet available to estimate the diffusion coefficient based on material specifications. Moreover, a new generation of models need to be developed for the prediction of the stepwise degradation initiated by chloride ingress and accelerated by cracking until the ultimate collapse of the reinforcement and concrete members.

6.4 PROBABILISTIC ASPECTS

A complete framework for the probabilistic analysis and design (level II and III) of vertical breakwaters and similar monolithic structures subject to wave action has been developed and implemented on a set of five representative structures. This also includes the related probabilistic tools, prediction models and associated methods to quantify the uncertainties involved in the models and input parameters. An optimisation procedure has also been developed to determine the target (or acceptable) probability of failure when loss of life is excluded and guidance is provided when human injury and loss of life are to be taken into account. Moreover, a partial safety factor system (PSFS) has been developed and calibrated on the basis of the full probabilistic approach. Nevertheless, there are still a number of topics candidate for further research:

(i) Extension of the probabilistic approach to include local morphologic changes (Fig. 6-2) when design formulae for scour will be available (prospective results from MAS3-CT97-0097).

(ii) Although Human and Organisation Errors (HOE) generally represent a major part of the uncertainties involved in the design process they have been omitted in PROVERBS, because they are too problem specific and because they can be accounted for afterwards in the probabilistic analysis. It would be useful to implement the suggested technique for this purpose and demonstrate how it works on a specific case study. The main HOE sources to be considered are given in Figure 6-3.

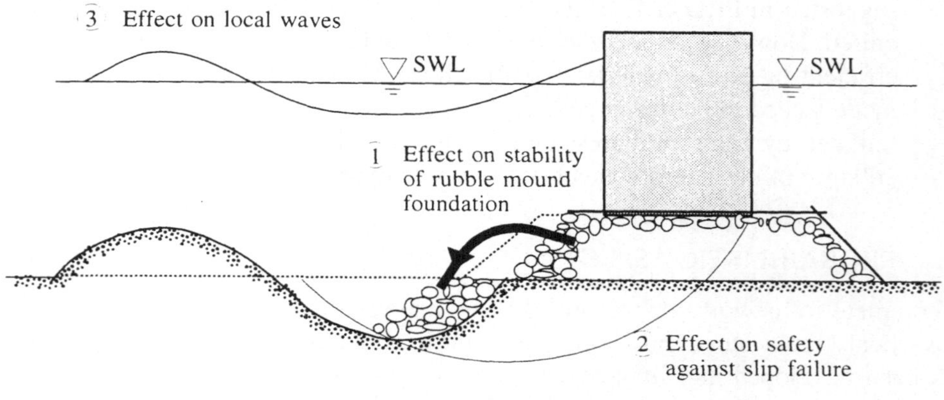

Figure 6-2. Effect of local morphological changes on breakwater stability.

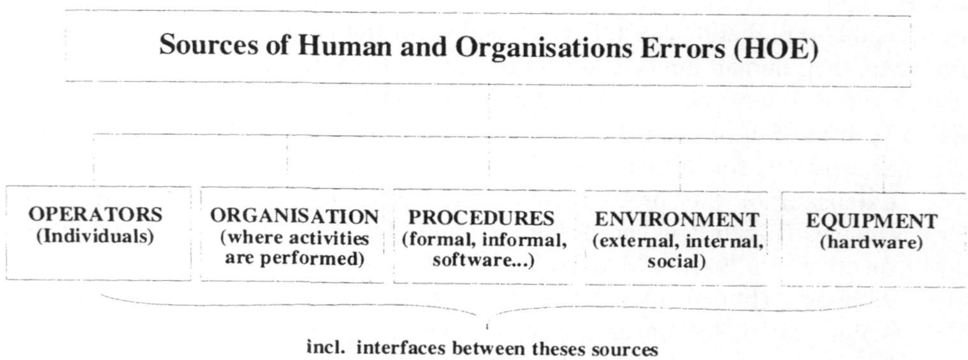

Figure 6-3. Sources of human and organisation errors.

(iii) The prediction and probabilistic analysis of degradation and stepwise failures is of utmost importance to set up optimal programmes and strategies for monitoring, repair and maintenance as part of the design process. Further probabilistic tools based on Bayesian decision methods need to be developed in order to specify the required monitoring instrumentation and procedure and to obtain the probability of the initiation of repair/maintenance as a function of service time.

(iv) The reliability tools can be extended and further refined to address in more detail not only the design phase, but also other life cycles of the breakwater, including the construction phase, the operation phase and the maintenance phase and where necessary also the emergency phase. These extensions and improvements are particularly required as a departure basis for a further more important ultimate step: the probabilistic risk analysis and risk management framework (Fig. 6-4).

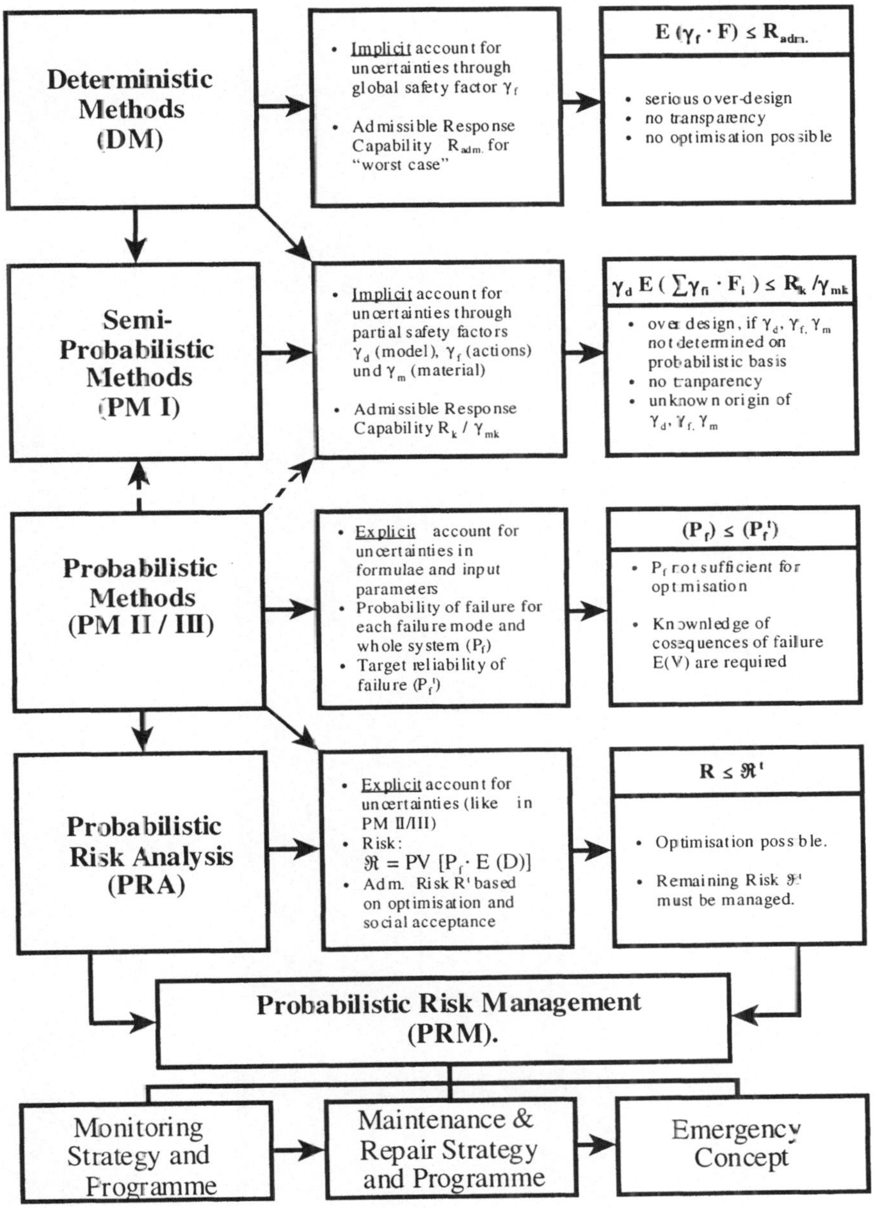

Figure 6-4. Probabilistic risk analysis and risk management framework (PRM).

ANNEX 1

Notations

GENERAL SYNTAX

In this annex a list of notations and symbols is given as they are used within the project and this book. Symbols and notations are in an alphabetical order as follows:

- Latin characters
- Greek characters

Since the analysis of all model tests and prototype measurements are performed on computers a second column was added showing the corresponding computer notation of the parameter. Small and capital letters are used despite the fact that many computer systems ignore these cases. To meet this requirement repetition of parameters (small and capital characters) is avoided. Furthermore the syntax of the computer notation is kept close to the non ASCII list in order to ensure readability and to avoid too many cross references to the list. The main purpose of this second column is to easily exchange data even if simple ASCII tables are used. It furthermore facilitates the use of simple headers in different charts or data bases used for analysis.

There is a wide range of parameters and this list will most probably not cover the whole range. However, some of the parameters can easily be extended by adding characters and without changing their general meaning. For example, the horizontal force at the breakwater F_h can be varied by adding "G" or "T" for representing calculated forces by Goda or Takahashi, respectively. Accelerations and displacements are very often measured at specific loactions at the breakwater, so these locations should be added to the respective symbol eg "$d_{h,\underline{b}}$" for horizontal displacement at the back of the structure.

As far as parameters describe stochastic parameters they should be underlined. Estimators for the variable x should be denoted as $\hat{x}$.

All dimensions of the parameters (third column) are given in SI units.

LIST OF NOTATIONS AND SYMBOLS

Param.	**Comp.**	**Dim.**	**Description**
A	Acc		Accidental action
A	A	m^2	Area
A_0	A0 .	m^2	Initial specimen cross section
A_c	Ac	m	Armour crest freeboard
A_c	Ac	m^2	Consolidated specimen cross section
A_e	Ae	m^2	Erosion area on cross-section
A_{tr}	Atr		Area under curve over time t_r
A_{td}	Atd		Area under curve over time t_d
A_{tt}	Att		Area under curve over time t_t
a	a	m	Distance, geometrical data, empirical coefficient
a_h	ah	m/s^2	Horizontal acceleration of structure
a_r	ar	m/s^2	Rotational acceleration of structure
a_v	av	m/s^2	Vertical acceleration of structure
$B_{1,2}$	B1,2	m	Structure width parameters
B_b	Bb	m	Width of rubble berm, at toe of wall (Fig. 2)
B_c	Bc	m	Width of caisson / structure in x-direction (Fig. 2)
B_{cw}	Bcw	m	Width of crown wall
$B_{c,eff}$	Bceff	m	Width of caisson effective for foundation
B_{eq}	Beq	m	Width of rubble berm, averaged over height of berm (h_b)
B_n	Bn	-	Bulk number of structure cross-section, defined A_t/D_{n50}
B_{wl}	Bwl	m	Structure width at static water level
b	b	m	Width, empirical coefficient
C, C_i,	C		Empirical Coefficients
C_c	Cc	-	Coefficient of Curvature $(D_{30})^2/(D_{60}xD_{10})$
COG	COG	-	Centre of gravity of caisson
C_u	Cu	-	Coefficient of Uniformity (D_{60}/D_{10})
C_r	Cr	-	Coefficient of reflection
$C_r(f)$	Cr_f	-	Reflection coefficient function

c'	cs	kN/m^2	Cohesion
c_0	c0	m/s	Wave celerity in deep water
c_g	cg	m/s	Group velocity
c_u	cu	kN/m^2	Undrained shear strength
c_v	cv	m^2/s	Consolidation coefficient
c_{vp}	cvp	m^2/s	Consolidation coefficient due to compressibility of pore water
c_{vs}	cvs	m^2/s	Consolidation coefficient due to compressibility of skeleton
D	Dd	m	Particle size or typical diameter
D_{10}	D10	mm	10% value of sieve curve (mean particle size at which 10% of the soil is finer)
D_{100}	D100	mm	Maximum particle size
D_{15}	D15	mm	15% value of sieve curve
D_{50}	D50	mm	Diameter of stone which exceeds the 50% value of sieve curve (mean particle size)
D_{85}	D85	mm	85% value of sieve curve
D_{85}/D_{15}	D85_15	-	Armour grading parameter
D_{air}	Dair	-	Damping coefficient of entrapped air oscillations
D_e	De	m	Effective particle diameter
D_n	Dn	m	Nominal particle diameter, defined $(M/\rho_r)^{1/3}$) for rock and $(M/\rho_c)^{1/3}$ for concrete armour
D_{n50}	Dn50	m	Nominal particle diameter calculated from the median particle mass M_{50}
D_r	Dr	-	Relative Density
DSS	DSS	-	Direct simple shear test
d	d	m	Water depth over berm in front of wall; diameter, also used as empirical coefficient
d_c	dc	m	Depth of structure in foundation (Fig. 2)
d_h	dh	mm	Horizontal displacement of structure
d_r	dr	mm	Rotational displacement of structure
dt	dt	s	Time increment (s. also Δt)
dt_{ah}	dtah	s	Time of occurrence of a_{hmax} relative to t_0

dt_{ar}	dtar	s	Time of occurrence of a_{rmax} relative to t_0
dt_{Fh}	dtFh	s	Time of $F_{h,max}$ relative to t_0 (Fig. 4)
dt_{Fhmin}	dtFhmin	s	Time of $F_{h,min}$ relative to t_0
dt_{Feff}	dtFeff	s	Time of $F_{eff,max}$ relative to t_0
dt_{Ftot}	dtFtot	s	Time of F_{tot} relative to t_0
dt_{Fu}	dtFu	s	Time of $F_{u,max}$ relative to t_0 (Fig. 4)
dt_{Fumin}	dtFumin	s	Time of $F_{u,min}$ relative to t_0
$dt_{Fu,q}$	dtFuq	s	Time of $F_{u,q}$ relative to t_0 (Fig. 4)
dtp_{MWLmax}	dtpMWLmax	s	Time of p_{MWLmax} relative to t_0 (Fig. 4)
dt_{pi}	dtpi	s	Time of $p_{i,max}$ relative to t_0
dt_{pu}	dtpu	s	Time of $p_{u,max}$ relative to t_0
dt_{Mh}	dtMh	s	Time of M_{hmax} relative to t_0
dt_{Mu}	dtMu	s	Time of M_{umax} relative to t_0
dt_{Mt}	dtMt	s	Time of M_{tmax} relative to t_0
dx_{pu}	dxpu	m	Location of pressure transducer recording maximum uplift pressure from front face of caisson
d_v	dv	mm	Vertical displacement of structure
dz_{pA}/dx_{PA}	dzpA/dxpA	m	Location of pA, if not at point A
dz_{pi}	dzpi	m	Location of pressure transducer recording maximum horizontal pressure, relative to base of caisson
dz_{pMWL}	dzpMWL	m	Location of pressure transducer above MWL, relative to MWL
E	E	MN/m^2	Elasticity modulus
E	Eff		Effect of action
	Err		Random error
E_d	Ed		Absorbed or dissipated wave energy
E_i	Ei		Incident wave energy
E_r	Er		Reflected wave energy
E_t	Et		Transmitted wave energy
e	e	-	Void ratio
e	e	m	Eccentricity of foundation load
e_{max}	emax	-	Maximum void ratio
e_{min}	emin	-	Minimum void ratio
F	Fac		Action
F	Fl	m	Fetch length, used in wave generation calculations

F_{eff}	Feff	kN/m	Effective force acting on caisson or crown wall element
F_G	FG	kN/m	Gravity force (weight) of caisson or crown wall element
F_h	Fh	kN/m	Horizontal force on caisson or crown wall element
$F_{h,99.9\%}$	Fh999	kN/m	Horizontal force at 99.9% non-exceedance level
$F_{h,99.8\%}$	Fh998	kN/m	Horizontal force at 99.8% non-exceedance level
$F_{h,99.0\%}$	Fh99	kN/m	Horizontal force at 99.0% non-exceedance level
$F_{h,98.0\%}$	Fh98	kN/m	Horizontal force at 98.0% non-exceedance level
$F_{h,95.0\%}$	Fh95	kN/m	Horizontal force at 95.0% non-exceedance level
$F_{h,1/250}$	Fh1250	kN/m	Mean of highest 1/250 horizontal wave forces
$F_{h,base}$	Fhbase	kN/m	Horizontal load of foundation per unit length at caisson-foundation interface
$F_{h,max}$	Fhmax	kN/m	Maximum horizontal wave force (Fig. 4)
$F_{h,min}$	Fhmin	kN/m	Minimum horizontal wave force (Fig. 4)
$F_{h,q}$	Fhq	kN/m	Quasi-static horizontal force (Fig. 4)
$F_{h,sub}$	Fhsub	kN/m	Horizontal load of subsoil per unit length at bedding layer-subsoil interface
F_{Hs}	FHs	-	Uncertainty of H_s
F_S	Fs	-	Factor of safety
F_{tot}	Ftot	kN/m	Maximum total force at back of structure
F_u	Fu	kN/m	Uplift force on caisson or crown wall element
$F_{u,99.9\%}$	Fu999	kN/m	Uplift force at 99.9% non-exceedance level
$F_{u,99.8\%}$	Fu998	kN/m	Uplift force at 99.8% non-exceedance level
$F_{u,99.0\%}$	Fu99	kN/m	Uplift force at 99.0% non-exceedance level

$F_{u,98.0\%}$	Fu98	kN/m	Uplift force at 98.0% non-exceedance level
$F_{u,95.0\%}$	Fu95	kN/m	Uplift force at 95.0% non-exceedance level
$F_{u,1/250}$	Fu1250	kN/m	Mean of highest 1/250 uplift wave forces
$F_{u,max}$	Fumax	kN/m	Maximum uplift wave force (Fig. 4)
$F_{u,min}$	Fumin	kN/m	Minimum uplift wave force (Fig. 4)
$F_{u,q}$	Fuq	kN/m	Quasi-static uplift force (Fig. 4)
$F_{u,Fhq}$	FuFhq	kN/m	Uplift force at the same time of occurrence of $F_{h,q}$ (Fig. 4)
$F_{v,base}$	Fvbase	kN/m	Vertical load of foundation per unit length at caisson-foundation interface
$F_{v,sub}$	Fvsub	kN/m	Vertical load of subsoil per unit length at bedding layer-subsoil interface
F*	FRca	-	Dimensionless freeboard, defined $(R_c/H_s)s_m/2\pi)^{1/2}$
F'	FRcb	-	Dimensionless freeboard, defined $R_c/(H_s^2L_{ps})^{1/3}$
f	f	Hz	Wave frequency, general
f	fc, fs	kN/m^2	Strength of a material
f_{air}	fair	Hz	Frequency of entrapped air oscillations
f_m or f_p	fm or fp	Hz	Frequency of peak of wave energy spectrum
f_{sam}	fsam	Hz	Sampling frequency
f(H)	f_H		Density function of H
G	G	MN/m^2	Shear modulus
G	Gac		Permanent action
G(x)	G_x		Design function, design condition satisfied when G≥0
GF	GF	-	Groupiness factor
G_{max}	Gmax	MN/m^2	Initial shear modulus
G_S	GS	-	Specific gravity
g	ga	m/s^2	Gravitational acceleration
g	gf		Failure function
H	H	m	Wave height, from trough to crest

H_0	H0	m	Deep water wave height
$H_{1/10}$	H110	m	Average of the highest 1/10 wave heights in a record
$H_{1/3}$	H13	m	Mean height of highest 1/3 of waves in a record
$H_{2\%}$	H2	m	Wave height exceeded by 2% of waves in a record
H_{bc}	Hbc	m	Local critical breaking wave height which describe the transition from pulsating to impact conditions at the structure
H_{bs}	Hbs	m	Local maximum wave height on depth h_s
H_c	Hc	m	Consolidated specimen height
H_i	Hi	m	Initial specimen height
H_{max}	Hmax	m	Maximum wave height in a record
H_{m0}	Hm0	m	Significant wave height from spectral analysis, defined $4.0m_0^{0.5}$
H_{nom}	Hnom	m	Nominal wave height, to be created at wave paddle
H_o	Ho	m	Offshore wave height, un-affected by shallow water processes
H_{rms}	Hrms	m	Root mean square wave height
H_s	Hs	m	Significant wave height, average of highest one-third of wave heights
H_s'	Hss	m	Threshold value of H_s
$H_{s,100}$	Hs100	m	Significant wave height reached once in 100 years
$\hat{H}_s^{3T}$	Hs3T	m	Central estimate of H_s exceeded once in 3T years
H_{si}	Hsi	m	Significant inshore wave height, average of highest one-third of wave heights
H_{so}	Hso	m	Significant offshore wave height, average of highest one-third of wave heights, un-affected by shallow water
$\hat{H}_s^{T}$	HsT	m	Central estimate of H_s exceeded once in T years
$\hat{H}_s^{3T}$	Hs3T	m	Central estimate of H_s exceeded once in 3T years

$\hat{H}_s^{TPf}$	HsTpf	m	Central estimate of H_s corresponding to an equivalent return period T_{Pf}
$h_{...}$	h...	m	Water depth in front of toe of structure; height
h_0	h0	m	Water depth in deep water
h_1, h_2	h1, h2	m	Water depth over toe armour, over other points on structure cross-section
h_b	hb	m	Height of berm above sea bed (Fig. 2)
h_c	hc	m	Height of caisson / structure (Fig. 2)
h_f	hf	m	Exposed height on caisson / crown wall over which wave pressures act (Fig. 2)
h_r	hr	m	Depth of rubble core beneath caisson to sea bed (Fig. 2)
h_s	hs	m	Water depth at toe of structure / mound (Fig. 2)
h_u	hu	m	Setup heights within a permeable mound
h_{umax}	humax	m	maximum setup heights within a permeable mound
h'	hd	m	Water depth from base of caisson
I	I_{xx}, I_{yy} etc	m^4	2nd moment of area
I_D	ID	-	Density index
I_h	Ih	kNs/m	Impulse of horizontal force at front face of breakwater
I_{hr}	Ihr	kNs/m	Impulse of horizontal force at front face of breakwater (integrated over t_r)
I_{hd}	Ihd	kNs/m	Impulse of horizontal force at front face of breakwater (integrated over t_d)
I_{hq}	Ihq	kNs/m	Impulse of horizontal force at front face of breakwater (integrated over t_q)
I_{ht}	Iht	kNs/m	Impulse of horizontal force at front face of breakwater (integrated over t_t)

I_r	Ir	-	Iribarren or surf similarity number = $\tan \alpha / s^{1/2}$
I_{ur}	Iur	kNs/m	Impulse of uplift force on base of breakwater (integrated over t_{rFu})
I_{ud}	Iud	kNs/m	Impulse of uplift force on base of breakwater (integrated over t_{dFu})
I_{uq}	Iuq	kNs/m	Impulse of uplift force on base of breakwater (integrated over t_{qFu})
I_{ut}	Iut	kNs/m	Impulse of uplift force on base of breakwater (integrated over t_{tFu})
i	i	-	Hydraulic gradient; radius of gyration
K	K	MN/m^2	Compressibility modulus of soil skeleton
K_{11}	K11	MN/m	Spring coefficient of foundation for horizontal deformation
K_{22}	K22	MNm	Spring coefficient of foundation for rotation deformation
K'_0	Ks0	-	Coefficient of lateral earth pressure at rest
K_c	Kc	-	Stress ratio during consolidation ($\sigma'_{1c}/\sigma'_{3c}$)
K_D	KD		Empirical damage coefficient used in Hudson equation
K_{DZ}	KDZ		Coefficient used in Hudson's equation for 'zero damage'
K_R	KR	-	Resistance or strength coefficient
K_{RR}	KRR		Damage coefficient used in Hudson's equation for rip-rap armour
K_s	Ks	-	Load coefficient
K_w	Kw	MN/m^2	Compressibility modulus of pore water
k	k	-	Wave number = $2\pi/L$
k	k	m/s	Darcy permeability
k_b	kb	-	Empirical factor for the influence of relative berm length, B_{eq}/d, on breaking wave heights
k_α, k_β			Failure mode coefficients

$L_{...}$	L...	m	Wave length, in the direction of wave propagation
L_m	Lm	m	Deep water wave length related to mean (T_m) period
L_o	Lo	m	Deep water wave length - $gT^2/2\pi$
L_p	Lp	m	Deep water wave length related to peak (T_p) period
L_{pi}	Lpi	m	Local inshore wave length related to peak period at structure, given approximately by $(gT_p^2/2\pi)$ $[\tanh(4\pi^2 h_s/gT_p^2]^{1/2}$
L^*	Ls	m	Penetration length used in set-up calculations
l	l	m	Length, span
l*	ls	m	Penetration length used in set-up calculations
l_c	lc	m	Length of the caisson in y-direction
l_{Fh}	lFh	m	Lever arm of F_h, related to bottom of structure (Fig. 3)
l_{Fu}	lFu	m	Lever arm of F_u, related to shoreward side of structure (Fig. 3)
$M_{...}$	M...	kNm/m	Overturning or bending moment
M_{11}	M11	kg	Element of mass matrix in mass-spring-dashpot model (mass of caisson + added mass)
M_{12}	M12	kgm	Element of mass matrix in mass-spring-dashpot model
M_{22}	M22	kgm^2	Element of mass matrix in mass-spring-dashpot model
M_{50}	M50	t	Median mass or armour unit derived from the mass distribution curve
M_{base}	Mbase	kNm/m	Moment load on foundation per unit length at caisson-foundation interface
M_{hmax}	Mh	kNm/m	Maximum horizontal moment
M_{hFhq}	MhFhq	kNm/m	Horizontal moment at time of maximum horizontal quasi-static force

M_{sub}	Msub	kNm/m	Moment load on subsoil per unit length at bedding layer-subsoil interface
M_{cmax}	Mc	kNm/m	Maximum overturning moment due to vertical force on crest
M_{tmax}	Mt	kNm/m	Maximum total moment due to all wave loads
M_{umax}	Mu	kNm/m	Maximum uplift moment
M_{uFhqx}	MuFhq	kNm/m	Uplift moment at time of maximum horizontal quasi-static force
M_{tFhq}	MtFhq	kNm/m	Total moment at time of maximum horizontal quasi-static force
m	m	-	Cotangent of bed slope; if not uniform, then average over length $5L_{op}$ from structure;
m	m	t	mass
m_0	m0		Zeroth moment of the wave energy density spectrum
m_2	m2		Second moment of the wave energy density spectrum
m_a	ma	t	Mass of armour unit
m_{cai}	mcai	kg/m	mass of caisson (per unit length)
$N_{...}$	N...	-	Number of values
N	Nax	kN	Axial force
N_a	Na	-	Total number of armour units in area considered
N_d	Nd	-	Number of armour units displaced, usually by more than D
N_{eqv}	Neqv	-	Equivalent number of cycles
N_f	Nf	-	Number of cycles to failure
N_H	NH	-	Number of events; used as number of H_s values in analysis of extreme values
N_{od}	Nod	-	Number of armour units displaced per D_n width
$N_{\%d}$	Npd	-	Number of armour units displaced, expressed as a % of total number of units in area studied - (N_d/N_a) x 100%
N_r	Nr	-	Number of armour units rocking

N_s	Ns	-	Stability number, $H_s/\Delta D_n = (K_D \cot\alpha)^{1/3}$
N_s*	Nss	-	Spectral stability number - $H_s/\Delta D_n(1/s_p)^{1/3}$
N_T	NT	-	Number of storms in the observation period
N_{wo}	Nwo	-	Number of waves overtopping expressed as proportion or % of total incident
N_z	Nz	-	Number of zero-crossing waves in a record = T_R/T_m
n	n	-	Porosity
n_{max}	nmax	-	Maximum porosity
n_{min}	nmin	-	Minimum porosity
n_v	nv	-	Volumetric porosity, volume of voids expressed as proportion of total volume
OCR	OCR	-	Overconsolidation ratio
$P_{...}$	P...	-	Encounter probability, also notional permeability factor used in calculation of armour stability
P	PS	kN	Pre-stressing force
P_{as}	Pas	-	Notional permeability factor used in calculation of armour stability
$P_{b\%}$	Pbp	%	Percentage of waves which are breaking or broken at the structure
P_f	Pf	-	Target probability of failure
P(x)	Pf_x	-	Probability function
$P_{i\%}$	Pip	%	Percentage of waves which may break directly onto the structure
POT	POT		Peak-over-threshold analysis
p	p	kPa	Pressure (eg pore pressure or wave pressure)
p'	ps	kN/m²	Average effective stress, $(\sigma'_1 + \sigma'_3)/2$
p'_o	po	kN/m²	Effective overburden stress
p1	p1	kPa	Pressure at the front face of the structure at MWL (Fig. 3)
p2	p2	kPa	Extrapolated pressure below the core beneath the structure (Fig. 3)

p3	p3	kPa	Pressure at the bottom of the structure (Fig. 3)
p4	p4	kPa	Pressure at the crest of the structure (Fig. 3)
p_a	pa	kPa	Atmospheric pressure
p_{Amax}	pAmax	kPa	Maximum pressure at or near point A, at the lowest point on the exposed face of the caisson
p_c	pc	kN/m^2	Pre-consolidation stress
$p_{i,max}$	pimax	kPa	Maximum impact pressure at the front face of the breakwater
p_{MWLmax}	pMWLmax	kPa	Impact pressure at the front face of the breakwater close to mean water level
$p_{u,max}$	pumax	kPa	Maximum uplift pressure at the bottom of the breakwater
p(x)	pdf_x	-	Probability density function
Q	Q		Mean overtopping discharge, per unit length of structure
Q	Qvar		Variable action
Q*	Qsa	-	Dimensionless overtopping discharge, defined $q/(T_m g H_s)$
Q#	Qss	-	Dimensionless overtopping discharge, defined $q/(gH_s^3)^{0.5}$
Q_o	Qo	-	Overtopping coefficient, having dimensions of q
Q_{bs}	Qbs	-	Beach sediment transport rate
Qo	Qo	m^3/m	Overtopping coefficient, having dimensions of q; also used as beach sediment transport rate
Q_p	Qp		Peakedness factor
q	q	kN/m^2	Deviator stress, $(\sigma'_1 - \sigma'_3)/2$
q_c	qc	MN/m^2	Cone resistance
q_o	qo	m^3/m	Volume of overtopping per wave and per unit length of structure
q_s	qs		Superficial velocity; or specific discharge, discharge per unit area, usually through a porous matrix
q_v	qv		Discharge velocity in porous flow

R	R	-	General strength or resistance of the system in reliability analysis
$R_{...}$	R...		Return period
R_c	Rc	m	Crest freeboard, level of crest less static water level
R_{ch}	Rch		Characteristic strength
R_f	Rf		Roughness of rockfill
R_u	Ru	m	Run-up level, relative to static water level
R_{us}	Rus	m	Run-up level of significant wave
$R_{u2\%}$	Ru2	m	Run-up level exceeded by 2% of run-up crests
R^*	Rcs	-	Dimensionless freeboard, defined in terms of the steepness of the mean wave period, $R^*_m = (R_c/H_s)(s_m/2\pi)^{1/2}$
R^*_p	Rcp	-	Dimensionless crest height for low-crest or reef breakwaters, defined $(R_c/H_s)(s_p/2\pi)^{1/2}$
$R_{d98\%}$	Rd98		Run-down level, below which only 2% pass
r	r	m	Radius
r	rr	-	Roughness or run-up reduction coefficient, usually relative to smooth slopes
r	rcc	-	Correlation coefficient
S	S	kN/kNm	Internal forces and moments
S_d	Sd	-	Damage number for (rock) armoured slopes $= A_e/D_{n50}^2$; also used as a general load or surcharge on the system in reliability analysis
S_{ch}	Sch		Characteristic load or surcharge
$S(f)$	S_f		Spectral density
S_r	Sr		Rock strength
S_R	SRr		Degree of saturation
S	s	-	Wave steepness $= 2\pi H / gT^2$
s_m	sm	-	Deep water wave steepness related to mean wave period $= 2\pi H/gT_m^2$
s_p	sp	-	Deep water wave steepness related to peak wave period $= 2\pi H/gT_p^2$

$T_{...}$	T	s	(Regular) wave period
T	Tdl		Structural, economic, or design life-time (in years)
T	Tors	kNm	Torsional moment
T_{H110}	TH110	s	Wave period associated with $H_{1/10}$
T_{H13}	TH13	s	Wave period associated with $H_{1/3}$
T_{Hm}	THm	s	Wave period associated with H_m
T_{Hmax}	THmax	s	Wave period associated with H_{max}
T_m	Tm	s	Mean wave period in time domain or mean wave period from spectrum
T_{nom}	Tnom	s	Nominal wave period, to be created at wave paddle
T_N	TN	s	Natural period of caisson-foundation system
$T_{N,R}$	TNR	s	Natural period of caisson-foundation system for roll motion
$T_{N,S}$	TNS	s	Natural period of caisson-foundation system for sway motion
T_{Pf}	Tpf	-	Return period = $(1 - (1 - P_f)^{1/T})^{-1}$
T_p	Tp	s	Wave period of spectral peak, inverse of peak frequency
T_R	TR	s	Length of wave record, duration of sea state
T_s	Ts	s	Wave period associated with H_s, not statistically significant
T_{si}	Tsi	s	Wave period associated with H_{si}
T_{so}	Tso	s	Wave period associated with H_{so}
Tx	Tx		Triaxial test
t	t	s	Time
t	th	m	Thickness
t_{pimax}	tpi	s	Time of maximum impact pressure $p_{i,max}$
t_{pMWL}	tpMWL	s	Time of impact pressure at MWL p_{MWL}
t_{pumax}	tpu	s	Time of maximum uplift pressure $p_{u,max}$
t_r	tr	s	Rise time, generally for horizontal force F_h (Fig. 4)
t_{rFu}	trFu	s	Rise time for uplift force F_u

t_d	td	s	Duration of horizontal force impact, relative to t_0 (Fig. 4)
t_{dFu}	tdFu	s	Duration of uplift force impact, relative to t_{0Fu} (Fig. 4)
t_0	t0	s	Index time of start of (horizontal force) event, relative to test duration (Fig. 4)
t_{0Fu}	t0Fu	s	Index time of start of uplift event, relative to test duration (Fig. 4)
t_q	tq	s	Time of maximum horizontal quasi-static force, relative to t_0 (Fig. 4)
t_{qFu}	tqFu	s	Time of maximum uplift quasi-static force, relative to t_{0Fu} (Fig. 4)
t_t	tt	s	Total duration of event, relative to t_0 (Fig. 4)
t_{tFu}	ttFu	s	Total duration of uplift event, relative to t_{0Fu} (Fig. 4)
U_{10}	U10	m/s	Wind speed, particularly at 10m above water surface
u	u	kN/m^2	Excess pore water pressure
u	u	m/s	Component of velocity along x axis
u_a	ua	kN/m^2	Average pore pressure, $(u_{max} + u_{min})/2$
u_{cy}	ucy	kN/m^2	Cyclic pore pressure, $(u_{max} - u_{min})/2$
u_p	up	kN/m^2	Permanent pore water pressure. This corresponds to the pore pressure at the end of a complete load cycle, when the shear stress returns to the original shear stress ($\tau = \tau_a$)
u_{max}	umax	kN/m^2	Maximum pore water pressure
u_{min}	umin	kN/m^2	Minimum pore water pressure
V	V	kN	Shear force
V_o	Vo	m^3	Initial specimen volume
V_c	Vc	m^3	Consolidated specimen volume
v	v	m	Component of displacement of a point
v	v	m/s	Component of velocity along y axis
$W_{...}$	W...		Armour unit weight
W	Wsec		Section modulus

W_{50}	W50	tm^2/s^2	Median armour unit weight
w	w	m/s	Component of velocity along z axis
w	w	%	Water content
X	X		Value of a property of a material
X, X_1, X_2	X, X1, X2		Variable, example values of X
	xmean		Mean value of variable
x	x		x-axis, perpendicular to front face of caisson, x = 0 at the face (Fig. 1)
x_{COG}	xCOG	m	x-value of centre of gravity
Y, Y_1, Y_2	Y, Y1, Y2		Variable, example values of Y; also X+E
y	y		y-axis, longitudinal direction of the breakwater, y = 0 in the middle of one section (Fig. 1)
Z	ZPf		Stochastic variable signifying reliability of failure function, also reliability function = R-S
z	z	m	z-axis, vertical direction, positive upwards, z = 0 at seabed (Fig. 1)
z_{COG}	zCOG	m	z-value of centre of gravity
α	alpfs	°	Structure front slope angle to horizontal; also used as a coefficient
	alpd	-	Distribution parameter
	alpPM	-	Coefficient in PM spectrum
α_c	alpc	-	Initial breakwater response slope, defined $A_t / h_c'^2$
β	BETA	°	Direction of wave propagation relative to normal to breakwater alignment
β	betd		Distribution parameter
β	betr		Reliability index
β_T	betT		Target reliability index
Γ	GAM		Gamma function, also Poisson ratio

γ	gam		Partial safety coefficient, also shear strain
γ_1	gam1		Partial coefficient related to characteristic value of X_i
γ_a	gama	-	Average shear strain = $(\gamma_{max} + \gamma_{min})/2$
γ_c	gamc	kN/m^3	Weight density of caisson
γ_{cy}	gamcy	-	Cyclic shear strain = $(\gamma_{max} - \gamma_{min})/2$
γ_{Hs}	gamHs		Partial coefficients on the wave height H_s
γ_{JS}	gamJS	-	Peak factor of JONSWAP spectrum
γ_{max}	gammax	-	Maximum shear strain within a cycle
γ_{min}	gammin	-	Minimum shear strain within a cycle
γ_p	gamp	-	Permanent shear strain. This corresponds to shear strain at the end of a complete load cycle, when the shear stress returns to the original shear stress ($\gamma = \gamma_a$)
γ_R	gamR	kN/m^3	Weight density of rubble
γ_S	gamS	kN/m^3	Weight density of subsoil
γ_W	gamW	kN/m^3	Weight density of water
γ_Z	gamZ		Reliability function for the formula adopted Z
Δ	DEL	-	Reduced relative density, eg. $(\rho_r/\rho_w)-1$
Δf	DELf	Hz	Frequency increment
ΔH	DELH	m	Change of specimen height
Δt	DELt	s	Time interval between samples
ε	eps	-	Linear strain
ε_a	epsa	-	Axial strain
ε_{vol}	epsvol	-	Volumetric strain
$\eta(t)$	eta_t	m	Surface elevation function referred to the mean water level (MWL)
η_c	etac	m	Wave crest elevation referred to MWL
η_{max}	etamax	m	Maximum wave crest elevation in a record referred to MWL
η_{min}	etamin	m	Minimum wave crest elevation in a record referred to MWL

η_O	etaO	-	Safety coefficient related to over-turning calculated from equilibrium of moments
η_s	etaS	-	Safety coefficient related to sliding calculated from equilibrium of forces
η_t	etat	m	Wave trough elevation referred to MWL
θ	THE	°	Direction of wave propagation relative to north
λ	lam	-	Model / prototype scale ratio; slenderness ratio
λ_a	lama	-	Fraction of air (aeration)
λ_{Hs}	lamHs	-	Average number of H_s data values per year, H/T
μ	mu	-	Coefficient of friction, particularly between concrete elements and rock
μ_x	mux		Mean of value ($\mu(x)$)
ν_L	nuL	-	Dynamic load factor
ν_d	nud	-	Dynamic response factor or dynamic resistance factor
ξ	xi	-	Iribarren number or surf similarity parameter, $= \tan\alpha/s^{1/2}$
ξ_m	xim	-	Iribarren number calculated in terms of s_m
ξ_p	xip	-	Iribarren number calculated in terms of s_p
ξ_c	xic	-	Critical value of Iribarren number distinguishing plunging from surging breakers
ρ	rho	t/m^3	Mass density, usually of fresh water
ρ_a	rhoa	t/m^3	Mass density of armour units
ρ_c	rhoc	t/m^3	Mass density of caisson or concrete
ρ_r	rhor	t/m^3	Mass density of rock
ρ_s	rhos	t/m^3	Mass density of soil
ρ_{sp}	rhosp	t/m^3	Mass density of soil particles

ρ_w	rhow	t/m^3	Mass density of sea water
σ	sig	kN/m^2	Normal stress
σ'	sigs	-	Normalised standard deviation σ/μ, also effective normal soil stress
σ'_{1c}	sig1c	kN/m^2	Major effective principal stress during consolidation
σ'_{3c}	sig3c	kN/m^2	Major effective principal stress during consolidation
σ'_a	siga	kN/m^2	Axial effective stress during consolidation
σ'_{vc}	sigvc	kN/m^2	Vertical effective stress during consolidation
σ'_{oct}	sigoct	kN/m^2	Octahedral effective normal stress
σ'_h	sigh	kN/m^2	Horizontal effective stress
σ'_r	sigr	kN/m^2	Radial effective stress
σ_1	sig1	kN/m^2	Major principal stress
σ_{FHs}	sigFHs	-	Standard deviation of F_{Hs}
σ_{xx}	sigxx	kN/m^2	Normal stress in x-plane
σ_{yy}	sigyy	kN/m^2	Normal stress in y-plane
σ_{zz}	sigzz	kN/m^2	Normal stress in z-plane
$\sigma(x)$	sig_x	-	Standard deviation of x
τ	tau	kN/m^2	Shear stress of rock mound or soil
τ_a	taua	kN/m^2	Average shear stress
τ_{cy}	taucy	kN/m^2	Cyclic shear stress (single amplitude)
τ_f	tauf	kN/m^2	Shear stress at failure
τ_{fcy}	taufcy	kN/m^2	Cyclic shear stress, $(\tau + \tau_{cy})_f$
τ_h	tauh	kN/m^2	Shear stress on horizontal plane
τ_{max}	taumax	kN/m^2	Maximum shear stress
τ_{min}	taumin	kN/m^2	Minimum shear stress
τ_o	tauo	kN/m^2	Initial shear stress
τ_{ref}	tauref	kN/m^2	Reference stress
Φ_N	PHIN		Normal distribution
$\Phi_B(-\beta)$	PHIN_b		Distribution function of the normal distribution
φ	phi	°	Angle of internal friction of rock or soil

φ	phi	rad	Rotational motion of the caisson / structure (Fig. 2)
φ_R	phiR	°	Angle of internal friction of rubble
φ_S	phiS	°	Angle of internal friction of subsoil
φ'_{peak}	phip	°	Angle of internal friction at peak stress
υ	ups	-	Damage, $= N_d/N_a$
Ψ	psi	°	Dilation angle (note - upper case) Factors defining representative value of variable actions 0 for combination values; 1 for frequent values; 2 for quasi-permanent values.
ω	ome	rad/s	Angular frequency, $= 2\pi f$

Subscripts for chapter 4:

a	structural steel
c	concrete
c	compression
cr (or crit)	critical
d	design
dst	destabilising
dir	direct
eff	effective
ext	external
f	flange
F (or P)	action
g (or G)	permanent action
h	high; higher
ind	indirect
inf	inferior; lower
int	internal
k	characteristic
l	lower; low
m (or M)	material

m	bending
m	mean
max	maximum
min	minimum
nom	nominal
p (or P)	pre-stressing force
pl	plastic
ps	pre-stressing steel
q (or Q)	variable action
R	resistance
rep	representative
s	reinforcing steel
S	internal moments and forces
stb	stabilising
sup	superior; upper
t (or ten)	tension
t (or tor)	torsion
u	ultimate
v	shear
w	web
x, y, z	co-ordinates
y	yield

FIGURES

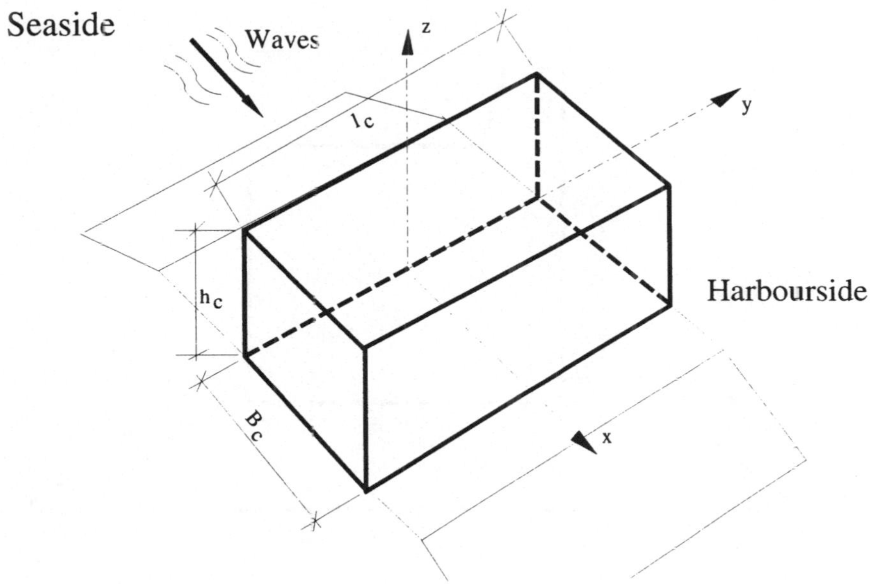

Figure 1. Coordinate system.

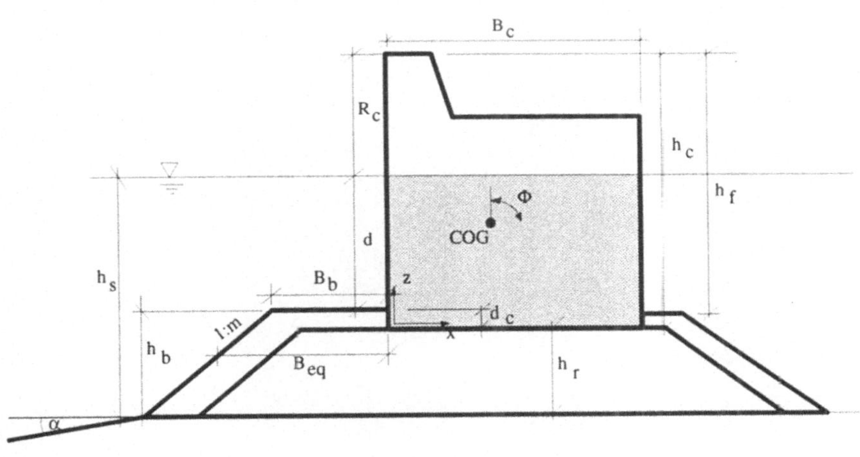

Figure 2. Definition of geometric parameters.

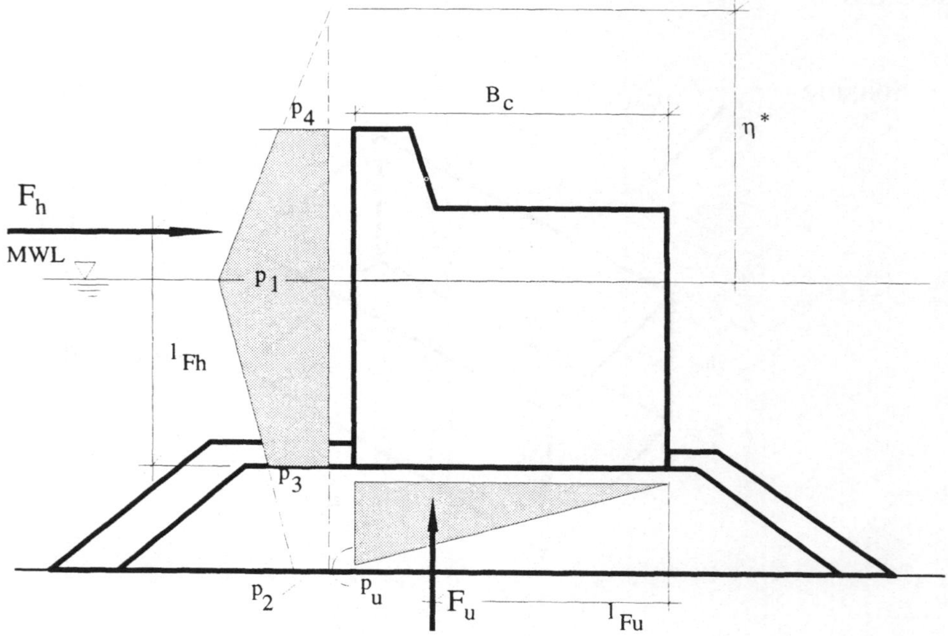

Figure 3. Definition of loading of structure.

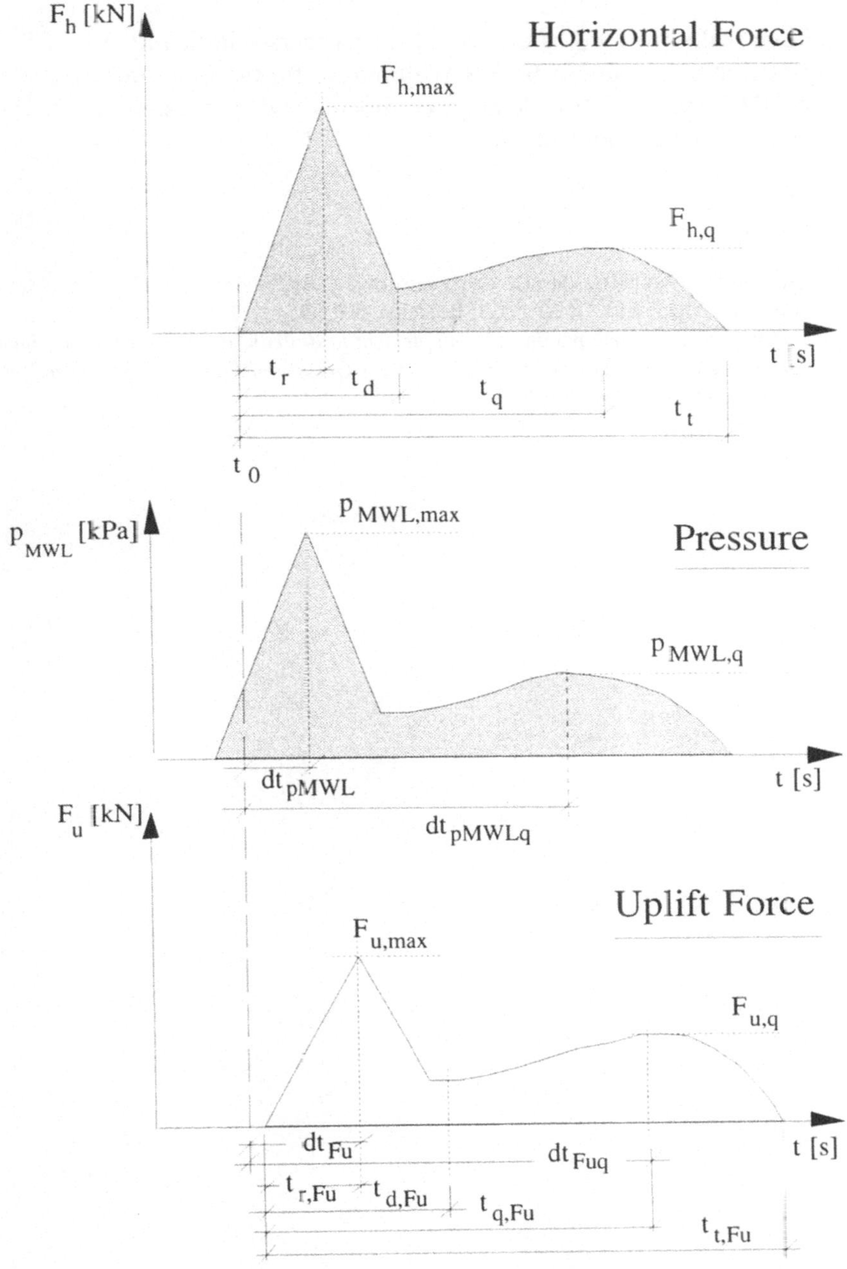

Figure 4. Definition of time series analysis.

ACKNOWLEDGEMENTS

This annex results from collaboration of all partners within PROVERBS. However, significant contributions by HR Wallingford for the first draft of this annex; DG, UoS, DUT and NGI for detailed comments and discussions of the various drafts are gratefully acknowledged.

REFERENCES

CIRIA/CUR (1991): Manual on the use of rock in coastal and shoreline engineering. Rotterdam, The Netherlands / Brookfield, USA: A.A. Balkema, 607 pp.

PIANC (1986): List of sea state parameters. *Supplement to Bulletin. Permanent International Association of Navigation Congresses, IAHR - Working Group on Wave Generation and Analysis,* Brussels, Belgium, vol. 52, 23 pp.

ANNEX 2

Full addresses of contact persons from partner institutes

01	**LWI**	**Prof. H. Oumeraci** TU Braunschweig Abteilung Hydromechanik und Küsteningenieurwesen Beethovenstr. 51a DE-38106 Braunschweig / GERMANY Tel: 0531/391-3930 Fax: 0531/391-8217 email: h.oumeraci@tu-bs.de
02	**HR**	**Prof. N.W.H. Allsop** HR Wallingford Ltd. - Howbery Park GB-Wallingford OX10 8BA / U.K. Tel: +44 1491 822 230 Fax: +44 1491 825 539 email: nwha@hrwallingford.co.uk
03	**DG**	**M.B. De Groot** GeoDelft (previously Delft Geotechnics) P.O. Box 69 NL-2600 AB Delft / THE NETHERLANDS Tel: +31 15 2693 787 Fax: +31 15 2610 821 email: dgo@geodelft.nl
04	**UoS**	**Prof. R.S. Crouch** University of Sheffield Sir Frederick Mappin Building Mappin Street GB-Sheffield S1 3JD / U.K. Tel: +44 114 222 5716 Fax: +44 114 222 5700 email: r.crouch@sheffield.ac.uk

05 **DUT** **Prof. J.K. Vrijling**
Delft University of Technology
Hydraulic and Offshore Engineering Section
Stevinweg 1
NL-2628 CN Delft / THE NETHERLANDS
Tel: +31 15 278-5278
Fax: +31 15 278-5124
email: a.ramdjan@ct.tudelft.nl

06 **AU** **Prof. H.F. Burcharth**
Aalborg University
Sohngaardsholmsvej 57
DK-9000 Aalborg / DENMARK
Tel: +45 9635 8482
Fax: +45 9814 2555
email: burcharth@civil.auc.dk

07 **BrU** **Prof. D.H. Peregrine**
University of Bristol
University Walk
GB-Bristol BS8 1TW / U.K.
Tel: +44 117 928-7971
Fax: +44 117 928-7999
email: d.h.peregrine@bristol.ac.uk

08 **CEP** **B.G. Madrigal**
Centro De Estudios De Puertos Y Costas (CEPYC)
c./ Antonio Lopéz 81
ES-28026 Madrid / SPAIN
Tel: +34 1 335 7625
Fax: +34 1 335 7622
email: braulio.g.madrigal@cedex.es

09 **DH** **Dr. M.R.A. Van Gent**
Delft Hydraulics
P.O. Box 177
NL-2600 MH Delft / THE NETHERLANDS
Tel: +31 15 285 8846
Fax: +31 15 285 8582
email: marcel.vangent@wldelft.nl

10 **PM** **Prof. L. Franco**
Politecnico di Milano
32 Pzz. Leonardo da Vinci
IT-20133 Milano / ITALY
Tel: +39 02 2399 6297
Fax: +39 02 2399 6298
email: leofranc@fenice.dsic.uniroma3.it

11 **UoP** **P.J. Hewson**
University of Plymouth
Palace Street
GB-Plymouth PL1 2DE / U.K.
Tel: +44 1752 233652
Fax: +44 1752 233658
email: phewson@civ.plym.ac.uk

12 **UoB** **Prof. A. Lamberti**
Università degli Studi di Bologna
Viale del Risorgimento 2
IT-40136 Bologna / ITALY
Tel: +39 051 644-3749
Fax: +39 051 644-8346
email: alberto@idraulica.ing.unibo.it

13 **UGE** **Prof. W. Richwien**
Universität Gesamthochschule Essen
Universitätsstr. 15
DE-45117 Essen / GERMANY
Tel: +49 201 183-2857
Fax: +49 201 183-2870
email: igb010@sp2.power.uni-essen.de

14 **ULH** **Prof. M. Bélorgey**
Université de Caen
Laboratoire de Mécanique
Rue de tilleuls 24
FR-14000 Caen / FRANCE
Tel: +33 231 565 711
Fax: +33 231 565 757
email: belorgey@meca.unicaen.fr

15 **SOG** **A. Martinez**
Sogreah Ingenierie SNC
Rue de Lorraine 6
FR-38130 Echirolles / FRANCE
Tel: +33 47633 4079
Fax: +33 47633 4296
email: martinez@sogreah.fr

16 **NGI** **T.J. Kvalstad**
Norwegian Geotechnical Institute
P.O. Box 3930 Ullevaal Hageby
N-0806 Oslo / NORWAY
Tel: +47 220 23046
Fax: +47 222 30448
email: tk@ngi.no

17 **UE** **T. Bruce**
University of Edinburgh
The King's Buildings
GB-Edinburgh EH9 3JL / U.K.
Tel: +44 131 650 8701
Fax: +44 131 667 3677
email: tom.bruce@ed.ac.uk

18 **CU** **Dr. I.J. Losada**
Universidad de Cantabria
Grupo de Ingenieria Oceanografica y de Costas
Avda. de los Castros S/N
ES-39005 Santander / SPAIN
Tel: +34 42 201810
Fax: +34 42 201860
email: inigo@puer.unican.es

19 **QuB** **G. Müller**
Queens University of Belfast
Stranmills Road
Belfast BT7 1NN / NORTHERN IRELAND
Tel: +44 1232 274517
Fax: +44 1232 663754
email: g.muller@qub.ac.uk

20 UoN **Prof. E. Benassai**
Università degli Studi di Napoli "Frederico II"
Via Claudio n. 21
IT-80125 Napoli / ITALY
Tel: +39 081 7683444
Fax: +39 081 5938936
email: benassai@ds.unina.it

21 BV **D. Berdin**
Bureau Veritas
17 bis Place des Reflets
FR-92400 Courbevoie / FRANCE
Tel: +33 1 4291 5291
Fax: +33 1 4291 5345
email: dberdin@bureauveritas.com

22 STC **J.-B. Kovarik**
Service Technique Central
B.P. 53 - boulevard Gambetta 2
FR-60321 Compiègne / FRANCE
Tel: +33 344 926027
Fax: +33 344 200675
email: jb.kovarik@stcpmvn.equipement.gouv.fr

23 ENEL **M. De Gerloni**
ENEL Società per Azioni
Centro di Ricerca Idraulica e Strutturale
Via Ornato 90/14
IT-20162 Milano / ITALY
Tel: +39 02 7224 3634
Fax: +39 02 7224 3530
email: degerloni@cris.enel.it

ANNEX 3

Full contents list of Volume II of PROVERBS

This annex contains the full list of all papers which are part of the so-called Volume II report of PROVERBS. These papers are authored by various researchers from the participating institutions and altogether build the basis for the results presented in the chapters of this book. All reports listed below can be purchased from the coordinator of the project (see no. 1 in Annex 2) or from the Task Leaders of PROVERBS (see nos. 2 – 5 in Annex 2).

VOLUME IIa – HYDRAULIC ASPECTS

Chapter 1: ALLSOP, N.W.H. (1999): Introduction.

Chapter 2.1: ALLSOP, N.W.H.; DURAND, N. (1999): Influence of steep seabed slopes on breaking waves for structure design. 28 pp.

Chapter 2.2: MCCONNELL, K.J. (1999): Derivation, validation and use of parameter map. 5 pp.

Chapter 2.3: CALABRESE, M.; VICINANZA, D. (1999): Estimation of proportion of impacts. 15 pp.

Chapter 3.1: VOORTMAN, H.G.; HEIJN, K.M. (1999): Wave transmission over vertical breakwaters. 9 pp.

Chapter 3.2:ALLSOP, N.W.H.; BESLEY, P.; FRANCO, L. (1999): Wave overtopping discharges. 8 pp.

Chapter 3.3:ALLSOP, N.W.H. (1999): Wave reflections. 13 pp.

Chapter 4.1: VOORTMAN, H.G.; VAN GELDER, P.H.A.J.M.; VRIJLING, J.K. (1999): The Goda model for pulsating wave forces. 5 pp.

Chapter 4.2: FLOHR, H.; MCCONNELL, K.J.; ALLSOP, N.W.H. (1999): Negative or suction forces on caissons: development of improved prediction methods. 17 pp., 1 Annex.

Chapter 4.3: BURCHARTH, H.F.; LUI, Z. (1999): Force reduction of short-crested non-breaking waves on caissons. 17 pp., 3 Annexes.

Chapter 4.4: VRIJLING, J.K.; VAN GELDER, P.H.A.J.M. (1999): Uncertainty analysis of non breaking waves. 12 pp.

Chapter 4.5: VAN GENT, M.R.A.; TORENBEEK, R.V.; PETIT, H.A.H. (1999): VOF model for wave interaction with vertical breakwaters. 11 pp.

Chapter 4.6: LÖFFLER, A.; KORTENHAUS, A. (1999): Non breaking waves - pressures on berms. 23 pp.

Chapter 5.1: KORTENHAUS, A.; OUMERACI, H.; ALLSOP, N.W.H.; MCCONNELL, K.J.; VAN GELDER, P.H.A.J.M.; HEWSON, P.J. ET AL. (1999): Wave impact loads - pressures and forces. 39 pp.

Chapter 5.2: WALKDEN, M.; WOOD, D.J.; BRUCE, T.; PEREGRINE, D.H. (1999): Seaward impact forces. 25 pp.

Chapter 5.3: ALLSOP, N.W.H.; CALABRESE, M. (1999): Impact loadings on vertical walls in directional seas. 19 pp.

Chapter 5.4: VAN GELDER, P.H.A.J.M.; VRIJLING, J.K.; HEWSON, P.J. (1999): Uncertainty analysis of impact waves and scale corrections due to aeration. 12 pp.

Chapter 5.5: LÖFFLER, A.; KORTENHAUS, A.; WOOD, D.J. (1999): Wave impact loads - pressures on a berm. 22 pp.

Chapter 6.1: WALKDEN, M.; MÜLLER, G. (1999): Strongly depth limited waves. 4 pp.

Chapter 6.2: MARTíN, F.L.; LOSADA, M.A. (1999): Wave loads on crown walls. 36 pp.

Chapter 6.3: MUTTRAY, M.; OUMERACI, H. (1999): Wave loads on caisson on high mounds. 28 pp.

Chapter 7: CRAWFORD, A.R.; HEWSON, P.J. (1999): Field measurements and database. 4 pp.

Chapter 8.1: DE GERLONI, M.; COLOMBO, D.; BÉLORGEY, M.; BERGMANN, H.; FRANCO, L.; PASSONI, G.; ROUSSET, J.-M.; TABET-AOUL, E.H. (1999): Alternative low reflective structures - perforated vertical walls. 41 pp.

Chapter 8.2: KORTENHAUS, A.; OUMERACI, H. (1999): Alternative low reflective structures - other type of structures. 29 pp.

VOLUME IIb –GEOTECHNICAL ASPECTS

Chapter 1: DE GROOT, M.B. (1999): Introduction.

Chapter 2: KVALSTAD, T.J. (1999): Soil investigations and soil parameters. 20 pp.

Chapter 3: LAMBERTI, A.; MARTINELLI, L.; DE GROOT, M.B. (1999): Dynamics. 56 pp.

Chapter 4: DE GROOT, M.B. (1999): Instantaneous pore pressures and uplift forces. 38 pp.

Chapter 5: KVALSTAD, T.J. (1999): Degradation and residual pore pressures. 37 pp.

Chapter 6: IBSEN, L.B.; JAKOBSEN, K.P. (1999a): Limit state equations for stability and deformation. 20 pp., 3 Annexes.

Chapter 6, Annex B: IBSEN, L.B.; JAKOBSEN, K.P. (1999b): Permanent deformations due to impact loading. 9 pp.

Chapter 6, Annex A: JAKOBSEN, K.P.; SØRENSEN, J.D.; BUCHARTH, H.F.; IBSEN, L.B. (1999): Failure modes - limit state equations for stability. 26 pp.

Chapter 6, Annex C: LAMBERTI, A. (1999): Combined effect of dilatancy in rubble mound and caisson inertia. 9 pp.

Chapter 7: KVALSTAD, T.J.; DE GROOT, M.B. (1999): Uncertainties. 30 pp.

Chapter 8: LESNY, K.; PERAU, E.; RICHWIEN, W. (1999): Influence of design parameters - stability analysis on feasibility level. 31 pp.

Chapter 9: KVALSTAD, T.J. (1999): Alternative foundations. 18 pp.

VOLUME IIc –STRUCTURAL ASPECTS

Chapter 1: CROUCH, R.S. (1999): Introduction.

Chapter 2: MARTINEZ, A.; KOVARIK, J.-B.; BERDIN, D. (1999): Structural design of vertical breakwaters - limitations of current practice and existing design codes. 37 pp.

Chapter 3: VROUWENVELDER, A.W.C.M.; BIELECKI, M. (1999): Caisson reliability during transport and placing. 36 pp.

Chapter 4: CROUCH, R.S. (1999a): In-service behaviour of cellular reinforced concrete caissons under severe wave impact. 39 pp.

Chapter 5: CROUCH, R.S. (1999b): Some observations on the durability and repair of concrete structures in a marine environment. 30 pp.

VOLUME IId –PROBABILISTIC ASPECTS

Chapter 1: VRIJLING, J.K. (1999): Introduction.

Chapter 2: VRIJLING, J.K. (1999): Fault tree analysis of a vertical breakwater. 9 pp.

Chapter 3: SØRENSEN, J.D.; BURCHARTH, H.F. (1999): Limit state equations including uncertainties. 26 pp., 1 Annex.

Chapter 4.1: VOORTMAN, H.G.; KUIJPER, H.K.T.; VRIJLING, J.K. (1999): Economic optimal design of vertical breakwaters. 17 pp.

Chapter 4.2: SØRENSEN, J.D.; BURCHARTH, H.F. (1999): Partial safety factor system. 24 pp.

Chapter 5.1 LAMBERTI, A.; MARTINELLI, L.; DE GROOT, M.B.; GOLÜCKE, K.; VAN HOVEN, A.; ZWANENBURG, C. (1999): Hazard analysis of Genoa Voltri breakwater. 40 pp.

Chapter 5.2: VOORTMAN, H.G.; VRIJLING, J.K. (1999): Reliability analysis of the Easchel breakwater. 29 pp.

Chapter 5.3: SØRENSEN, J.D.; BURCHARTH, H.F. (1999): Other representative structures: Mutsu-Ogawara, Niigata East and West. 19 pp.

ANNEX 4

Summary of final results of PROVERBS

Outcome and End Users

PROVERBS has developed probability based methods for the analysis and design of a wide range of harbour and coastal monolithic structures subject to wave attack. These techniques will be used by coastal / harbour engineers working in central government, port authorities, coastal or local government authorities, consulting engineers, contractors, material suppliers, and further related specialists.

Main benefits

The use of the new methods will directly:

- Improve the efficiency of analysis / design for breakwaters / seawalls;
- Improve the competitiveness of this class of structures;
- Improve the consumer choice within Europe.

These improvements will then:

- Reduce costs of construction and ownership;
- Reduce risks of structural failure or under-performance;
- Increase competitiveness of European coastal / harbour engineers working world-wide.

Less direct, but still important benefits will accrue from:

- Increased skill levels and knowledge within Europe;
- Increased business exposure for European coastal / harbour engineers.

Hydrodynamic key results

The key results within Task 1 of PROVERBS (hydrodynamic aspects) are summarised in Table 1:

Table 1. Summary of selected key results in Task 1 (Hydrodynamic Aspects).

	KEY RESULTS	DESCRIPTION, IMPORTANCE & FURTHER REMARKS
1	Development of a parameter map to identify the possible loading cases of waves attacking the front face of a vertical breakwater based on structure geometry, water depth and wave conditions in the nearfield, including a method to assess the occurrence probability of breaking wave impacts.	• Developed from laboratory testing at HR Wallingford, Leichtweiß-Institut and further laboratories. • The parameter map allows to identify the impact loading or further relevant wave loading based only on three simple non-dimensional parameters: - relative berm height h_b/h_s which decides about structure type - relative wave height H_{si}/d and - relative berm width B_{eq}/L_{pi} which both decide about the loading case. • A method based on new developed breaking criteria which accounts for the reflection properties of the structure has been developed to predict the probability of occurrence of impact loading.
2	Development of a research strategy and new formulae to predict impact loading (horizontal and uplift forces) as a function of load duration based on theoretical, numerical and experimental studies, together with field measurement.	• The new procedure includes the following three steps which are similar for both horizontal and uplift loads: + From (statistical) GEV-Distribution the relevant peak force $F_{max}^* = F_{max}/\rho g H_b^2$ is selected for the intended design purpose + From the new formula $F_{max}^* =$ f (relative rise time $t_r/\sqrt{gd_b}$) and a further relationship for the impact load duration $t_d = f(t_r)$, the corresponding impact rise time t_r and load duration t_d are obtained by assuming a simplified triangular shape of the load history + Based on laboratory results a simplified parametric pressure distribution (4 parameters for horizontal load and 2 for uplift) is derived • The new prediction tools now allow to apply dynamic analyses for the design and to explain observed prototype failures.
3	Development of improved scaling rules to transfer laboratory results of breaking wave impact loads to prototype conditions based on a better physical understanding of the processes by means of theoretical/numerical, small- and large-scale model studies together with prototype measurements	Detailed insight in the individual processes involved in the whole load history including dynamics of entrained / entrapped air • Based on this detailed insight a separation method is developed to individually scale up each of the components of the load history by Froude or Cauchy law, depending on whether gravity or compressibility is dominant for the related process. • A more pragmatic procedure based on corrective factors (when only Froude scaling is used) is also provided which is based on a purely empirical relationship providing the level of aeration as a function of the number of breaking waves within a given time interval • The achieved results provide the engineer with improved tools for scaling and the researcher with a sound basis towards more improvements in this field.

	KEY RESULTS	DESCRIPTION, IMPORTANCE & FURTHER REMARKS
4	Prediction of the reduction of impact forces as a function of caisson length, wave obliquity and short-crestedness	• New experimental results of 3D-tests conducted at three different laboratories confirmed and extended the previous investigations within the Mast II/MCS-Project for the reduction of pulsating wave forces as a function of wave obliquity and short-crestedness • A new prediction formula is developed for the reduction factor of impact forces as a function fo the relative caisson length (L_c/L_{op}) and wave obliquity β for long-crested and short-crested sea • The new result will allow to avoid overdesign by substantially reducing the impact forces (up to 25 % for caisson lengths of 0,2 L_{op}) as used by the present conservative design approaches.
5	Seaward impact forces induced by overtopping waves plunging into the harbour basin	• Based on small-scale model tests and numerical modelling using the pressure-impulse theory the generation mechanisms of impact loading directed seaward have been identified. The impact loading on the rear face and the bottom slab of the caisson breakwater is induced by the plunging plume of overtopping waves into the harbour basin. The entrapped air under the plume plays a dominant role in the generation mechanisms • The results will allow to achieve a better design with respect to the seaward tilting failures which have often been observed by low vertical breakwaters.
6	Development and implementation of Artificial Neural Networks (ANNs) as a complementary prediction tool for horizontal pulsating forces, including a new technique to quantify the reliability intervals of the prediction	• Based on the integrated use of data sets from several model studies performed at different institutes, ANN-Modelling has been performed leading to a better prediction of the pulsating forces than the standard GODA method • With the new developed technique to determine the uncertainties of the predicted forces, ANN-Modelling can also be applied as a predictive tool in probabilistic design.
7	Development of new prediction formulae for pulsating wave forces on perforated caisson breakwaters	• Based on prototype measurements at two sites in Italy and France, and on small and large-scale model tests, new formulae to predict total wave forces on perforated structures have been developed. Three different approaches have been used for this purpose. • The new formulae have shown that a reduction of horizontal wave forces in the order of 30% and 50% can be achieved by using perforated caissons with a single chamber and with three chambers, respectively. • An improved physical understanding and better prediction models have also been achieved with respect to the hydraulic performance, building an important departure basis for the development of more innovative structures.

	KEY RESULTS	DESCRIPTION, IMPORTANCE & FURTHER REMARKS
8	Development of wave load formulae for perforated superstructures of high-mound composite breakwaters	• Based on large-scale model tests on new perforated structures on a high rubble foundation new prediction formulae for the associated wave loads have been developed • The improved understanding of the overall hydraulic performance which has been achieved by these results will also allow to improve the constructional features of these structures and to develop further innovative crown walls.
9	Development of new wave load formulae for crown walls	• Based on prototype measurement, small and large-scale modelling new formulae have been developed to predict the wave pressure distributing on the vertical front and the bottom slab of the crown wall. The new method makes use of the run-up height and two empirical parameters which depends on the incident wave steepness and the characteristics of the berm. • The results build a sound scientific basis for the development of more complete final design tools for this type of structures

Geotechnical key results

The key results within Task 2 of PROVERBS (geotechnical aspects) are summarised in Table 2:

Table 2. Summary of selected key results in Task 2 (Geotechnical Aspects).

	KEY RESULTS	DESCRIPTION, IMPORTANCE & FURTHER REMARKS
1	Development of soil data base, including strategy for soil investigations	• Soil data base developed by collection and analysis of available data at various laboratories as well as by performing new experimental investigations within PROVERBS, including related uncertainties. • Guidance for soil investigations required for feasibility and preliminary/detailed design levels, including diagrams and methods for interpretation of soil tests (Fig. 1-24). • The results will enhance the opportunity for a unified procedure for foundation design of breakwaters.
2	Development and improvement of engineering models for the simulation of the dynamic response of the breakwater to wave impact loading	• Various engineering tools have been developed/refined to simulate the swaying and rocking oscillatory motions (elastic models) and the permanent displacements related to sliding over the rubble foundation and the bearing capacity failures (plastic models). The required dynamic input parameters and the validation of the models have been obtained by means of large scale model tests, centrifuge testing and sophisticated finite element simulations and full-scale tests on breakwaters. • The elastic models are useful as they provide a first insight into the load transmission mechanisms to the foundation and can predict the dynamic load factor as well as the natural periods of oscillation of the structure-foundation system.

	KEY RESULTS	DESCRIPTION, IMPORTANCE & FURTHER REMARKS
3	Understanding and prediction of instantaneous pore pressure in rubble foundation (uplift forces) and in subsoil induced *directly* by cyclic wave pressure along seabed and rubble foundation boundary and *indirectly* by the rocking motions of the structure subject to wave loading	• Through combined hydraulic model testing, theoretical analysis, centrifuge testing and prototype measurements, supplemented by FEM-modelling a physical understanding of the processes involved in the structure-foundation-interaction have been achieved. • *For the pressure in the rubble foundation* various solutions have been suggested to explain the observed deviations from the commonly used triangular uplift pressure distribution along the bottom slab for both quasi-stationary (non-impact) and non-stationary (impact) conditions. Possible maximum values of these deviations are also provided. • *For the pressure in the subsoil* the results have shown that in many cases undrained conditions can be assumed and that they may affect only a thin top layer of the foundation. Furthermore gas content in the subsoil is considerably important and the most critical situation is expected to occur only at the seaward and shoreward edge of the rocking caisson breakwater.
4	Understanding and prediction of the degradation (cohesive subsoils) and "residual" pore pressure (non-cohesive subsoils) induced by cyclic wave loading including their effect (reduction) on soil stiffness and shear strength	• As in the case of instantaneous pressure an improved physical understanding of the most relevant phenomena has been achieved. Furthermore, design diagrams based on centrifuge modelling and FEM-simulations have been developed to predict the reduction of soil stiffness (shear modulus) and shear strength which are induced by excess pore pressure generated by cyclic wave loading. • *For sandy subsoils* the results have shown that there is no significant shear strength reduction for dense sands while this is critical only for relatively loose sand layers. A prediction formula to quantify this effect is therefore provided. • *For clayey subsoils* it is shown that the reduction in shear strength can be predicted by using the results of monotonic (static) undrained tests with a proper correction factor to account for cyclic wave loading. A design diagram to determine this correction factors as a function of the over-consolidation ratio and the loading conditions has been developed. • In both sandy and clayey subsoils the shear modulus may be reduced by a factor of 2 as compared to its initial value.
5	Formulation of Limit State Equations (LSEs) for the main foundation failure modes including among others: (i) sliding over rubble foundation (ii) bearing capacity in rubble foundation (iii) bearing capacity in subsoil	• Simplified LSEs for relevant sliding and bearing capacity failure modes, including limitations (prescribed rupture surface, explicit neglecting of dilation, limited number of rupture surfaces) • More sophisticated LSEs for detailed design which can identify the most unfavourable rupture surface and guidance for the use of numerical modelling, including a warning notice on the limitations of the models (Bishop slip circle analysis and FE-models). • Assessment of conservatism of the commonly used 2D-approach of rupture surface in bearing capacity analysis

	KEY RESULTS	DESCRIPTION, IMPORTANCE & FURTHER REMARKS
		as compared to actual 3D-rupture surfaces (up to 40%). • Solution alternatives for stability analysis, including cumulative effects and stepwise failure.
6	Quantification of uncertainties in soil data and models	• Development of "operational fault tree" in order to provide an overview of all uncertainties associated with the input (soil) parameters, the models and the output parameters. • Uncertainties of soil parameters based on the various steps of the determination of these parameters are quantified for feasibility design levels, guidance is given for further design levels. • Model uncertainties are evaluated for the models describing the dynamic response, the instantaneous pore pressure, the degradation/residual pore pressure and the stability by performing several hindcasts of large scale model tests and prototype tests. • The quantified uncertainties can readily be used for feasibility studies.
7	Relative influence of design parameters on failure models (sensitivity analysis), including guidance for design improvement and suggestions for non-conventional foundation alternatives	• Development of a combined parameter map for most relevant loading cases/failure modes combinations as related to design parameters for an example structure. • Development of general tendencies with respect to the most dominant parameters (eccentricity of dead load, internal friction angle etc) and to the most relevant failure modes (sliding along base, bearing capacity in subsoil etc) and verification of their validity for further structures. • The results have allowed to suggest several practical solutions for design improvements, including non conventional alternatives.

Structural key results

The key results within Task 3 of PROVERBS (structural aspects) are summarised in Table 3:

Table 3. Summary of selected key results in Task 3 (Structural Aspects).

	KEY RESULTS	DESCRIPTION, IMPORTANCE & FURTHER REMARKS
1	Identification of deficiencies of existing codes of practice and development of new approach and framework for future codes related to structural design of cellular caissons	<u>Deficiencies</u>: (i) Lack of guidance for wave loads & load combinations for ULS & SLS, (ii) No guidance on tolerable deformations/displacements, (iii) Lack of harmonisation for safety levels, (iv) Safety factors only based on experience or/and semi-probabilistic approach <u>Implications</u>: Since no single existing code can readily be applied for the structural design of caisson breakwaters and a peacemeal approach by assembling parts of various codes is not appropriate, a new approach to account for maritime aspects in the existing codes has to be developed.

	KEY RESULTS	DESCRIPTION, IMPORTANCE & FURTHER REMARKS
		Future codes: Most focus on the specification of wave loading according to Eurocode framework (characteristic, design and combined values associated with frequent, quasi-permanent and accidental conditions) and on a unified framework to avoid inconsistencies between codes. This is achieved through the introduction at the last stage of the analysis of a single model factor for both actions and resistances which depends on model uncertainties, required safety level and design life.
2	Structural behaviour of floating and towed caissons during pre-service conditions (transport & placing)	Floating and towing phase: 3 failure modes: (i) instabilities due to severe motions, (ii) cracking of concrete (durability) and (iii) plate collapse. Sinking and placing phase: 2 failure modes: cracking and plate collapse, both due to uneven foundation. The full structural probabilistic analysis developed for each failure modes, incl. associated target reliabilities builds a framework for further research and for drafting design guidelines.
3	Identification of most relevant in-service failure modes and formulation of associated limit state equations	• Analysis of role of all structural caisson members with respect to load transmission up to the foundation • The results have shown that: -Multi-cell caissons are highly redundant structures with many load paths -simple models based on structural behaviour of individual members can and must be used for preliminary design before using more refined full 3D-dynamic analysis.
4	Development and use of more refined models for in-service structural behaviour to assess uncertainties and perform full structural dynamic analysis.	Development & comparison of results of three hierarchical refined models for a case study (Genoa Voltri Breakwater): • linear 3DoF model (three degrees of freedom) • non-linear shell FEM (layered shell finite element model) • 3D-FEM (full three dimensional continuum finite element model), incl. caisson pre-processor The results have shown among others that: • prior to performing non-linear analysis, linear models should first be used to provide a first insight into load transmission mechanisms. • Depending on structural model used the bending moments induced by wave impacts on a front wall may differ by a factor of two. • 3D-FEM analysis is recommended for calculation of direct stress, bending moment and shear forces in a member. It also well illustrates the progressive transmission of pressure pulse through the structure into the foundation.
5	Durability of reinforced concrete caissons in the marine environment	• Analysis of possible consequences of cracking, onset of corrosion and progressive material degradation • Formulation and implementation of limit state functions for cracking during pre-service and in service conditions and for the onset of corrosion. • The results are also relevant for the development of monitoring, inspection and maintenance programmes.

Probabilistic key results

The key results within Task 4 of PROVERBS (probabilistic aspects) are summarised in Table 4:

Table 4. Summary of selected key results in Task 4 (Probabilistic Design Tools).

	KEY RESULTS	DESCRIPTION; IMPORTANCE & FURTHER REMARKS
1	Development of methods for uncertainty evaluation and statistical description	After the various sources and types of uncertainties have been identified and classified, including wave and soil parameters, concrete properties as well as the models for the load incl. implementation of Artificial Neural Networks for wave loading and stability calculations, methods have been implemented to account for these uncertainties in the LSEs. The adopted procedure can be extended later to account for Human and Organisation Errors (HOE).
2	Evaluation of methods to select target reliability	Since the definition of a target reliability represents one of the most decisive and difficult tasks in probabilistic design, all the methods and techniques available are reviewed and analysed, including code calibration and cost optimisation over life time. The most suitable methods for the class of structure, the limit state and the safety class considered are then identified and adapted to PROVERBS.
3	Development of methods for economic optimisation	A probabilistic method for economic optimisation which account for ultimate and serviceability limit states and maintenance over the life time of the structure has been developed and implemented to optimise caisson dimensions and height of rubble foundation. The developed procedure also allows to select the optimal target reliability on a rational and economic basis.
4	Formulation of LSEs incl. uncertainties and fault tree analysis for most relevant failure modes	The Limit State Equations (LSEs) have been formulated for hydraulic, geotechnical and structural failure modes (including cracking and outset of corrosion as well as transport and placing of caissons). The modular structure allows an update of the various models and uncertainties involved, as well as of the associated knowledge related to the failure mechanisms and the interaction between the failure modes.
5	Development of a Partial Safety Factor System (PSFS)	A PSFS for the design of vertical breakwaters calibrated on the basis of the established full probabilistic framework has been developed and implemented for example structures. The PSFS accounts for the uncertainties involved in the parameters and models associated with the considered Limit States, for the safety class, the type of limit state, the structure life time and the amount of information and control.
6	Application to representative structures	A fictitious breakwater with a sandy bed layer (Easchel Breakwater), Genoa Voltri Breakwater as well as further three Japanese breakwaters were used as example representative structures to implement the developed probabilistic procedure.